T0253751

EVIDENCE AND EVOLUTION

How should the concept of evidence be understood? And how does the concept of evidence apply to the controversy about creationism as well as to work in evolutionary biology about natural selection and common ancestry? In this rich and wide-ranging book, Elliott Sober investigates general questions about probability and evidence and shows how the answers he develops to those questions apply to the specifics of evolutionary biology. Drawing on a set of fascinating examples, he analyzes whether claims about intelligent design are untestable; whether they are discredited by the fact that many adaptations are imperfect; how evidence bears on whether present species trace back to common ancestors; how hypotheses about natural selection can be tested, and many other issues. His book will interest all readers who want to understand philosophical questions about evidence and evolution, as they arise both in Darwin's work and in contemporary biological research.

ELLIOTT SOBER is Hans Reichenbach Professor and William Vilas Research Professor in the Department of Philosophy, University of Wisconsin-Madison. His many publications include *Philosophy of Biology, 2nd Edition* (1999) and *Unto Others: The Evolution and Psychology of Unselfish Behavior* (1998) which he co-authored with David Sloan Wilson.

EVIDENCE AND EVOLUTION

The logic behind the science

ELLIOTT SOBER

CAMBRIDGE
UNIVERSITY PRESS

CAMBRIDGE
UNIVERSITY PRESS

University Printing House, Cambridge CB2 8BS, United Kingdom

One Liberty Plaza, 20th Floor, New York, NY 10006, USA

477 Williamstown Road, Port Melbourne, VIC 3207, Australia

314-321, 3rd Floor, Plot 3, Splendor Forum, Jasola District Centre, New Delhi - 110025, India

79 Anson Road, #06-04/06, Singapore 079906

Cambridge University Press is part of the University of Cambridge.

It furthers the University's mission by disseminating knowledge in the pursuit of education, learning and research at the highest international levels of excellence.

www.cambridge.org
Information on this title: www.cambridge.org/9780521692748

First published 2008

A catalogue record for this publication is available from the British Library

Library of Congress Cataloging in Publication data
Sober, Elliott.
Evidence and evolution : the logic behind the science / Elliott Sober.
p. cm.
Includes bibliographical references.
ISBN 978-0-521-87188-4 (hardback : alk. paper) – ISBN 978-0-521-69274-8 (pbk. : alk. paper)
1. Evolution (Biology)–Philosophy. 2. Natural selection–Philosophy. 3. Evidence.
4. Probabilities–Philosophy. I. Title.
QH360.5.S625 2008
576.8–dc22
2007051438

ISBN 978-0-521-87188-4 Hardback
ISBN 978-0-521-69274-8 Paperback

In memory of my friend Berent Enç (1938–2003)

Contents

List of figures   *page* ix
Preface   xv
Acknowledgements   xix

1 Evidence   1

   1.1 Royall's three questions   3
   1.2 The ABCs of Bayesianism   8
   1.3 Likelihoodism   32
   1.4 Frequentism I: Significance tests and probabilistic *modus tollens*   48
   1.5 Frequentism II: Neyman–Pearson hypothesis testing   58
   1.6 A test case: Stopping rules   72
   1.7 Frequentism III: Model-selection theory   78
   1.8 A second test case: Reasoning about coincidences   104
   1.9 Concluding comments   107

2 Intelligent design   109

   2.1 Darwin and intelligent design   109
   2.2 Design arguments and the birth of probability theory   113
   2.3 William Paley: The stone, the watch, and the eye   118
   2.4 From probabilities to likelihoods   120
   2.5 Epicureanism and Darwin's theory   122
   2.6 Three reactions to Paley's design argument   125
   2.7 The no-designer-worth-his-salt objection to the hypothesis
of intelligent design   126
   2.8 Popper's criterion of falsifiability   129
   2.9 Sharpening the likelihood argument   131
   2.10 The principle of total evidence   136
   2.11 Some strengths of the likelihood formulation of the design argument   139

2.12 The Achilles heel of the likelihood argument   141
2.13 Paley's stone   147
2.14 Testability   148
2.15 The relationship of the organismic design argument to Darwinism   154
2.16 The relationship of Paley's design argument to contemporary
      intelligent-design theory   154
2.17 The relationship of the design argument to the argument from evil   164
2.18 The design argument as an inductive sampling argument   167
2.19 Model selection and intelligent design   177
2.20 The politics and legal status of the intelligent-design hypothesis   184
2.21 Darwinism, theism, and religion   186
2.22 A prediction   188

3   Natural selection   189

3.1   Selection plus drift (SPD) versus pure drift (PD)   192
3.2   Comparing the likelihoods of the SPD and PD hypotheses   199
3.3   Filling in the blanks   201
3.4   What if the fitness function of the SPD hypothesis contains a valley? 212
3.5   Selection versus drift for a dichotomous character   215
3.6   A breath of fresh air: Change the *explanandum*   219
3.7   Model selection and unification   226
3.8   Reichenbach's principle of the common cause   230
3.9   Testing selection against drift with molecular data   235
3.10 Selection versus phylogenetic inertia   243
3.11 The chronological test   253
3.12 Concluding comments   261

4   Common ancestry   264

4.1   *Modus Darwin*   265
4.2   What the common ancestry hypothesis asserts   268
4.3   A Bayesian decomposition   275
4.4   A single character: Species matching and species mismatching   277
4.5   More than one character   294
4.6   Concluding comments on the evidential significance of similarity   310
4.7   Evidence other than similarity   314
4.8   Phylogenetic inference: The contest between likelihood and cladistic
      parsimony   332

Conclusion   353
Bibliography   368
Index   385

Figures

1.1 Present evidence and its downstream consequences. *page* 4
1.2 Three possible distributions of longevities. 20
1.3 A flat prior density distribution for p and the non-flat posterior density occasioned by observing one head in four tosses. 22
1.4 When the coin lands heads in five of twenty tosses, the maximum likelihood estimate of $p = Pr$(the coin lands heads | the coin is tossed) is $p = \frac{1}{4}$. 23
1.5 When two independent and reliable witnesses each report on whether proposition p is true, two yeses provide stronger evidence for p than one, and one yes provides stronger evidence than zero. 43
1.6 Smith and Jones differ in their inclinations to place different orders for breakfast. 44
1.7 A new set of breakfast inclinations for Smith and Jones. 44
1.8 S either has tuberculosis or does not, and you, the physician, must decide whether to accept or reject the hypothesis H that S has tuberculosis. 58
1.9 If $p = \frac{1}{4}$ is the null hypothesis and $p = \frac{3}{4}$ is the alternative to the null, and $\alpha = 0.05$ is chosen, the Neyman–Pearson theory says that the null hypothesis should be rejected if and only if twelve or more heads occur in thirty tosses of the coin. 61
1.10 Each of the observations can be represented by a data point. L(LIN) is the straight line that fits the data best; L(PAR) is the parabola that fits best. 67
1.11 L(LIN) is the straight line that is closest to the data; the LIN model postulates an error distribution around this line. 68
1.12 If a coin lands tails on the first two tosses and heads on the third, this outcome might be the result of two different experiments. 73

1.13 A fixed-length experiment in which a coin is tossed twenty
      times and a flexible-length experiment in which a coin is
      tossed until six heads occur.   74
1.14 The prediction problem that Akaike considered.   83
2.1   Two theistic hypotheses.   110
2.2   Creationism and theistic evolutionism.   112
2.3   If A individuals have a fitness of 0.6 and B individuals
      have a fitness of 0.2, no other evolutionary forces impinge,
      and the population is infinitely large, trait A must
      evolve to 100-percent representation.   157
2.4   A trait that evolves from a value of 10 to a value of 20
      by the process of Darwinian gradualism in an infinite
      population must have a fitness function that
      monotonically increases from 10 to 20.   158
2.5   If there are n parts to an eye, how fit are organisms that
      have 0, or 1, or 2,..., or $(n-1)$, or all n?   159
2.6   An arch surmounted by a keystone satisfies the
      definition of irreducible complexity.   161
2.7   Hypothetical example of epistatic fitness relationships.   163
2.8   If we accept the bridge principle $q \approx p$, we can estimate
      the value of p by observing the frequency f.   173
2.9   The (One) model unifies the 20 million observations;
      the (20 Million) model treats each toss of each coin as
      a separate problem and is therefore more disunified.   183
2.10 Evolutionary biology proposes a unified model of the
      features that organisms have. Intelligent-design theory
      proposes a disunified model.   183
3.1   The pure-drift (PD) hypothesis can be thought of as a
      random walk on a line. The selection-plus-drift (SPD)
      hypothesis can be represented as a biased walk, influenced
      by a probabilistic attractor, the optimal phenotype.   193
3.2   Three fitness functions that have the same optimum
      ($\theta = 12$).   196
3.3   According to the SPD hypothesis, a population that has a
      given trait value at t_0 can be expected to move in the
      direction of O, the optimal trait value.   197
3.4   According to the PD hypothesis, a population that has a
      given trait value at time t_0 has that initial state as its
      expected value at all subsequent times, though the
      uncertainty surrounding that expected value increases.   198

3.5 The likelihoods of the SPD and the PD hypotheses.   199

3.6 The population must evolve from its ancestral state A to
     its present state P.   200

3.7 The body size of ancestors of current polar bears (S)
     can be (a) observed, or inferred from (b) fossilized relatives
     (FR_1 and FR_2), or from (c) extant relatives (ER_1 and ER_2).   204

3.8 The solid curve represents Cook and Cockrell's (1978)
     estimate of how the amount of food (f) a ladybird obtains
     from eating an aphid depends on the amount of time (t)
     spent feeding on it.   206

3.9 Given the trait values of present-day polar bears and their
     relatives, the principle of parsimony provides estimates
     of the character states of the ancestors A_1 and A_2.   208

3.10 If $P = a$ is the present trait value and the lineage has
     experienced pure drift, the maximum likelihood estimate
     of the trait value of the ancestor is $A = a$.   209

3.11 If $P = a$ is the present trait value and selection has been
     pushing the lineage towards the optimal value O, the
     maximum likelihood estimate of the trait value of the
     ancestor is not $A = a$.   210

3.12 A fitness function for the camera, cup, and compound eye
     that has a valley.   214

3.13 The SPD and PD hypotheses differ in the probabilities
     they specify for a lineage's ending in the state $P = 1$.   217

3.14 The observed fur lengths for different bear species show a
     downward trend and are closely clustered around an
     independently motivated optimality line.   220

3.15 Two scenarios in which selection causes bear lineages to
     evolve in the direction of an optimality line.   220

3.16 The ancestors A_1 and A_2 both have optimal trait values
     and their environments get colder.   221

3.17 If two descendant lineages stem from a common ancestor
     A and then evolve in the direction of an optimality line
     that has a negative slope, the expectation is that a line
     through D_1 and D_2 will also have a negative slope,
     if the trait's heritability is approximately the same in
     the two lineages.   222

3.18 If two descendant lineages stem from a common
     ancestor A and then evolve by drift, the expectation is
     that a line through D_1 and D_2 will have zero slope.   223

3.19 If two descendant lineages stem from a common ancestor *A* and then overshoot the optimality line postulated by the adaptive hypothesis, does this count as evidence favoring drift over selection? 224

3.20 Survival ratios and male care of offspring in anthropoid primates. 228

3.21 Possible explanations of patterns of variation, all for hypothetical data. 230

3.22 Although the principle of the common cause is sometimes described as saying that an "observed correlation" entails a causal connection, it is better to divide the inference into two steps. 232

3.23 Given the phylogeny, the neutral theory entails that the expected difference between 1 and 3 equals the expected difference between 2 and 3. 238

3.24 The number of nonsynonymous and synonymous differences that exist within and between three Drosophila species at the Adh locus. 241

3.25 The relative rate test and the McDonald–Kreitman test focus on different events in this tree. 242

3.26 Selection for character state 1 raises the probability that the descendant *D* will exhibit that character state. 247

3.27 If smoking causally contributes to lung cancer, smoking should raise the probability of lung cancer for people who have the same degree of asbestos exposure. 248

3.28 To test for phylogenetic inertia, lineages alike in their selective regimes must be compared. 249

3.29 The fact that species have common ancestors permits phylogenetic inertia and selection to each be tested by means of controlled comparisons without estimating ancestral trait values. 251

3.30 When the principle of parsimony is used to reconstruct the character states of ancestors in this phylogenetic tree, the conclusion is that trait *T* and trait *W* each evolved once. 254

3.31 The probability of the data (the trait values of tip species) is affected by the character states assigned to ancestors A_1, A_2, and A_3. 256

3.32 Two reconstructions of ancestral character states. 256

3.33 The two reconstructions of ancestral character states depicted in Figure 3.32 assign different events to branches *a–e*. 257

3.34 Two hypotheses about events in the lineage leading to land vertebrates that make different predictions about the trait combinations that land vertebrates and their relatives should exhibit. 259

4.1 Two competing genealogical hypotheses about the phylogeny of human beings (*H*), chimpanzees (*C*), and gorillas (*G*). 265

4.2 If you are a diploid organism with one chromosome pair, two of your four grandparents must have failed to make any genetic contribution to your genome. 270

4.3 Hypothesis (a), that there was a LUCA, is denied by both (b) and (c), which disagree as to how much relatedness there is among the *n* organisms and fossils ($S_1, \ldots, S_n$) that exist now. 271

4.4 A CA_1 and a CA_3 genealogy for Bacteria (B), Archaea (A), and Eukaryotes (E), both of which involve rampant lateral gene transfer in early life. 273

4.5 Three scenarios under which organisms *X* and *Y* share a trait because it was transmitted to them from an earlier organism *O*. 275

4.6 The common-ancestry and separate-ancestry hypotheses. 279

4.7 Two possible transformation series for a trait *T* that has n states. 285

4.8 When *X* and *Y* are scored for whether they match on a dichotomous trait *T*, there are two possible observations. 293

4.9 Three fitness functions: (a) frequency independent selection for trait A; (b) drift; (c) frequency dependent selection for the majority trait. 299

4.10 Four likelihood ratios, two of which depend on the amount of time between ancestor and descendants. 304

4.11 Two character distributions for the two species *X* and *Y*. 308

4.12 Two alternatives to the hypotheses that all the traits of the taxa *W*, *X*, *Y*, and *Z* stem from a single common ancestor. 317

4.13 If the evolutionary process is gradual, the CA hypothesis predicts the existence of ancestors that had intermediate forms, regardless of the character state of the common ancestor *Z*. 319

4.14 Either X and Y have a common ancestor or they do not (SA). Cells represent probabilities of the form $Pr(\pm\text{intermediate} \mid \pm \text{ CA})$. Gradualism is assumed. 320

4.15 Either X and Y have a common ancestor or they do not (SA). Cells represent the probability that we have observed an intermediate, or that we have not, conditional on CA and conditional on SA. 321

4.16 Observing an intermediate fossil favors CA over SA, and failing to so observe favors SA over CA, if $a > 0$ and $q < 1$. 322

4.17 These dated fossils form an intermediate series between the extant species X and Y. 323

4.18 H, C and G are each temporally extended lineages; time slices drawn at random from H and from C can be expected to be temporally more proximate to each other that time slices drawn at random from H and from G (or from C and from G). 327

4.19 The genealogy of X, Y, and Z is $(XY)Z$. 328

4.20 Each of the dichotomous traits A and B can experience two changes and each kind of change can occur on each of the two branches. 335

4.21 Models are more complex the larger the number of adjustable parameters they contain. 336

4.22 Two sites in two aligned sequences that come from different branches of a phylogenetic tree. 337

4.23 Four models of molecular evolution and their logical relationships. 338

4.24 Conjunctions of the form "tree topology & process model" containing adjustable parameters; these are nuisance parameters in the context of making inferences about topologies. 340

4.25 The example described in Felsenstein (1978) in which parsimony can converge on the incorrect tree as more and more data are consulted. 347

4.26 The tree in Figure 4.25 is in the "Felsenstein zone" when $p \gg q$. 348

Preface

Biologists study living things, but what do philosophers of biology study? A cynic might say "their own navels," but I am no cynic. A better answer is that philosophers of biology, and philosophers of science generally, study science. Ours is a second-order, not a first-order, subject. In this respect, philosophy of science is similar to history and sociology of science. A difference may be found in the fact that historians and sociologists study science as it is, whereas philosophers of science study science as it ought to be. Philosophy of science is a *normative* discipline, its goal being to distinguish good science from bad, better scientific practices from worse. This evaluative endeavor may sound like the height of hubris. How dare we tell scientists what they ought to do! Science does not need philosopher kings or philosophical police. The problem with this dismissive comment is that it assumes that normative philosophy of science ignores the practice of science. In fact, philosophers of science recognize that ignoring science is a recipe for disaster. Science itself is a normative enterprise, full of directives concerning how nature ought to be studied. Biologists don't just describe living things; they constantly evaluate each other's work. Normative philosophy of science is continuous with the normative discourse that is ongoing within science itself. Discussions of these normative issues should be judged by their quality, not by the union cards that discussants happen to hold.

Pronouncements on "the scientific method" all too often give the impression that this venerable object is settled and fixed – that it is an Archimedean point from which the whole world of scientific knowledge can be levered forward. The fact of the matter is that a thorough grasp of scientific inference is a goal, not a given. Like our current understanding of nature, our present grasp of the nature of scientific inference is fragmentary and a work in progress. Scientists themselves disagree about the methods of inference that should be used, and so do statisticians and philosophers. For this reason, the first chapter of this book, on the concept of evidence, is not a report on a complacent consensus. The position I develop on what evidence means in science is controversial. It is an intervention in the long-standing disagreement between frequentists and Bayesians. I wrote this chapter for neophytes, not sophisticates. No prior understanding of probability is presupposed; I try to build from the ground up.

The methods of inference used in science take two forms. Some are entirely general, in the sense that they apply no matter what the subject matter is. These are the sorts of procedures described in texts on deductive logic and statistics. A method for estimating the average blood pressure in a population of robins is also supposed to apply to the problem of estimating the average weight in a pile of rocks. The different sciences also include methods that are narrower in scope; these methods are tailor-made to apply to a specific subject matter. For example, in evolutionary biology, a concept of parsimony has been developed that underwrites inferences about phylogenetic trees; this method is not general in its subject matter, it applies only to hypotheses about genealogies of a certain sort. The usefulness of this concept of parsimony has been controversial in evolutionary biology. When I consider the role of parsimony considerations in evolutionary biology in Chapters 3 and 4, I again will be intervening in a methodological dispute that is alive within science itself.

When scientists disagree about which of several competing inference methods they should use, it often is fairly obvious that there is a philosophical dimension to their dispute. But philosophical questions also can be raised when there is a thoroughgoing scientific consensus. No competent biologist now doubts that human beings and chimps have a common ancestor. The detailed similarities that unite these two species are overwhelming. It takes a philosopher to see a question in the background – why does detailed similarity provide evidence of common ancestry? Philosophers can ask this question without doubting the good judgment of the scientific community. They want to uncover the assumptions that need to be true for this inference from similarity to common ancestry to make sense. Analyzing inferences that seem to be obviously correct has long been a favorite project for philosophers.

Two grand ideas animate the Darwinian theory of evolution, both in the form that Darwin gave it and also in the form that modern Darwinians endorse. These are the ideas of common ancestry and natural selection. In each case, we can think of Darwinian ideas as competing with alternatives. The hypothesis that the species we now observe trace back to a common ancestor competes with the hypothesis that they originated separately and independently. The hypothesis that a trait in a species – say, the long fur that polar bears now have – evolved by natural selection competes with the hypothesis that it evolved by random genetic drift and with other hypotheses that describe other possible causes of character change and stasis. Most of Chapters 3 and 4 is devoted to understanding how the Darwinian position can be tested against its competitors. But I also spend time exploring how ideas about natural selection and common ancestry interact with each other. Biologists use information about common ancestry to test hypotheses about natural selection. And inferences about ancestry often rely on information about how various traits have evolved. The two parts of the Darwinian picture are *logically independent* of each other, but they are *methodologically interdependent*.

This book is aimed at philosophers of science and evolutionary biologists. Both tend to have little patience with creationism, so I want to explain why I devote Chapter 2 to its evaluation. I do not think that "intelligent design" is a substantive scientific theory, but I am not satisfied with the standard reasons that have been offered to explain why this is so. For example, Karl Popper's ideas on falsifiability are often used in this context, but philosophers of science have long realized that there are serious problems with Popper's solution to the demarcation problem – the problem of separating science from nonscience. In Chapter 2, I try to develop a better account of testability that clarifies what is wrong with the hypothesis of intelligent design. Another standard critique of creationism begins with the fact that many of the adaptations we find in nature are highly imperfect. It is claimed that an intelligent designer would never have produced such arrangements. I explain in Chapter 2 why I find this criticism of creationism problematic. Although it isn't true that every word of Chapter 2 matters to the material in Chapters 3 and 4, there nonetheless is a through-line from Chapter 1 to Chapters 3 and 4 that passes through Chapter 2. The Duhem–Quine thesis about scientific testing is introduced in Chapter 2 and so is the concept of a fitness function; both play important roles in what comes after.

Chapter 3 begins where Chapter 2 leaves off, by asking whether hypotheses about natural selection are in any better shape than hypotheses about intelligent design. It is no fair switching standards – setting the bar impossibly high when evaluating creationism, but lowering the bar when evolutionary hypotheses are assessed. I begin with the apparently simple problem of explaining why polar bears now have (let us assume) fur that is, on average, 10 centimeters long. Which is the more plausible explanation: that the trait evolved by natural selection or that it evolved by drift? In the first few sections of Chapter 3, I describe what needs to be known if one wishes to test these hypotheses against each other. The result is a catalog of difficulties. I then argue that the situation is transformed if we take up a different problem: Rather than trying to explain why polar bears have an average fur length of 10 centimeters, we might try to explain why bears in cold climates have longer fur than bears in warm ones. This new problem is easier to solve, and the fact that bears have a common ancestor plays a role in solving it. The rest of Chapter 3 discusses some of the methods that biologists have used to test hypotheses about natural selection; for example, they use DNA sequence data and they also infer the chronological order of the novelties that evolve in a phylogenetic tree.

Chapter 4 addresses a question I mentioned before: Why, or in what circumstances, is the similarity of two species evidence that they have a common ancestor? After developing an answer to this question that is based on the concept of evidence described in Chapter 1, I explore Darwin's idea that similarities that are useless to the organisms that have them provide stronger evidence for common ancestry than adaptive similarities do. Although Darwin's suggestion is right for a large class of adaptive similarities, it emerges that that there is a type of adaptive similarity for which the situation is precisely the reverse. I then consider how intermediate fossils and biogeographical distribution provide evidence concerning common ancestry. The chapter concludes with a discussion of two conflicting methods for inferring phylogenetic trees.

The title of this book may be a little misleading, but I hope that the subtitle corrects a misapprehension that the title may encourage. The title perhaps suggests that this is a book that describes the evidence *for* evolution. There are many good books that do this; they are works of *biology*. The book before you is not a member of that species; rather, it is a work of *philosophy*. My goal in what follows is not to pile up facts that support this or that proposition in evolutionary biology. Rather, I want to describe the tools that ought to be used to assess the evidence that bears on evolutionary ideas. Scientists, ever eager to draw conclusions about nature, reach for patterns of reasoning that seem sensible, but they rarely linger over why the procedures they use make sense. Although this book is not a work of science, I hope that scientists will find that some of the thoughts developed here are worth pondering. I also hope that the philosophers who read this book will be intrigued by the evolutionary setting of various epistemological problems.

Acknowledgements

I have been lucky in my collaborators, both philosophical and biological. Some of these coauthors will find that some of the ideas in this book are drawn from papers we have written together (citations indicate where the extractions and insertions occurred); others will find a connection to work we have done together that is less direct, but I hope they will see that it is tangible nonetheless. This book would be very different or would not exist at all (depending on how you define "the same book"), had it not been for my interactions with these talented people: Martin Barrett, Ellery Eells (whom I miss very much), Branden Fitelson, Malcolm Forster, Christopher Hitchcock, Christopher Lang, Richard Lewontin, Gregory Mougin, Steven Orzack, Larry Shapiro, Mike Steel, Christopher Stephens, Karen Strier, and David Sloan Wilson.

I also have been lucky that many philosophers and biologists read parts of this book and reacted with criticisms and suggestions. Some even read the whole thing. Let me mention first the dauntless souls who plowed through the entire manuscript and gave me valuable comments: Martin Barrett, Juan Comesaña, James Crow, Malcolm Forster, Thomas Hansen, Daniel Hausman, Steven Leeds, Richard Lewontin, Peter Vranas, and Nigel Yoccoz. They read, as far as I know, of their own free will. I'm not sure I can say the same of the students who took seminars with me in which the manuscript was discussed, but their comments have been no less helpful. My thanks to Craig Anderson, Mark Anderson, Matthew Barker, John Basl, Ed Ellesson, Joshua Filler, Patrick Forber, Michael Goldsby, Casey Helgeson, John Koolage, Matthew Kopec, Hallie Liberto, Deborah Mower, Peter Nichols, Angela Potochnik, Ken Riesman, Susanna Rinard, Michael Roche, Armin Schulz, Shannon Spaulding, Tod van Gunten, Joel Velasco, Jason Walker, and Brynn Welch. Matthew Barker and Casey Helgeson also helped me with the references, John Basl with the figures, and Joel Velasco with the corrections.

I next want to thank the people who read portions of the manuscript and sent me comments or who responded to questions that came up as I wrote; at times I felt I was being helped by an army of experts. For this I am grateful to Yuichi Amitani, Eric Bapteste, Gillian Barker, David Baum, John Beatty, Ken Burnham, David Christensen, Eric Cyr Desjardins, Ford Doolittle, John Earman, Anthony Edwards, Branden Fitelson, Steven Frank, Richard Healey, Jonathan Hodge, Dan Hartl, Edward Holmes, John Huelsenbeck, James Justus, Bret Larget, Paul Lewis, William Mann, Sandra Mitchell, John Norton, Ronald Numbers, Samir Okasha, Roderick Page, Bret Payseur, Will Provine, Alirio Rosales, Bruce Russell, Larry Shapiro, Mike Steel, Christopher Stephens, Scott Thurow, and Carl Woese.

I am deeply indebted to the Vilas Trust at the University of Wisconsin; were it not for the research support provided by my William Vilas Professorship, I would not have been able to work so long and hard on this project. I also am grateful to the Rockefeller Foundation for the month's stay I had during May–June 2006 at their research center, the Villa Serbeloni in Bellagio, Italy. This is where I wrote a draft of Chapter 1 in delightful circumstances that still make me smile each time I think of them. Finally, I want to thank Sandra Mitchell and John Norton at the University of Pittsburgh's Center for Philosophy of Science for organizing a workshop on my book manuscript that took place in March 2007; I learned a lot during this event and the book is better because of it.

Evidence

Scientists and philosophers of science often emphasize that science is a fallible enterprise. The evidence that scientists have for their theories does not render those theories certain. This point about *evidence* is often represented by citing a fact about *logic*: The evidence we have at hand does not deductively entail that our theories must be true. In a *deductively valid argument*, the conclusion must be true if the premises are. Consider the following old saw:

> All human beings are mortal.
>
> Socrates is a human being.
> _____
> Socrates is mortal.

If the premises are true, you cannot go wrong in believing the conclusion. The standard point about science's fallibility is that the relationship of evidence to theory is *not* like this. The correctness of this point is most obvious when the theories in question are far more *general* than the evidence we can bring to bear on them. For example, theories in physics such as the general theory of relativity and quantum mechanics make claims about what is true at *all* places and *all* times in the entire universe. Our observations, however, are limited to a very small portion of that immense totality. What happens here and now (and in the vicinity thereof) does not deductively entail what happens in distant places and at times remote from our own.

If the evidence that science assembles does not provide certainty about which theories are true, what, then, does the evidence tell us? It seems entirely natural to say that science uses the evidence at hand to say which theories are *probably* true. This statement leaves room for science to be fallible and for the scientific picture of the world to change when new evidence rolls in. As sensible as this position sounds, it is deeply controversial. The controversy I have in mind is not between science and

1

nonscience; I do not mean that scientists view themselves as assessing how probable theories are while postmodernists and religious zealots debunk science and seek to undermine its authority. No, the controversy I have in mind is alive *within* science. For the past seventy years, there has been a dispute in the foundations of statistics between Bayesians and frequentists. They disagree about many issues, but perhaps their most basic disagreement concerns whether science is in a position to judge which theories are probably true. Bayesians think that the answer is *yes* while frequentists emphatically disagree. This controversy is not confined to a question that statisticians and philosophers of science address; scientists use the methods that statisticians make available, and so scientists in all fields must choose which model of scientific reasoning they will adopt.

The debate between Bayesians and frequentists has come to resemble the trench warfare of World War I. Both sides have dug in well; they have their standard arguments, which they lob like grenades across the no-man's-land that divides the two armies. The arguments have become familiar and so have the responses. Neither side views the situation as a stalemate, since each regards its own arguments as compelling. And yet the warfare continues. Fortunately, the debate has not brought science to a standstill, since scientists frequently find themselves in the convenient situation of not having to care which of the two approaches they should use. Often, when a Bayesian and a frequentist consider a biological theory in the light of a body of evidence, they both give the theory high marks. This allows biologists to walk away happy; they've got their answer to the biological question of interest and don't need to worry whether Bayesianism or frequentism is the better statistical philosophy. Biologists care about making discoveries about *organisms*; the *nature of reasoning* is not their subject, and they are usually content to leave such "philosophical" disputes for statisticians and philosophers to ponder. Scientists are *consumers* of statistical methods, and their attitude towards methodology often resembles the attitude that most of us have towards consumer products like cars and computers. We read *Consumer Reports* and other magazines to get expert advice on what to buy, but we rarely delve deeply into what makes cars and computers tick. Empirical scientists often use statisticians, and the "canned" statistical packages they provide, in the same way that consumers use *Consumer Reports*. This is why the trench warfare just described is not something in which most biologists feel themselves to be engulfed. They live, or try to live, in neutral Switzerland; the Battle of the Marne (they hope) involves others, far from home.

This book is about the concept of evidence as it applies in evolutionary biology; the present chapter concerns general issues about evidence that will be relevant in subsequent chapters. I do not aim here to provide anything like a complete treatment of the debate between Bayesianism and frequentism, nor is my aim to end the trench warfare that has persisted for so long. Rather, I hope to help the reader to understand what the shooting has been about. I intend to start at the beginning, to not use jargon, and to make the main points clear by way of simple examples. There are depths that I will not attempt to plumb. Even so, my treatment will not be neutral; in fact, it is apt to irritate both of the entrenched armies. I will argue that Bayesianism makes excellent sense for many scientific inferences. However, I do agree with frequentists that applying Bayesian methods in other contexts is highly problematic. But, unlike many frequentists, I do not want to throw out the Bayesian baby with the bathwater. I also will argue that some standard frequentist ideas are flawed but that others are more promising. With respect to frequentism as well, I feel the need to pick and choose. My approach will be "eclectic"; no single unified account of all scientific inference will be defended here, much as I would like there to be a grand unified theory.

One further comment before we begin: I have contrasted Bayesianism and frequentism and will return to this dichotomy in what follows. However, there are different varieties of Bayesianism, and the same is true of frequentism. In addition, there is a third alternative, likelihoodism (though frequentists often see Bayesianism and likelihoodism as two sides of the same deplorable coin). We will separate these inferential philosophies more carefully in what follows. But for now we begin with a stark contrast: Bayesians attempt to assess how probable different scientific theories are, or, more modestly, they try to say which theories are more probable and which are less. Frequentists hold that this is not what the game of science is about. But what do frequentists regard as an attainable goal? Hold that question in mind; we will return to it.

1.1 ROYALL'S THREE QUESTIONS

The statistician Richard Royall begins his excellent book on the concept of evidence (Royall 1997: 4) by distinguishing three questions:

(1) What does the present evidence say?
(2) What should you believe?
(3) What should you do?

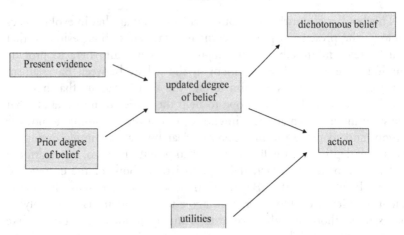

Figure 1.1   Present evidence and its downstream consequences.

If you are rational, you form your beliefs by consulting the evidence
you have just gained, and when you decide what to do (which actions to
perform), you should take account of what you believe. But answering
question (2) requires more than an answer to (1), and answering question
(3) requires more than an answer to (2). The extra elements needed are
depicted in Figure 1.1.

Suppose you are a physician and you are talking to the patient in your
office about the result of his tuberculosis test. The report from the lab says
"positive." This is your present evidence. Should you conclude that the
patient has tuberculosis? You want to take the lab report into account, but
you have other information besides. For example, you previously had
conducted a physical exam. Before you looked at the test report, you had
some opinion about whether your patient has tuberculosis. The lab report
may modify how certain you are about this. You update your degree of
belief by integrating the new evidence with your prior information. This
may lead you say to him "your probability of tuberculosis is 0.999."

If your patient is a philosopher who enjoys perverse conversation, he
may reply, "but tell me, doctor, do I have tuberculosis, or not?" He
doesn't want to know how *probable* it is that he has tuberculosis; he wants
to know *whether* he has the disease – *yes or no*. This raises the question of
whether a proposition's having a probability of 0.999 suffices for one to
believe it, where belief is conceptualized as a dichotomous category:
Either you believe the proposition or you do not. It may seem that a high
degree of belief suffices for believing a proposition (even if it does not

suffice for being certain that the proposition is true), but there are complications. Consider Kyburg's (1970) lottery paradox. Suppose 1,000 lottery tickets are sold and the lottery is fair. *Fair* means that one ticket will win and each has the same chance of winning. If high probability suffices for belief, you are entitled to believe that ticket no. 1 will not win, since the probability of ticket 1's not winning is $\frac{999}{1000}$. The same is true of ticket no. 2; you should believe that it won't win. And so on, for each of the 1,000 tickets. But if you put these 1,000 beliefs (each of the form *ticket i will not win*) together with the rest of what you believe, your beliefs have become contradictory: You believe that some ticket will win (since you believe the lottery is fair), and you have just accepted the proposition that no ticket will win. Kyburg's solution to this puzzle is to say that acceptance does not obey a rule of conjunction; you can accept A and accept B without having to accept the conjunction $A\&B$.[1] This may be the best one can do for the concept of dichotomous belief, but it raises the question of whether we really need such a concept. After all, our everyday thought is littered with dichotomies that, upon reflection, seem to be crudely grafted to an underlying continuum. For example, we speak of people being *bald*, but we know that there is no threshold number of hairs that marks the boundary.[2] We are happy to abandon these crude categories when we need to, but we return to them when they are convenient and harmless.

If it makes sense to talk about rational acceptance and rational rejection, those concepts must bear the following relation to the concept of evidence:

If learning that E is true justifies you in *rejecting* (i.e., disbelieving) the proposition P, and you were not justified in rejecting P before you gained this information, then E must be evidence *against P*.

If learning that E is true justifies you in *accepting* (i.e., believing) the proposition P, and you were not justified in accepting P before you gained this information, then E must be evidence *for P*.

A theory of rational acceptance and rejection must provide more than this modest principle, which may seem like a mere crumb, hardly worth

---

[1] See Kaplan (1996) for a theory of rational acceptance that, unlike Kyburg's, obeys the conjunction principle.

[2] I say we "know" this, but Williamson (1994) and Sorenson (2001) have argued that in each use of a vague term, there is a cutoff, even if speakers are not aware of what it is. Their position is counterintuitive, but it cannot be dismissed without attending to their arguments (which we won't do here).

mentioning at all. But, in fact, it *is* worth stating, since later in this chapter it will do some important philosophical work.[3]

Even if this modest principle linking evidence and rational acceptance seems obvious, there is an old philosophical reason for pausing to ponder it. In the seventeenth century, Blaise Pascal sketched an argument that came to be called *Pascal's wager*. Earlier proofs of the existence of God had tried to demonstrate that there is evidence that God exists; Pascal endeavored to show that one ought to believe in God even if all the evidence one has is evidence *against*. The rough idea is this: If there is a God, you'll go to Heaven if you're a believer and go to Hell if you're not; on the other hand, if there is no God, it won't much affect your well-being whether or not you believe. Pascal wrote when probability theory was just starting to take its modern mathematical form, and his argument is a nice illustration of ideas that came to be assembled in *decision theory*. Though there is room to dispute the details of this argument (on which see Mougin and Sober 1994), the wager is of interest here because it appears to challenge the "modest" principle just enunciated. The wager purports to provide a reason for accepting the proposition that God exists even though it does not cite any evidence that there is a God. It is easy to think of nontheological arguments that pose the same challenge. Suppose I promise to give you $1,000,000 if you can get yourself to believe that the President is now juggling candy bars. If I am trustworthy, I have given you a reason to believe the proposition though I have not provided any evidence that it is true.

Commentators on Pascal's wager often distinguish two types of rational acceptance. The *act of accepting* a proposition can make good prudential sense, but that does not mean that *the proposition accepted* is well supported by evidence. When acceptance is driven by the costs and benefits that attach to the act of believing, I'll call this "prudential acceptance." When it is driven by the bearing of evidence on the proposition believed, I'll use the term "evidential acceptance." The modest principle linking evidence and "acceptance" really pertains to *evidential* acceptance. The principle, modified in this way, is true; in fact, it may even be true *by definition*. However, this does not settle whether it is ever permissible to

---

[3] It is interesting that the concept of evidence relates pairs of propositions to each other, while the concepts of acceptance and rejection relate propositions to persons. Smoke is evidence for fire, regardless of whether any agent takes this fact to heart. However, rational acceptance (or rejection) means that a person is justified in accepting (or rejecting) some proposition. The present disciplinary divide between philosophers of science and epistemologists coincides to a considerable degree with this distinction between questions concerning how propositions are related to each other and questions concerning how propositions are related to persons.

indulge in prudential acceptance. William James (1897) defends the right to believe when the evidence is silent in his essay "The Will to Believe." W. K. Clifford (1999) replies, in "The Ethics of Belief," that it is always wrong "to believe upon insufficient evidence." I will not try to adjudicate between these two positions. Suffice it to say that the modest principle stated earlier is binding on those who commit to having evidence control what they believe.

It may seem a long jump from Pascal's seventeenth-century theology to the hard edges of twentieth-century statistics, but Pascal's concept of prudential acceptance lives on in frequentism. The following remark by Neyman and Pearson (1933: 291) has often been quoted:

No test based upon the theory of probability can by itself provide any valuable evidence of the truth or falsehood of [an] hypothesis [...] But we may look at the purpose of tests from another viewpoint. Without hoping to know whether each separate hypothesis is true or false, we may search for rules to govern our behavior with regard to them, in following which we insure that, in the long run of experience, we shall not be too often wrong.

Neyman and Pearson think of acceptance and rejection as *behaviors*, which should be regulated by prudential considerations, not by "evidence," which, for them, is a will o' the wisp. The prudential considerations they have in mind do not involve going to Heaven or Hell, but rather pertain to having true beliefs or false ones. There is no such thing as allowing "evidence" to regulate what we believe. Rather, we must embrace a policy and stick to it. If we do so, we can be certain (or, at least, it is overwhelmingly probable) that the percentage of false beliefs we accumulate over the long run will be held below some predesignated minimum. Not that present-day frequentists are all so dismissive of the concept of evidence (§1.4). But frequentists, early and late, have often embraced the idea of *prudential* belief.

Let us return to Figure 1.1. Suppose you, the physician, are 99.9 percent certain that your patient has tuberculosis, this degree of belief being based on the present tuberculosis test result and on other information you had from before. The thing to notice next is that your degree of belief does not, by itself, dictate what you should *say* or *do*. Should you tell your patient what you think? Should you remain silent? Should you lie? Should you hand him the pink pills you have in your desk? A rational decision about what to do requires more than the evidence you have and more than the degree of belief you have; a choice of action requires the input of values (which economists call *utilities*).

1.2   THE ABCS OF BAYESIANISM

Bayesianism is an answer to Royall's question (2): What should you believe? Bayesianism refines this question, substituting the concept of degree of belief for the dichotomous concept of believing or not believing a proposition. In our running example, Bayesianism addresses the question of how certain you should be that your patient has tuberculosis, given that his tuberculosis test came back positive.

Bayes' theorem

Bayesianism is based on Bayes' theorem, but the two are different. Bayes' theorem is a result in mathematics.[4] It is called a theorem because it is derivable from the axioms of probability theory (in fact, from a standard definition of conditional probability). As a piece of mathematics, the theorem is not controversial. Bayesianism, on the other hand, is a philosophical theory – it is an epistemology. It proposes that the mathematics of probability theory can be put to work in a certain way to explicate various concepts connected with issues about evidence, inference, and rationality.

Here is the rough idea of how Bayesianism uses Bayes' theorem: Before you make an observation, you assign a probability to the hypothesis H; this probability may be high, medium, or low (all probabilities by definition must be between 0 and 1, inclusive). After you make the observation, thereby learning that some observation statement O is true, you update the probability you assigned to H to take account of what you just learned. The probability that H has before the observation is called its *prior probability*; it is represented by $Pr(H)$. The word "prior" just means *before*; it doesn't mean that you know its value a priori (i.e., without any empirical input at all). The probability that H has in the light of the evidence O is called H's *posterior probability*; it is represented by the conditional probability $Pr(H \mid O)$; read this as "the probability of H, given O." Bayes' theorem shows how the prior and the posterior probability are related.

Now for the derivation of the theorem. Forget for just a moment that H means hypothesis and O means observation. Just regard them as any two

---

[4] A special case of the theorem was derived by Thomas Bayes and was published posthumously in the *Proceedings of the Royal Society* for 1764. Bayes' derivation was laborious and not fully general, very unlike the now-standard streamlined derivation I'll describe here.

propositions. Kolmogorov's (1950) definition of conditional probability is this:

$$Pr(H \mid O) = \frac{Pr(H \ \& \ O)}{Pr(O)}.$$

The definition is intuitive. For example, what is the probability that a card drawn at random from a standard deck is a heart, given that it is red? According to the Kolmogorov definition, this conditional probability has the same value as the ratio $Pr(\text{heart} \ \& \ \text{red})/Pr(\text{red})$. The denominator has a value of $\frac{1}{2}$. The proposition in the numerator, *heart & red*, is equivalent to *heart*, so the value for the numerator is $\frac{1}{4}$. Hence, the conditional probability has a value of $\frac{1}{2}$. By switching Hs and Os with each other in the Kolmogorov definition, you can see that it also is true that

$$Pr(O \mid H) = \frac{Pr(O \ \& \ H)}{Pr(H)}.$$

This means that the probability of the conjunction $H\&O$ can be expressed in two different ways:

$$Pr(H \ \& \ O) = Pr(H \mid O) \ Pr(O) = Pr(O \mid H)Pr(H).$$

From the second equality in the previous line, we obtain

$$\text{Bayes' theorem: } Pr(H \mid O) = \frac{Pr(O \mid H)Pr(H)}{Pr(O)}.$$

Here is some more terminology. I've already mentioned the *posterior probability* and the *prior probability* that appear in Bayes' theorem, but two other quantities are also mentioned. $Pr(O)$ is the *unconditional probability of the observations*. And R. A. Fisher dubbed $Pr(O \mid H)$ the *likelihood of H*. Because Fisher's terminology has become standard in statistics, I will use it here. However, this terminology is confusing, since in ordinary English, "likely" and "probably" are synonymous. So, beware! You need to remember that "likelihood" is a technical term. The likelihood of H, $Pr(O \mid H)$, and the posterior probability of H, $Pr(H \mid O)$, are different quantities and they can have different values. The likelihood of H is the probability that H confers on O, not the probability that O confers on H. Suppose you hear a noise coming from the attic of your house. You consider the hypothesis that there are gremlins up there bowling. The likelihood of this hypothesis is very high, since if there are gremlins bowling in the attic, there probably will be noise. But surely you don't think that the noise makes it very probable that there are gremlins up there bowling. In this example, $Pr(O|H)$ is high and $Pr(H|O)$ is low. The gremlin hypothesis has a high likelihood (in the technical sense) but a low probability.

Let me add two more details that underscore the distinction between H's probability and its likelihood.

$$Pr(H) + Pr(notH) = 1$$

and

$$Pr(H|O) + Pr(notH|O) = 1$$

as well. The probability of a proposition and the probability of its negation sum to one; this is true for prior and also for posterior probabilities. But likelihoods need not sum to one; $Pr(O|H) + Pr(O|notH)$ can be less than 1, or more. Suppose you observe that Sue is a millionaire and wonder whether she won her wealth in last week's lottery. Your observation is very improbable under the hypothesis that she bought a ticket in the lottery and also under the hypothesis that she did not. To summarize this point: If you know the probability of H, you thereby know the probability of $notH$; but knowing the likelihood of H leaves the likelihood of $notH$ completely open.

Another difference between likelihoods and probabilities concerns the difference between logically stronger and logically weaker hypotheses. Consider the following two hypotheses about the next card you'll be dealt from a standard deck:

$$H_1 = \text{It's a heart.}$$

$$H_2 = \text{It's the Ace of Hearts.}$$

The hypothesis H_2 is *logically stronger* than H_1; this means that H_2 entails H_1, but not conversely. Suppose the dealer is careless and you catch a glimpse of the card before it is dealt; you observe O = the card is red. Notice that H_1 has the higher posterior probability; $Pr(H_1|O) = \frac{1}{2}$ while

$Pr(H_2 | O) = \frac{1}{26}$. But the two hypotheses have identical likelihoods, since $Pr(O | H_1) = Pr(O | H_2) = 1$. It is a theorem of probability theory that

If proposition X entails proposition Y, then $Pr(X) \leq Pr(Y)$, and $Pr(X | \text{data}) \leq Pr(Y | \text{data})$ no matter what the data are.

Logically stronger hypotheses can't have higher probabilities than logically weaker hypotheses, but they can have higher likelihoods. This point about likelihoods is illustrated by the relationship of H_1 and H_2 to the observation $O' = $ the card is an ace.

A rule for updating

The different quantities used in Bayes' theorem are all available *before* you find out whether the statement O is true. You can know the value of $Pr(H | O)$ without knowing whether O is true, just as you can know that a conditional (an if/then statement) is true without knowing whether its antecedent (the if part) is true. All Bayes' theorem tells you is how the different probabilities it mentions, all assigned values at the same time, must be related. The theorem is, so to speak, a *synchronic* statement. But, as mentioned, Bayesianism provides advice about how you should change your degree of belief as you acquire new evidence. Bayes' theorem, therefore, must be supplemented by a rule for updating: This rule describes how probabilities should be related *diachronically*.

The rule of updating by strict conditionalization says that if O is the totality of the new information you have acquired, your *new* probability for H should be equal to your *old* value for $Pr(H | O)$. In other words: $Pr_{now}(H) = Pr_{then}(H | O)$, if O is all the evidence you acquired between then and now.

Before the result of the tuberculosis test is placed before you, you know the value of $Pr(S$ has tuberculosis | the test is positive) and $Pr(S$ has tuberculosis | the test is negative). These are your old posterior probabilities. When you learn that the test turned out positive, your new degree of belief for the proposition that S has tuberculosis is the one you assigned to the first of these conditional probabilities.

When I say that this rule for updating applies to "your" probability, does this mean that the Bayesian framework concerns only subjective degrees of belief? No – it is more general than this. You can think of this rule as giving normative advice to agents on how they should adjust the amount of certainty they have. But a rule for updating also provides advice concerning what you should think the objective probability of a proposition is. If you think that the objective prior probability of drawing the Ace of Hearts from a normal deck is $\frac{1}{52}$, and you think that the objective posterior probability of the card's being the Ace of Hearts, given that it is red, is $\frac{1}{26}$, and you learn (just) that the next card drawn will be red, then your new objective probability for the card's being the Ace of Hearts should be $\frac{1}{26}$. It is useful to keep Bayesianism's *epistemological* advice about how probabilities should be assigned and manipulated separate from the *semantic* question of what probability statements mean. Not that interesting connections can't be drawn between the two issues. But first things first.

Strict conditionalization involves the idealization that an act of observation has the result that you find out that an observation statement is true or that it is false. What you learn isn't just that O is *probably true*; you learn that O is *true*. You then use this information to modify the degree of belief you have for some other proposition H. Bayesianism with strict conditionalization is a kind of hybrid philosophy, in which you accept or reject O but you do not apply the concept of dichotomous belief to H. Richard Jeffrey (1965) proposed a rule for updating in which you acquire only a degree of belief in O; the concept of dichotomous belief is thoroughly abandoned. Jeffrey's *probability kinematics* describes how your newly acquired degree of belief in O should affect your degree of belief in H.[5] For the purposes of this book, we can ignore Jeffrey's refinement and think of Bayesianism in terms of the idea of strict conditionalization. In what follows, I won't go to the trouble of distinguishing old probability assignments from new ones. Since I'll be focusing on the version of Bayesianism that uses the rule of strict conditionalization, I'll treat the posterior probability $Pr(H \mid O)$ as representing your updated degree belief once you learn that O is true (provided that O is *all* you learned).

Notice that the rule for updating by strict conditionalization addresses the case in which you *now* have a probability for proposition H, and you also had a (conditional) probability for that proposition *earlier*. It therefore fails to apply to cases of conceptual innovation in which H involves concepts that you just formulated. You didn't have a conditional

---

[5] Although Jeffrey's conditionalization is more realistic than strict conditionalization in terms of its characterization of the input, it has a logical oddity that strict conditionalization avoids. The *order* in which new evidence arrives can affect the final degree of belief in Jeffrey's conditionalization, but not in strict.

probability for *H* earlier because *H* uses concepts you didn't have available back then. This is an especially important feature of some scientific innovations; scientists often work within the confines of a fixed stock of concepts, but every so often they break out. Evolutionists sometimes draw a distinction between micro- and macroevolution (§2.19); the former describes changes that occur within an enduring species whereas the latter describes changes that result in the appearance of new species. Kuhn's (1962) distinction between normal science and revolutionary science is similar; there is science pursued within an existing "paradigm" and science that results in the formation of new paradigms. Bayesian updating by strict conditionalization makes more sense in connection with the micro-changes that occur within normal science; it is controversial whether it can represent the macro-changes that occur in scientific revolutions.[6]

Posterior probabilities, likelihoods, and priors

Let's apply Bayes' theorem to the running example that you are a doctor and your patient has a positive tuberculosis test result. You want to use this new information to figure out how certain you should be that he has tuberculosis. Bayes' theorem says that

$$(4)\ Pr(\text{tuberculosis} \mid +\text{result}) = \frac{Pr(+\text{result} \mid \text{tuberculosis}) Pr(\text{tuberculosis})}{Pr(+\text{result})}.$$

Bayes' theorem also can be stated for the hypothesis that *S* does *not* have tuberculosis:

$$(5) \quad Pr(\text{no tuberculosis} \mid +\text{result})$$
$$= \frac{Pr(+\text{result} \mid \text{no tuberculosis}) Pr(\text{no tuberculosis})}{Pr(+\text{result})}.$$

Combining (4) and (5) yields the following equality of ratios:

$$(6) \quad \frac{Pr(\text{tuberculosis} \mid +\text{result})}{Pr(\text{no tuberculosis} \mid +\text{result})}$$
$$= \frac{Pr(+\text{result} \mid \text{tuberculosis})}{Pr(+\text{result} \mid \text{no tuberculosis})} \times \frac{Pr(\text{tuberculosis})}{Pr(\text{no tuberculosis})}.$$

---

[6] See Eells (1985) and Earman (1992) for discussion of the closely related problem of old evidence. The problem described above is located in what Earman calls "the problem of new theories."

Notice that the quantity $Pr(+ \text{ result})$, the unconditional probability of the observations, which is present in both (4) and (5), now has disappeared. Proposition (6) says that the ratio of posterior probabilities equals the ratio of likelihoods times the ratio of priors.

Before you observe the test result, you have your two prior probabilities; these must sum to one, but their ratio may of course be greater than unity, or less. Will your observation of the positive test result lead you to change your degrees of belief? They cannot if the two likelihoods are the same. If

$$Pr(+ \text{ result} \mid \text{tuberculosis}) = Pr(+ \text{ result} \mid \text{no tuberculosis}),$$

the ratio of the posterior probabilities will be the same as the ratio of priors. In this case, the observation is uninformative. In fact, you needn't even bother to check how the test came out. On the other hand, if

$$Pr(+ \text{ result} \mid \text{tuberculosis}) > Pr(+ \text{ result} \mid \text{no tuberculosis}),$$

your observation makes a difference. A positive test result will increase your confidence that S has tuberculosis (and reduce your confidence that he does not). In this case, the observation has the effect of making the ratio of posterior probabilities larger than the ratio of priors. The likelihood ratio, the first product term on the right-hand side of (6), is *the* pathway by which the test result can lead you to revise your degree of belief in whether S has tuberculosis. For Bayesianism, there is no other.

Another way to see this point is to delve more deeply into the instance of Bayes' theorem given in (4). What does "the unconditional probability of the observation" mean? A positive test result can occur when S has tuberculosis, but it also can occur when S does not (in which case the test result is mistaken). Both these possibilities are represented in the unconditional probability of the observations:

(7) $Pr(+ \text{ result}) = Pr(+ \text{ result} \mid \text{tuberculosis}) Pr(\text{tuberculosis})$

$\qquad\qquad + Pr(+ \text{ result} \mid \text{no tuberculosis}) Pr(\text{no tuberculosis}).$

The unconditional probability of the observation is the *average* probability that the observation has under the two alternative hypotheses, where the average is taken by using weighting terms supplied by the prior probabilities; in other words, $Pr(+ \text{ result})$ is a weighted average of the two likelihoods. If we use (7) to rewrite (4), we obtain:

(8) $Pr(\text{tuberculosis} \mid + \text{ result})$

$$= \frac{Pr(+ \text{ result} \mid \text{tuberculosis}) Pr(\text{tuberculosis})}{Pr(+ \text{ result} \mid \text{tuberculosis}) Pr(\text{tuberculosis}) + Pr(+ \text{ result} \mid \text{no tuberculosis}) Pr(\text{no tuberculosis})}.$$

If $Pr(+ \text{ result} \mid \text{tuberculosis}) = Pr(+ \text{ result} \mid \text{no tuberculosis})$, the denominator in (8) is equal to $Pr(+ \text{ result} \mid \text{tuberculosis})$, in which case (8) simplifies to

$$Pr(\text{tuberculosis} \mid + \text{ result}) = Pr(\text{tuberculosis}).$$

Without a difference in likelihoods, the posterior probability must have the same value as the prior; the observation has not affected your degree of belief.

Confirmation

As mentioned earlier, Bayesianism is more than Bayes' theorem. The philosophy goes beyond the mathematics because the philosophy proposes definitions of key epistemological concepts. For example, Bayesianism defines confirmation as probability-raising and disconfirmation as probability-lowering:

(Qual) O confirms H if and only if $Pr(H \mid O) > Pr(H)$.

   O disconfirms H if and only if $Pr(H \mid O) < Pr(H)$.

   O is confirmationally irrelevant to H if and only if
   $$Pr(H \mid O) = Pr(H).$$

Confirmation does not mean *proving true* and disconfirmation does not mean *proving false*; confirmation and disconfirmation mean only that an observation should increase or reduce your confidence that H is true. Thus, the observation that O is true can confirm H even though $Pr(H \mid O)$ is still low; the posterior probability just has to be higher than the prior. And O can disconfirm H even though $Pr(H \mid O)$ is still high; O just has to lower H's probability. Bayesian confirmation and disconfirmation involve *comparisons* of probabilities; they say nothing about the *absolute values* of any probability. Bayes' theorem allows an equivalent definition of Bayesian confirmation to be extracted from the one given above:

O confirms H if and only if $Pr(O \mid H) > Pr(O \mid notH)$.

To see whether O confirms H, don't ask whether H, if true, would lead you to expect that O is true. Rather, ask whether H makes O more probable than *notH* does.

The definitions stated in (Qual) characterize a *qualitative* concept of confirmation. They do not provide a measure of *degree* of confirmation; (Qual) doesn't say *how much* O confirms H. How might a *quantitative* concept be defined? Here are some candidates to consider, where $DoC(H,O)$ represents the degree to which O confirms H:

(Diff) $$DoC(H,O) = Pr(H \mid O) - Pr(H).$$

(Ratio) $$DoC(H,O) = \frac{Pr(H \mid O)}{Pr(H)}.$$

(L-Ratio) $$DoC(H,O) = \frac{Pr(O \mid H)}{Pr(O \mid notH)}.$$

All three of these definitions agree that (Qual) is true. However, they are not *ordinally equivalent*; they can disagree as to whether O_1 confirms H_1 more than O_2 confirms H_2. For example, suppose that

$$Pr(H_1 \mid O_1) = 0.9 \qquad Pr(H_1) = 0.5$$
$$Pr(H_2 \mid O_2) = 0.09 \qquad Pr(H_2) = 0.02.$$

According to (Diff), the difference measure, O_1 confirms H_1 more than O_2 confirms H_2, since $0.4 > 0.07$. But, according to the ratio measure, the reverse is true, since $\frac{9}{5} < \frac{9}{2}$. The fact that these and other measures sometimes disagree has given rise to a lively debate among Bayesians as to which measure is best (Fitelson 1999). Bayesians who despair of resolving this question try to restrict their discussion of confirmation to the qualitative definition (Qual).

Do we need to measure degree of confirmation? Perhaps the qualitative notion is enough. After all, there seems to be little reason to compare how much the fossil record confirms the Darwinian theory of evolution with how much Eddington's observation of light bending during an eclipse confirms the GTR. True, but there are other scientific contexts in which quantitative questions about confirmation matter. For example, in Chapter 4 we'll consider the hypothesis that two or more species share a common ancestor, and we'll investigate whether the *adaptive* similarities that the species share or the *neutral* similarities that they share provide stronger evidence in favor of that hypothesis. Even if

$Pr(X$ and Y have a common ancestor $\mid X$ and Y share adaptive trait $T_1) > Pr(X$ and Y have a common ancestor) and $Pr(X$ and Y have a common ancestor $\mid X$ and Y share neutral trait $T_2) > Pr(X$ and Y have a common ancestor).

there is another question that remains to be addressed. If it makes sense to ask which kind of similarity provides stronger evidence for common ancestry, (Qual) is not enough.

Reliability

What does it mean to say that a tuberculosis test is "reliable"? Does it mean that what the test says has a high probability of being true? That is, does it mean that

(9) Pr(tuberculosis | + result) and Pr(no tuberculosis | − result)

are both large?

Or does it mean that when the person taking the test has tuberculosis (or not), the procedure can be relied upon to say what is true? That is, does it mean that

(10) Pr(+ result | tuberculosis) and Pr(− result | no tuberculosis)

are both large?

As emphasized earlier, it is important not to confuse $Pr(O \mid H)$ and $Pr(H \mid O)$. Recall the example about the gremlins. But what does the word "reliability" mean?

Here's how I think the term is used in ordinary English: When a witness is reliable, what he or she says is probably true. Witnesses who are apt to pick up on what is true might be said to be *sensitive*; if the proposition is true, they will probably notice that it is and tell you. In my view, ordinary usage pairs "reliable" with (9) and "sensitive" with (10). But whether or not this is how the terms are used in everyday discourse, *aficionados* of probability have come to use the term "reliability" to indicate that (10) is true, not that (9) is.[7] A reliable tuberculosis test procedure has a large likelihood ratio for each possible test outcome:

$$(R) \quad \frac{Pr(+ \text{ result} \mid \text{tuberculosis})}{Pr(+ \text{ result} \mid \text{no tuberculosis})} \gg 1.0 \quad \frac{Pr(- \text{ result} \mid \text{no tuberculosis})}{Pr(- \text{ result} \mid \text{tuberculosis})} \gg 1.0.$$

---

[7] Actually, the terminology is more varied. For example, a "reliable" method for ranking options given a set of data is sometimes defined as one that usually returns the same ranking across different data sets; a method that ignores the data and always imposes the same ranking would be perfectly "reliable" in this sense.

Given this meaning, your patient S can obtain a positive test result on the reliable tuberculosis test you gave him and still it is highly improbable that he has tuberculosis. This will be true if the prior probability of S's having tuberculosis is sufficiently low (imagine that S is drawn at random from a population in which tuberculosis is very rare and then is given the test). To verify that this can happen, have another look at the relationship of the three ratios described in proposition (6).

Why is the term "reliability" often used by probabilists with the meaning described in (R)? Is this sheer perversity on their part? In fact, there is reason to focus on (R) even though people take tuberculosis tests to find out if they (probably) have the disease. Imagine using the same test procedure in two populations. In the first, people frequently have tuberculosis; in the second, they rarely do. There is a useful sense of "reliability" in which the test procedure is equally reliable in the two populations. Yet, if people are sampled at random in the two populations and then take the test, Pr(tuberculosis) is higher in the first population than in the second. If the test is equally reliable in the two cases, Pr(tuberculosis $|$ + test outcome) will be higher in the first case than in the second. Tuberculosis tests are in this respect like a great many detectors and measurement procedures. Whether the test returns a positive or a negative verdict is determined just by facts specific to the person or thing taking the test; thermometers are related to ambient temperature in the same way, and pregnancy tests are related to pregnancy in that way as well. Whether the person has a common or a rare condition is irrelevant to what the test will say. To put the point abstractly, *likelihoods are often independent of priors*. But posterior probabilities depend on both likelihoods *and* priors. This feature that a test procedure has, which is stable across different applications in different populations, is worth noting; this is why the ratios described in (R) are important.

In saying that the posterior probability of tuberculosis "depends" on priors and likelihoods, but that the likelihoods are "independent" of priors and posteriors, I am describing the *physical* characteristics of test procedures, not the *mathematical* relationships characterized by Bayes' theorem. In Bayes' theorem, each of the quantities mentioned is a mathematical function of the other three; given any three values, you can calculate the fourth. However, this symmetry with respect to mathematical dependence is not present when we consider physical relationships. Whether a tuberculosis test is apt to yield a positive result depends on whether the person taking the test has tuberculosis, not on whether tuberculosis is common or rare.[8]

Expectation and expected value

It is often said that a baby born in the USA today can expect to live about seventy-eight years. What does this mean? The reality is that a baby not only might have a longer life than this, or a shorter one. Each possible lifespan has its own probability; p_1 is the probability of living exactly one year, p_2 is the probability of living exactly two, and so on. The figure of seventy-eight years is the mathematical expectation, a technical term:[9]

$$E(S\text{'s longevity} \mid S \text{ is born in the USA in 2008})$$
$$= 1(p_1) + 2(p_2) + \cdots + n(p_n) = \sum i(p_i) = 78 \text{ years.}$$

$E(x \mid y)$ represents the expected value of x given y; notice that x is a quantity and y is a proposition. Probabilities must fall between 0 and 1, but expected values need not. The expected value is an average; in fact, it is a *weighted* average, because the different possible longevities have different probabilities.

If seventy-eight years is the life expectancy, does that mean that you should expect a US newborn to live about seventy-eight years? That depends on how different possible longevities are distributed around this mean value. Figure 1.2 shows three hypothetical distributions. Each is symmetrical and is centered on seventy-eight years, so 78 is the average value according to each. It wouldn't make much sense to expect a baby to live about seventy-eight years if (a) were true. According to (a), a baby will probably live only a very short life or a very long one; it will be exceedingly rare for a baby to live about seventy-eight years. In (b), all lifespans from 0 to 156 years are equally probable, so here again it would not make sense to use the expected value as the value you should expect. In (c), not only is 78 the expected value, but it is highly probable that a US newborn will live about seventy-eight years. There is less variation around the mean value in (c) than there is in (a) and (b). In (c), it is sensible to use the expected value as the approximate value you'd expect.

---

[8] In §4.5, we'll examine a kind of evolutionary process, one that involves frequency dependent selection, in which priors and likelihoods do not exhibit this type of independence.
[9] To keep the example simple, I assume that lifespans come in whole numbers of years. This permits the expected value to be expressed as a summation over discrete quantities. If we take time to be a continuous quantity, the expectation will be an integral.

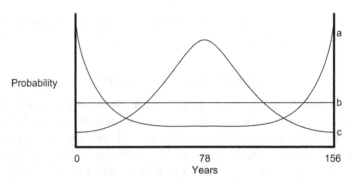

Figure 1.2 Three possible distributions of longevities. Each has the same expected
value, seventy-eight years.

Induction

One of the important contributions that Bayesianism has made to
understanding scientific reasoning is that it has thrown light on the
traditional idea of learning by induction. Induction, as I use the term,
means making an inference about a population based on a sample drawn
from it. The inference may concern what *the next object* sampled will
probably be like, or what *all the objects* in the population are probably
like. There is a lot more to scientific reasoning than inductive sampling,
but it is enlightening to see what induction looks like through Bayesian
lenses.

Here is a seemingly plausible principle of inductive reasoning that
Reichenbach (1938) called *the straight rule*:

If you toss a coin *n* times and *h* of those tosses come up heads, infer that *Pr*(the
coin lands heads | the coin is tossed) = *h*/*n*.

This rule is not the only one to consider. For example, Laplace (1820)
described a *rule of succession*:

If you toss a coin *n* times and *h* of those tosses come up heads, infer that *Pr*(the
coin lands heads | the coin is tossed) = $(h + 1)/(n + 2)$.

The two rules disagree (though they disagree less the more you toss).
Which is the right one to use? Reichenbach's rule looks simple and it
seems to "go by the evidence," while Laplace's seems to introduce a funny
correction to what the evidence is saying. Is this a good reason to prefer
Reichenbach to Laplace? Bayesianism provides a framework for answering

this question. But, more importantly, Bayesianism exposes a deficiency present in both rules; there is a kind of assumption that neither rule makes explicit but that needs to be in place if any such rule is to make sense. Notice that both rules draw a conclusion about the value of a posterior probability, based on the evidence at hand, but neither rule states values for any prior probability. Bayesianism asserts that this is *magical thinking*. The observations alone cannot give you a posterior probability; you need to have a prior probability as well. A central thesis of Bayesianism is: *no probabilities out without some probabilities in*.

Laplace was well aware of this point. He identified an assignment of prior probabilities that allowed him to *prove* that the rule of succession is correct. Let p be the probability of heads on each toss. We assume that tosses are independent of each other; results on earlier tosses don't affect the probability of heads on later ones. Laplace's assumptions about prior probabilities include the postulate that p has the same chance of falling between 0.1 and 0.2 as it has of falling between 0.8 and 0.9 and that its chance of falling between 0.3 and 0.6 is the same as its chance of falling between 0.4 and 0.7. Perhaps it sounds strange to assign a probability to a probability; if so, think of p as a physical property of the coin, perhaps one that concerns how symmetrical it is. In any event, to fully describe how Laplace conceived of the prior probabilities associated with p, we need to think about the fact that there are infinitely many values that p might have. This means that Laplace can't express his postulate about prior probabilities by saying that all point values of p have the same probability. If they all have a probability of zero, they sum to zero; and if they all have a positive value, they sum to infinity. What is required is that they sum to unity. The solution is to shift from talk of *probability* to talk of *probability density*, an idea depicted in Figure 1.3. Densities take values from zero to infinity. The prior density represented in the figure always has a value of 1, so the area under this density curve has a value of unity. Probabilities are areas under density curves. Laplace's assumption was that the prior density curve is flat. Each point value for p has a probability of zero and a probability density of 1.[10]

According to this prior density curve, the expected value of p is $\frac{1}{2}$. Notice that the curve is symmetrical around $p = \frac{1}{2}$. Imagine a factory that manufactures coins according to this prior density function. A tenth of

---

[10] Laplace thought that this assumption is justified by the principle of indifference, which we'll examine in the next section. Here we'll simply examine the assumption's consequences.

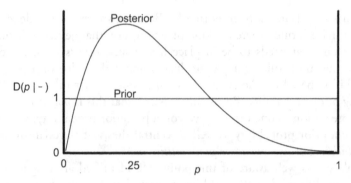

Figure 1.3 A flat prior density distribution for p and the non-flat posterior density occasioned by observing one head in four tosses. The prior expected value of p is 0.5; given this prior, the posterior expected value of p is 0.33.

the coins it produces have $0 < p < 0.1$, a tenth have $0.1 < p < 0.2$, and so on. So the average coin produced from this factory has a value of $p = \frac{1}{2}$. If you draw a coin at random from this prior distribution, and if you allow yourself to think of the expected value of p as the value you should expect p to have (thus setting aside the previous section's warning about how expected values should be interpreted), you can say that Laplace's assumption about priors entails that you should expect the coin to be fair before you have tossed it even once. This vindicates what the rule of succession says when $h = n = 0$; in this case, $\frac{(h+1)}{(n+2)} = \frac{1}{2}$. The next step is to understand what happens when you start tossing the coin. Does Laplace's rule give correct values for the expected value of p, conditional on the observations you have made? Surprisingly, the answer is *yes*.

We already know from the gremlins example that the hypothesis with the highest likelihood need not be the one with the highest posterior probability. The reason is that the prior probability is an "anchor"; the observations can lead the posterior probabilities to depart from the priors, but the priors still influence what values those posterior probabilities will have. If you obtain one head in four tosses, you have some evidence that the expected value of p is lower than $\frac{1}{2}$. But this does not permit you to ignore the prior expected value. This is why the posterior expectation moves away from the prior value of $\frac{1}{2}$ in the direction of $\frac{h}{n} = \frac{1}{4}$ and ends up somewhere in between, with a posterior expectation of $\frac{1}{3}$. How much of a shift the rule of succession tells you to make depends not just on the frequency of heads in the observations, but on the absolute number of tosses. Observing one head in four tosses occasions a smaller shift away from $\frac{1}{2}$ than observing 100 heads in 400 tosses. The posterior expectation in the former case, as just noted, is $\frac{1}{3}$, while that in the latter case is $\frac{101}{402}$.

Laplace's rule is correct if you start with a flat prior density and you think that the proper target of this inductive rule is to infer the expected value of p. Where does that leave Reichenbach? Perhaps there is another assignment of prior probabilities that justifies the straight rule. Let's investigate this question by initially changing the subject. Instead of thinking about the *probabilities* of hypotheses, let's think about their *likelihoods*. Suppose we observe five heads in twenty tosses of the coin. What value of $p = Pr$(the coin lands heads | the coin is tossed) will maximize the probability of the observations, again assuming that tosses are independent of each other? The maximum likelihood estimate of this parameter is $p = \frac{5}{20} = 0.25$. The likelihood of this hypothesis is depicted in Figure 1.4, relative to the observations we actually made (five heads in twenty tosses) and also with respect to other observations that could have occurred but did not. The figure also represents the likelihood of the hypothesis that $p = \frac{3}{4}$ relative to different possible data sets. Note that the hypothesis $p = \frac{1}{4}$ says that the actual observations were more probable than the hypothesis $p = \frac{3}{4}$ says they were. In fact, the $p = \frac{1}{4}$ hypothesis makes the data more probable than *any* assignment of a point value to p does; it provides the estimate of *maximum* likelihood. The maximum likelihood estimate of p is just the sample frequency; it doesn't matter

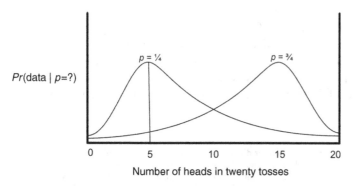

Figure 1.4 When the coin lands heads in five of twenty tosses, the maximum likelihood estimate of $p = Pr$(the coin lands heads | the coin is tossed) is $p = \frac{1}{4}$. The likelihood of the estimate $p = \frac{3}{4}$ is lower.

whether you observe one head in four tosses, or five in twenty, or 100 in 400 – the maximum likelihood estimate is the same.

The fact that the hypothesis $p = \frac{1}{4}$ has a higher *likelihood* than the hypothesis $p = \frac{3}{4}$ does not say anything about their *probabilities*. If those hypotheses are to have posterior probabilities, they must have priors. So what priors should we assign? More specifically, is there a prior density distribution of values for p that allows Reichenbach's rule to always generate the right value for the posterior expected value of p? Surprisingly, the answer is *no*. Notice that the straight rule pays no attention to the prior values; it simply goes by the maximum likelihood estimate. There is no prior distribution that legitimizes this policy.[11] The rule of succession is typical in this regard; it moves the estimate from the prior expected value of $\frac{1}{2}$ towards the maximum likelihood estimate of h/n, but does not go all the way there. The only case in which the rule of succession yields a value that is identical with the maximum likelihood estimate is when $h/n = 0.5$; in this case $(h + 1)/(n + 2)$ also equals 0.5. The general point is that *every* prior distribution will have a prior expected value, and this will always exert some influence on what the posterior expected value is. The straight rule cannot be given a Bayesian foundation.[12]

Trouble in Paradise

If all scientific inferences resembled the problem you face when your patient's tuberculosis test has a positive result, Bayesianism would be a thoroughly adequate philosophy of scientific inference. Before describing the fly in the ointment (in fact, there are two), let us examine some features of this example.

In the example of tuberculosis diagnosis, the two hypotheses are exclusive and exhaustive.[13] This is why $Pr(S$ has tuberculosis$) + Pr(S$ does not have tuberculosis$) = 1.0$. What is more, when you assign values to these prior probabilities, you are not merely reporting your subjective degree of certainty. You can point to frequency data concerning how

---

[11] Or, more precisely, no prior distribution that obeys the axioms of probability permits this. A flat *improper* prior (which goes outside the unit interval) can do so.

[12] Not that Reichenbach thought that the straight rule requires a Bayesian justification. Rather, he was impressed with the fact that the straight rule converges on the true value of p as the data set is made large without limit. This property, which statisticians call *statistical consistency*, will be discussed in §1.7 and §4.8.

[13] I assume here that your patient, S, exists and that this is not up for test.

often people have tuberculosis in the population to which S belongs. Of course, S belongs to many populations; for example, suppose that S lives in the USA, lives in Wisconsin, and lives in Madison, and that the frequencies of tuberculosis in these three populations differ. Philosophers often recommend considering the narrowest population on which you have frequency data, but I don't think that that is the only consideration. It matters whether you can regard S as being drawn at random from this or that population; if you can, the frequency data for that population provide a defensible prior. Although there are interesting issues here as to what the best assignment of value to the prior probability is, the point I want to emphasize is that frequency data are relevant and available.

The same virtue attaches to the values assigned to the likelihoods $Pr(+$ result $|$ tuberculosis$)$ and $Pr(+$ result $|$ no tuberculosis$)$. These are not numbers pulled from thin air, nor are they mere introspective reports about your attitudes. Rather, they too can be justified by pointing to frequency data. It is a familiar fact that scientific instruments, including the devices employed in medical diagnosis, are used to test hypotheses. The point of relevance here is that those devices are themselves tested. You can see how well a tuberculosis test performs by giving the test to a large number of people whom you know have tuberculosis and also to a large number whom you know do not. Frequencies within large samples provide a substantial justification for one assignment of values to the likelihoods rather than another.

In saying this, I am not denying the main lesson of the previous section. Frequency data do not by themselves *deductively entail* an assignment of value to a posterior probability. The fact that $p = h/n$ is the maximum likelihood estimate for a coin's probability of landing heads does not entail that this is the most probable value; still less does it entail that this is the true value. It is useful to think of the probability one is trying to estimate as a theoretical quantity; the evidence one uses to make this estimate is an observed frequency. The observations do not deductively entail the theory. However, with large samples, almost any prior probability will produce the same, or nearly the same, assignment of posterior probabilities. This is what Bayesians mean when they refer to the *swamping of priors*. Two agents can begin with different prior probabilities, but if they both update by using a sufficiently large data set, their posterior probabilities will be very close; the difference in priors has *washed out*. In this case, you will not go far wrong by ignoring whatever prior probabilities you start with and just using Reichenbach's straight rule. The rule is invalid, as noted, but the values it delivers will usually be sensible for large random samples.

It is important to recognize how important it is for prior probabilities to be grounded in evidence. We often calculate probabilities to resolve our own uncertainty or to persuade others with whom we disagree. It is no good assigning prior probabilities simply by asking that they reflect how certain we feel that this or that proposition is true. Rather, we need to be able to cite reasons for our degrees of belief. Frequency data are not the only source of such reasons, but they are one very important source. The other source is an empirically well-grounded theory. When a geneticist says that Pr(the offspring has genotype Aa | mom and dad both have the genotype Aa) $= \frac{1}{2}$, this is not just an autobiographical comment. Rather, it is a consequence of Mendelism, and the probability assignment has whatever authority the Mendelian theory has. That authority comes from empirical data.

I don't want to overstate my praise for the objectivity of the quantities that figure in the Bayesian answer to the question of whether your patient has tuberculosis. Skeptical questions can always be pursued back to a point where you do not know how to answer, or you "answer" by stamping your foot and insisting on the legitimacy of assumptions that cannot be further justified. This is true for any claim about knowledge or justification; the present context is no exception. But to insist that the Bayesian solution to the diagnostic problem is "purely subjective" is to mistake the part for the whole. The objective component is substantial and compelling.

There is a world of difference between this quotidian case of medical diagnosis and the use of Bayes' theorem in testing a deep and general scientific theory, such as Darwin's theory of evolution or Einstein's general theory of relativity. The difference may be, at the end of the day, a matter of degree, but still the difference is profound. When we assign prior probabilities to these theories, what evidence can we appeal to in justification? We have no frequency data as we do with respect to the question of whether S has tuberculosis. If God chose which theories to make true by drawing balls from an urn (each ball having a different theory written on it), the composition of the urn would provide an objective basis for assigning prior probabilities, if only we knew how the urn was composed. But we do not, and, in any event, no one thinks that these theories are made true or false by a process of this kind. As I mentioned, frequency data are not the only convincing justification that an assignment of prior probabilities can have. An empirical theory, like Mendelism, that is itself justified by observations can provide such probabilities. But this possibility does not bear fruit in the case of Darwin's theory or Einstein's; we have no empirically well-grounded theory of the processes by which theories like Darwin's or Einstein's are made true. In fact, maybe there is no such theory; perhaps Darwin's and Einstein's theories simply are true (or not), with no chance process leading to the one outcome or the other.

Although frequency data and a well-supported empirical theory can provide a basis for assigning prior probabilities, the principle of indifference cannot. This idea used to be a cornerstone of Bayesianism, but it is rare for contemporary Bayesians to have anything good to say about it. The principle says that if you are completely ignorant about which of a set of exclusive and exhaustive propositions is true, that you should assign them equal probabilities that sum to one. The problem with this principle is that there are multiple ways to slice the logical space into parts, which means that the same proposition can receive different prior probabilities depending on how the cake is sliced. It once was hoped that logic and language would somehow ground the principle of indifference, but this no longer seems even remotely plausible; logic and language do not furnish prior probabilities, at least not if prior probabilities are to have some authority in arguments in which people disagree. So do not fall into the trap of reasoning thus:

Either God exists or he does not.

$\overline{\text{Therefore, } Pr(\text{God exists}) = Pr(\text{God does not exist}) = \frac{1}{2}}$.

This is a trap because the pie can also be divided in three:

Either God exists and Christianity is true, God exists and Christianity is false, or there is no God.

$\overline{\text{Therefore, } Pr(\text{God exists and Christianity is true})} = Pr(\text{God exists}$ and Christianity is false) $= Pr(\text{God does not exist}) = \frac{1}{3}$.

If the principle of indifference licenses the first inference, why does it not license the second? And if it licenses both, it has lapsed into contradiction.

Laplace appealed to the principle of indifference to justify the prior density distribution he used to derive the rule of succession, so the dilemma of embracing either arbitrariness or contradiction arises in this context as well. Bertrand's paradox provides a nice illustration of how the principle of indifference goes wrong in the continuous case. Suppose I tell you that a cube manufactured by a certain factory has edges that are between 1 and 2 inches in length. If this is all you know about the cube, you might conclude that all possible lengths between 1 and 2 have the same prior density (= 1). This implies that

Pr(the length of an edge is between 1 and 1.5 inches)

$$= Pr\text{(the length of an edge is between 1.5 and 2 inches)} = \frac{1}{2}.$$

However, the information I gave you also allows you to see that each side of the cube has an area that is somewhere between 1 and 4 square inches, and this might suggest that all possible areas between 1 and 4 have the same prior densities (= 1). This entails that

Pr(the area of a side is between 1 and 2.5 square inches)

$$= Pr\text{(the area of a side is between 2.5 and 4 square inches)} = \frac{1}{2}.$$

The problem is that assigning equal priors to the lengths an edge might have contradicts assigning equal priors to the areas a side might have.

The questions just explored concerning the assignment of values to prior probabilities also attach to likelihoods, or rather they attach to some of them. In the case of S and whether he has tuberculosis, assignments of values to Pr(+ test result | S has tuberculosis) and to Pr(+ test result | S does not have tuberculosis) can be justified. The problem is that only half of this is true in many other testing situations. For example, when Arthur Stanley Eddington tested the general theory of relativity (GTR) by examining how much bend there was in starlight during a solar eclipse, he was able to ascertain a value for Pr(observation | GTR). But what value could he assign to Pr(observation | $notGTR$)? The negation of the GTR is what philosophers call a *catchall hypothesis*. There are many specific theories (T_1, T_2, ..., T_n) that are incompatible with the GTR. The likelihood of *notGTR* is the average likelihood of these specific alternatives, weighted by the probability they have conditional on the GTR being false:

$$Pr(\text{observation} \mid notGTR) = \sum\nolimits_i Pr(\text{observation} \mid T_i) Pr(T_i \mid notGTR).$$

Some alternatives to the GTR have not even been formulated yet, so it is hard to see how anyone can say what their likelihoods are. And what objective meaning could there be in saying that various alternatives have this or that probability of being true if the GTR is false? If the likelihood of the catchall hypothesis *notGTR* cannot be calculated, there is no saying whether Eddington's observation confirms the GTR, since

$$Pr(GTR \,|\, \text{observation}) > Pr(GTR) \text{ if and only if}$$
$$Pr(\text{observation} \,|\, GTR) > Pr(\text{observation} \,|\, notGTR).$$

As it happens, Eddington did not test the GTR against its negation; rather, he tested it against Newtonian theory, which made a concrete prediction about how much the light in the eclipse should bend. It turned out that

$$Pr(\text{observation} \,|\, \text{GTR}) \gg Pr(\text{observation} \,|\, \text{Newtonian theory}).$$

Unlike "*S* has tuberculosis" and "*S* does not have tuberculosis," the GTR and Newtonian theory are not exhaustive. Of course, if we think of the likelihoods as merely reflecting subjective degrees of confidence, someone might assert, as an autobiographical remark, that the GTR has a higher likelihood than its negation; but someone else, with equal autobiographical sincerity, could assert the opposite. And both would be right if the probabilities involved were merely subjective. In science, we want more than this.[14]

Let me comment, finally, on *Pr*(observation), the unconditional probability of the evidence. In the case of the tuberculosis test, the unconditional probability of a positive test result can be estimated empirically. You can estimate how often people have tuberculosis and how often not; and you can estimate how often people in each group who take the test have positive test results. This allows you to estimate the value of $Pr(+ \text{ test result})$, since this quantity is defined as $Pr(+ \text{ test result} \,|\, \text{tuberculosis})Pr(\text{tuberculosis}) + Pr(+ \text{ test result} \,|\, \text{no tuberculosis})Pr(\text{no tuberculosis})$. But what of the comparable quantity in Eddington's test? What is the unconditional probability that starlight bends a certain amount during an eclipse of the type that Eddington studied? It isn't true that the prior probabilities on *GTR* and *notGTR* are reflected in the fact that a given proportion of the physical systems that populate our universe

---

[14] Earman (1992: 117) uses the Eddington example to illustrate the problem of assigning likelihoods to catchalls.

are relativistic while the rest are not. We can't estimate Pr(observation) by seeing how often starlight bends during eclipses. This reveals, incidentally, why it can be misleading to say that Pr(observation) describes how "unsurprising" the observations are. Even if it is true that starlight *always* bends the same amount during eclipses of the type that Eddington observed, this does not mean that Pr(observations) ≈ 1. The relevant question is what the average probability is of this observation *under each hypothesis considered*, where the average is taken by using the *prior probabilities* of the hypotheses.

Philosophical Bayesianism, Bayesian statistics, and logic

Bayesian philosophers of science assign prior probabilities to scientific theories like the GTR and do not hesitate to assign likelihoods to catchall hypotheses – for example, to the GTR's negation. They concede that there is a subjective element in these assignments, though they hasten to note that there are numerous subjective elements in frequentism as well (we will examine these in due course). Bayesian philosophers think that it is a matter of intellectual honesty to acknowledge subjective elements when they intrude. They are inevitable. What could justify pretending that they are not there?

Bayesian statisticians in their professional work rarely assign prior probabilities to "big" theories like the GTR and they rarely assign likelihoods to catchalls like *notGTR*. But both these practices are standard in connection with hypotheses that are more modest. For example, when Bayesians consider the genealogical relationships that humans, chimps, and gorillas might bear to each other (§4.8), they often assign equal priors to the three competing hypotheses $(HC)G$, $H(CG)$, and $(HG)C$. Given the observed similarities and differences that those three species exhibit, it is possible to compute the likelihoods of the three hypotheses and then to compute their posterior probabilities. The effect of assigning equal priors is that all the real work is done by the likelihoods; if the priors are equal, the hypothesis of greatest likelihood must also be the hypothesis that has the greatest posterior probability. Bayesians might just as well say that what interests them here is the likelihoods and make no judgment at all about priors or posteriors. A similar comment applies when Bayesian statisticians perform *sensitivity analyses*; by examining various assignments of priors, they calculate how changing the priors affects the calculated posterior probabilities. Here again, what one is learning about are the likelihoods of the hypotheses under study; given the likelihood ratio of

H_1 to H_2, changing the ratio of priors will bring with it changes in the ratio of posterior probabilities. Describing these changes is just a way of describing the likelihood ratio.

Even though Bayesian statisticians often soft-pedal their assignments of prior probabilities to hypotheses, there is a deeper commitment on the part of Bayesians that concerns how likelihoods are sometimes computed. If a coin is tossed twenty times and seven heads are obtained, it is perfectly clear what the probability of that outcome is according to the hypothesis that the coin is fair (i.e., that $p = \frac{1}{2}$). But consider the hypothesis that the coin is *not* fair: i.e., that $p \neq \frac{1}{2}$. What is the probability of seven heads in twenty tosses according to this catchall? There are many ways the coin might fail to be fair, which correspond to different values of p, and these different values of p confer different probabilities on the observations. The likelihood of the hypothesis that $p \neq \frac{1}{2}$ is an *average* over the likelihoods of all the point values that p might have if it differs from $\frac{1}{2}$. This average takes the form of the following summation:

$$Pr\left(7 \text{ heads} \mid p \neq \frac{1}{2} \ \& \ 20 \text{ tosses}\right)$$
$$= \sum_i Pr(7 \text{ heads} \mid p = i \ \& \ 20 \text{ tosses})$$
$$\times Pr\left(p = i \mid p \neq \frac{1}{2} \ \& \ 20 \text{ tosses}\right).$$

The hypothesis that $p \neq \frac{1}{2}$ is, in this respect, just like the negation of the GTR. Notice that priors on different values of p do not occur in this expression, but something rather like them does. As we will see, frequentists also consider hypotheses like $p \neq \frac{1}{2}$, but they do not compute the *average* likelihoods of those hypotheses. The handling of such hypotheses (which statisticians call "composite") is a fundamental divide that separates Bayesians from frequentists.

For Bayesian philosophers, rationality does not require you to deny the subjective elements that inevitably intrude in inference; rather, the point is to regulate that subjectivity in the right way. For them, being rational has to do with how you *change* what you believe as new evidence arrives; your starting point is not something that Bayesian philosophers feel they need to address. Bayesian philosophers often see Bayesianism as analogous to deductive logic in this respect (Howson 2001). Deductive logic does not tell you what you should take your premises to be; logic is solely in the business of giving advice on what follows from them. So, the fact that priors and likelihoods are sometimes subjective is just a fact of life with which we all have to deal. Subjective Bayesians see themselves as facing these facts squarely in the face; they think their critics are ostriches burying their heads in the sand.

If Bayesianism is simply the logic that each of us should use to regulate our degrees of belief, the criticisms I have described of that philosophy do not apply. But an epistemology should do more than this. We need to identify which of our probability assignments can be justified interpersonally. And we also need to see if there are objective considerations that Bayesians ignore. The first of these tasks leads to likelihoodism; the second will lead us to consider frequentist ideas.

1.3 LIKELIHOODISM

Strength in modesty

The problems with Bayesianism just described suggest a fallback position that preserves much of what Bayesianism has to offer while abandoning the elements of the philosophy that are too subjective. This is likelihoodism. When prior probabilities can be defended empirically, and the values assigned to a hypothesis' likelihood and to the likelihood of its negation are also empirically defensible, you should be a Bayesian.[15] When priors and likelihoods do not have this feature, you should change the subject. In terms of Royall's three questions (§1.1), you should shift from question (2), which concerns what your degree of belief should be, to question (1), which asks what the evidence says. The likelihoodist does not answer this question by using the Bayesian concept of confirmation; you don't ask if the evidence raises, lowers, or leaves unchanged the hypothesis' probability. Rather, you compare only those hypotheses to each other that have determinate likelihoods. For example, instead of trying to compare the GTR to its own negation, you do what Eddington did: You compare the GTR with a specific alternative theory, Newtonian theory, and you use the law of likelihood (so named by Hacking 1965) to interpret the data:

Law of likelihood: The observations O favor hypothesis H_1 over hypothesis H_2 if and only if $Pr(O|H_1) > Pr(O|H_2)$. And the degree to which O favors H_1 over H_2 is given by the likelihood ratio $Pr(O|H_1)/Pr(O|H_2)$.

---

[15] Sometimes we can say what the value is of $Pr(O|H)$ without needing empirical information. For example, we know a priori (if we know anything a priori) that Pr(the next ball drawn will be green | 20 percent of the balls in the urn are green and the draw will be random) = 0.2.

The concept of *favoring* used in the law of likelihood involves a three-place relation that connects two hypotheses and a body of evidence. One also might call it the relation of *differential support*, although this terminology is apt to mislead; it may encourage the impression that the law of likelihood says that O supports H_1 to one degree, that O supports H_2 to another, and that the question is whether the first is greater than the second. This is not what the law means. According to likelihoodism, there is no such thing as the degree to which O supports a single hypothesis. Support is essentially *contrastive*.

The law of likelihood contains two ideas: a *qualitative assessment* of the bearing of the observations on the two hypotheses (expressed by an inequality) and a *quantitative measure* of how strongly or weakly the observations favor one hypothesis over the other (expressed by the likelihood ratio). The quantitative component goes beyond what the qualitative component says, just as the choice of a measure of degree of confirmation goes beyond the Bayesian definition of qualitative confirmation. And a similar question applies: even assuming that the qualitative law of likelihood is true, why should you use the likelihood *ratio* as your measure? The likelihoodist wants a measure of favoring that does not require any assignment of values to prior or posterior probabilities, or any assignment of values to the likelihoods of catchalls (if those values can't be defended by evidence), so that precludes using the possible definitions of degree of confirmation mentioned in §1.2. But why not define favoring in terms of the likelihood *difference*, $Pr(O \mid H_1) - Pr(O \mid H_2)$? One reason is suggested by a pattern that arises when there are multiple pieces of evidence that are independent of each other, conditional on each of the two hypotheses considered. Suppose, for example, that

$Pr(O_i \mid H_1) = 0.99$, for each of the 1,000 observations $O_1, \ldots, O_{1,000}$.

$Pr(O_i \mid H_2) = 0.3$, for each of the 1,000 observations $O_1, \ldots, O_{1,000}$.

With conditional independence, we have

$$Pr(O_1 \& \ldots \& O_{1,000} \mid H_1) = (0.99)^{1,000}$$
$$\text{and } Pr(O_1 \& \ldots \& O_{1,000} \mid H_2) = (0.3)^{1,000}.$$

The likelihood of each of these hypotheses, relative to the 1,000 observations, is very close to zero, so their difference is tiny; however, the ratio of the two likelihoods is $(33)^{1,000}$, which is huge. Since each of these 1,000 observations favors H_1 over H_2, the 1,000 observations should do so more powerfully than any of them does singly. This recommends the likelihood ratio over the likelihood difference as a measure of strength of evidence. Can it be shown that the likelihood ratio is the best of all possible measures? Perhaps a compelling argument for this stronger conclusion can be given, or perhaps this part of the law of likelihood should be regarded as a postulate to be judged by the intuitiveness and usefulness of its applications. In any event, there is a feature of this example that will come up later in this chapter and in subsequent chapters as well: A probabilistic hypothesis such as H_1 can do an excellent job predicting what happens in each of 1,000 experiments, in each case assigning a very high probability to the outcome that in fact takes place. Yet, the likelihood of the hypothesis goes down and down as one triumph is laid upon another. This underscores the fact that it is the *relationship* between the likelihoods of different hypotheses that matters, not the *absolute value* of any single hypothesis' likelihood.

Since likelihoodism agrees that Bayesianism makes sense in many cases, we can consider how the Bayesian concept of *confirmation* is related to the law of likelihood's qualitative notion of *favoring* when both uncontroversially apply (e.g., in the example of tuberculosis diagnosis discussed in §1.2). For O to confirm H_1, it must be true that $Pr(O|H_1) > Pr(O|notH_1)$. The observation provides Bayesian confirmation of H_1 precisely when H_1 has a higher likelihood than its negation. In contrast, the favoring relation posited by likelihoodism need not pit H_1 against its own negation; the question is whether H_1 has a higher likelihood than H_2, for some alternative hypothesis H_2 that is of interest. Here's a simple example that illustrates how O can provide Bayesian confirmation of H_1 without O's favoring H_1 over a hypothesis H_2 that is incompatible with H_1:

Example 1: Let O = the card is red, H_1 = the card is a heart, H_2 = the card is a diamond. Then $Pr(O|H_1) = 1$, $Pr(O|notH_1) = \frac{1}{3}$, and $Pr(O|H_2) = 1$.

And here's an example that exhibits the opposite pattern in which O does not provide Bayesianism confirmation of H_1 though it favors H_1 over a hypothesis H_2 that is incompatible with H_1:

Example 2: let O = the card is a 7, H_1 = the card is a heart, H_2 = the card is the Ace of Spades. Then $Pr(O|H_1) = \frac{1}{13} = Pr(O|notH_1)$, and $Pr(O|H_2) = 0$.

The likelihoodist concept of favoring describes what the evidence says about the competition between any two hypotheses that both probabilify the data at hand. The Bayesian concept of confirmation addresses a special case; it describes what the evidence says about the competition between a hypothesis and its own negation. Both questions are of interest from a Bayesian point of view. On the other hand, if Bayesianism has the problems described in §1.2, we need the concept of favoring for those problem cases, since Bayesian confirmation will not be able to do the needed work.[16]

Three objections to likelihoodism

The law of likelihood is a proposal; it is not a mathematical theorem (like Bayes' theorem). The law proposes that the informal concept of favoring (or differential support) be explicated in terms of the formal concept of likelihood comparison. To judge this proposal, we must determine how well it conforms to, and renders precise and systematic, our use of the informal concept. Our goal here, familiar from other projects of philosophical explication, is not to exactly mimic the everyday concept, which may contain various ambiguities, opacities, incoherences, indeterminacies, and even contradictions (Carnap 1947, 1950). The philosopher's job is not the same as the lexicographer's.

The previous paragraph conveys a formula that philosophers often offer that describes how the definitions they propose ought to be judged, and there is something to it. However, something more is needed with respect to the case at hand. Something important would be missed if the law of likelihood were judged solely on the basis of how it clarifies the meaning of the English word "likely." As already noted, Fisher's use of the term "likelihood" is radically at variance with ordinary usage. However, this is not an objection to Fisher's *idea*, just a comment on the infelicity of his choice of *label*. What matters about the law of likelihood is whether it isolates an epistemologically important concept. The same is true of the likelihoodist's use of terms like "favoring" and "support." A formal proposal that describes how an informal concept should be understood is to be judged by the light it throws on the informal concept, but it also should be judged by the light it throws, period.[17]

---

[16] There is more to likelihoodism than I have described here. For example, there is the likelihood principle. For discussion of what this principle means and how it is related to the law of likelihood, see Grossman (unpublished). One difference is that the law of likelihood describes the bearing of a single data set on two hypotheses while the likelihood principle says when two data sets are evidentially equivalent.

[17] A similar point was already visible in the discussion of what "reliability" means in §1.2.

The need to restrict the law of likelihood

Suppose you are Madison's top meteorologist. You gather data on the present weather configuration in the Midwest and (let us suppose) you have at hand a true theory of how weather systems change. Your job is to make a weather forecast. Based on the information you have, you infer that the probability of snow in Madison tomorrow is 0.9. It would be natural for you to express this by saying that your information *supports* the prediction that there will be snow; and it also would be natural to say that your information *favors* the hypothesis that it will snow over the hypothesis that it will not. But here the support and the favoring reflect facts about the *probabilities* of hypotheses not about their *likelihoods*. What your data and theory tell you is that

$$Pr(\text{snow tomorrow} \mid \text{present data \& theory}) = 0.9$$
$$> Pr(\text{no snow tomorrow} \mid \text{present data \& theory}) = 0.1.$$

You are not computing whether

$$Pr(\text{present data} \mid \text{snow tomorrow}) > Pr(\text{present data} \mid \text{no snow tomorrow}).$$

Your data and theory favor your weather prediction by making it probable, not by giving it a likelihood higher than that of some competing hypothesis.

An even starker example is provided by the following example. Suppose you want to predict whether the next card dealt to you will be a heart. The dealer looks at this card and, before he turns it over and places it in front of you, says, "This is the Ace of Hearts." You know that the dealer is truthful. What, then, is your epistemic situation? You're interested in ascertaining the truth value of the hypothesis $H =$ the next card is a heart. From what the dealer says, you know that proposition O is true where $O =$ the next card is the Ace of Hearts. Should you compute the likelihood of H or the probability of H? The likelihood of H is:

$$Pr(O \mid H) = \frac{1}{13}.$$

The probability of H is

$$Pr(H \mid O) = 1.0.$$

Surely you should focus on the probability. And it would not be an abuse of language to say that the dealer's comment *strongly supports* the hypothesis that the next card will be a heart; what the dealer says *favors* that hypothesis over the hypothesis, say, that the next card will be a spade.

These examples and others like them would be good objections to likelihoodism if likelihoodism were not a fallback position that applies only when Bayesianism does not.[18] The likelihoodist is happy to assign probabilities to hypotheses when the assignment of values to priors and likelihoods can be justified by appeal to empirical information. Likelihoodism emerges as a statistical philosophy distinct from Bayesianism only when this is not possible. The present examples therefore provide no objection to likelihoodism; we just need to recognize that the ordinary words "support" and "favoring" sometimes need to be understood within a Bayesian framework in which it is the probabilities of hypotheses that are under discussion; but sometimes this is not so. Eddington was not able to use his eclipse data to say how probable the GTR and Newtonian theory each are. Rather, he was able to ascertain how probable the data are, given each of these hypotheses. *That's* where likelihoodism finds its application.

How can a preposterous hypothesis be extremely likely?

The gremlin example invites the following objection to the law of likelihood: The hypothesis that there are gremlins bowling in the attic has a likelihood that is as high as a likelihood can be; it has a value of 1. So, the law of likelihood says that the gremlin hypothesis is very well supported. But this is silly. The noises we hear do not make it at all likely that there are gremlins up there bowling. This is not a well-supported hypothesis at all. Hence, the law of likelihood is false.

The complaint that the gremlin hypothesis can't be "likely" or "well supported" is easily explained by the fact that the speaker assigns the gremlin hypothesis a very low prior. Imagine that the objector has inspected thousands of attics and has never seen a gremlin and that reputable authorities have assured him that gremlins are a myth. When he arrives at your house, his prior that there are gremlins bowling in your attic is low; once he hears the noises, his probability that there are

---

[18] Fitelson (2007) uses this kind of problem to argue that the law of likelihood is false and should be modified to read as follows: O favors H_1 over H_2 if and only if $Pr(O \mid H_1) > Pr(O \mid H_2)$ and $Pr(O \mid notH_1) < Pr(O \mid notH_2)$. This principle does not follow from the Law (notice that both are biconditionals), though if the right-hand side of Fitelson's modified principle is true, so is the right-hand side of the law of likelihood. Notice also that using Fitelson's principle requires one to have likelihoods for catchall hypotheses, which likelihoodism maintains are often unavailable.

gremlins up there bowling remains low, though the Bayesian must concede that the observation increases the hypothesis' probability.[19] This is why the objector judges that the gremlin hypothesis is not "likely," by which he means that it is not very probable. Fair enough, but that is not an objection to the law of likelihood. As noted, we need to recognize that Fisher's terminology was not well chosen. The terms "likely" and "probably" are used interchangeably in ordinary English, but that is not an objection to the law of likelihood.

Although Bayesians sometimes make this objection to the law of likelihood, the fact of the matter is that Bayesianism is committed to the view that likelihoods are the one and only vehicle by which observations can change the probabilities we assign to hypotheses. This was the point I discussed in connection with proposition (6). Bayesians as well as likelihoodists need a word to use in describing the epistemological significance of the fact that $Pr(E|H) > Pr(E|notH)$. The law of likelihood uses the word "favoring," and "differential support" might be used here as well. Of course, the law of likelihood also applies this term in a wider context, namely when one is comparing H with an alternative hypothesis other than its own negation. But the point of this term is not to assess the overall plausibility of H but to describe what a particular observation says about the competition between H and some alternative hypothesis. The law of likelihood does not say that the gremlin hypothesis is rendered plausible by the noise you hear.

Edwards (1972) discusses the same sort of objection in connection with another example. You draw a card from a deck and it turns out to be the seven of spades. Now consider the hypothesis that each of the cards in the deck is a seven of spades; this hypothesis has a likelihood of 1.0. In contrast, the likelihood of the hypothesis that the deck is "normal" is only $\frac{1}{52}$. This leads the law of likelihood to conclude that the card you've observed favors the stacked hypothesis over the normal hypothesis. But surely, the objection concludes, the stacked hypothesis is not more plausible or better supported. I leave it to the reader to construct and evaluate the likelihoodist's reply.

Likelihoodism and the definition of conditional probability

Likelihoodists think they have a philosophy that comes into its own when no evidence is available to back up assignments of prior probabilities. But

---

[19] To see this, consider the following consequence of Bayes' theorem: If H entails E and $0 < Pr(E) < 1$ and $0 < Pr(H) < 1$, then $Pr(H|E) > Pr(H)$.

how can this be true, given the Kolmogorov definition of conditional probability (§1.2)? Recall that the definition says that

$$\text{(K)} \qquad Pr(O \,|\, H) = \frac{Pr(O \,\&\, H)}{Pr(H)}.$$

There, in the denominator on the right-hand side, a prior probability has popped up, just what likelihoodists say they can do without when they talk about likelihoods!

The answer to this challenge is that likelihoodists should think of the Kolmogorov definition as correct only when various unconditional probabilities are "well defined." When they are not, the concept of conditional probability can and should be taken to stand on its own; it does not need to be defined in terms of unconditional probabilities. There are good reasons for this approach that do not depend on any qualms one might have about Bayesianism. For example, consider the fact that Kolmogorov's (K) says that the conditional probability is undefined if $Pr(H) = 0$. But surely there are contexts in which a conditional probability has a value even though the conditioning proposition has a probability of zero. Suppose I make you the following promise: If the coin I am about to toss lands heads, I will buy you a ticket in a fair lottery in which 1,000 tickets are sold. If the coin fails to land heads, you will have no ticket, and so you can't win the lottery. You know that I am trustworthy, so you conclude that Pr(you win the lottery | the coin lands heads) $= \frac{1}{1,000}$. However, I then take measures to ensure that the coin *cannot* land heads. Maybe I bend the coin, or place it in a tossing device that ensures tails every time, or maybe I just lock it in a vault and thereby ensure that the coin can never be tossed. If you buy the Kolmogorov definition of conditional probability, the information that the coin can't land heads should lead you to say that the conditional probability just stated is not correct. The value is not $\frac{1}{1,000}$; rather, it is *not defined.* On the other hand, if conditional probability is a primitive concept, the conditional probability can have the value given even though the conditioning proposition has a probability of zero (Hajek 2003). This position has the additional virtue of allowing Pr(the coin lands heads | the coin lands heads) to have a value of *unity* instead of being *not defined.*

There is an epistemic point that is also worth considering. We often know the value of $Pr(O \,|\, H)$ even though we have no clue as to the value of $Pr(H)$. As mentioned in §1.2, we can estimate the value of $Pr(+$ test result | tuberculosis) by giving the test to thousands of people whom we know have tuberculosis. This procedure does not require that we know how common or rare tuberculosis is, and so we may be entirely in the dark as to the value of Pr(tuberculosis). The defender of Kolmogorov's definition is right to reply that proposition (K) is not a claim about *knowledge*; it does not say that to *know* the value of a conditional probability you first must *find out* the values of the two unconditional probabilities cited. (K) asserts a symmetric *mathematical* (or *logical*) dependence, not an asymmetric *epistemic* dependence. The right question to ask about Kolmogorov's (K) is whether there must exist unconditional probabilities for $H\&O$ and for H if there is such a thing as the conditional probability $Pr(H\,|\,O)$.

The answer depends on what we mean by probability and on the example we consider. Bayesians usually adopt the idealization that rational agents have degrees of belief for all the sentences of their language. The Bayesian framework is one in which a *complete probability function* is deployed over all the sentences in some language. If O_1, O_2, ... O_n, and H_1, H_2, ... H_m are all sentences in the language, then the probability function assigns a prior probability to each of those atomic sentences and to all Boolean combinations definable from them (e.g., to the negations of each and to all disjunctions and conjunctions constructed from this set). Posterior probabilities are definable from the relevant priors via proposition (K). This is not the best way to understand what likelihoodists are up to. According to likelihoodism, the language we speak is far more wide-ranging than the probability models we use. On a given occasion, we may specify a value for $Pr(O\,|\,H_1)$ and for $Pr(O\,|\,H_2)$, but none for $Pr(O\,|\,notH_1)$, and none for $Pr(H_1)$ or $Pr(H_2)$. We use this *partial* probability function to do the needed work. Not only don't we *know* the value of $Pr(O\,|\,notH_1)$, or of $Pr(H_1)$, or of $Pr(H_2)$; in addition, there may be no such values to know. The model we use does not include these even as unknown quantities.

What likelihoodists mean by probability is not simply that an agent has some degree of belief. For one thing, the concept of probability needs to be interpreted more normatively. $Pr(O\,|\,H)$ is the degree of belief you *ought* to have in O given that H is true. But likelihoodists also like to think of these conditional probabilities as reflecting objective matters of fact. If Pr(the card is the Ace of Hearts | the card is dealt from this deck) $= \frac{1}{52}$, this is because of the physical composition of the deck and the physical properties of the process of dealing. When likelihoodists insist that probabilities must be "objective," they mean that probabilities must be grounded in such physical details.[20] When the physical processes at

---

[20] The word "objective" used by likelihoodists does not mean what so-called objective Bayesians have meant by the term: that probabilities must be derivable from logical features of the language we speak.

work generate frequency data, these data provide evidence we can use to infer the values of the underlying probabilities.[21]

Is Kolmogorov's (K) the right way to think about conditional probability when probability is understood in the way that likelihoodists propose? If there exists a physical process that leads people with tuberculosis who are tested to have a positive test result with a certain frequency, is there also a physical process that leads some people, but not others, to have tuberculosis? Arguably so, in which case $Pr(+$ test result $|$ tuberculosis) and Pr(tuberculosis) will both figure in a useful model. But now consider Eddington. There was a physical process that led the light to bend during the eclipse; this is the process that the GTR purports to describe. But is there, in addition, a physical process whose result was that the GTR, or some competing theory, became true? Arguably not. If not, likelihoodists will not include $Pr(GTR)$ in their probability model. This is why your interpretation of probability should influence whether you regard Kolmogorov's (K) as a proper definition or just as a postulate that is true in favorable circumstances.

Kolmogorov's proposition (K), like Bayes' theorem, should be understood as having a certain rider attached. They do not assert that all the quantities they describe make sense. Rather, each of them should be understood in terms of the following preface: *in any model that uses the following quantities, here is how those quantities must be related.* When (K) is understood in this way, you can see that the following criticism is misguided: "If you assign a value to a hypothesis' likelihood, you are committed to saying that the hypothesis has a prior, whether you know its value or not."

The principle of total evidence

Bayesians and likelihoodists have their disagreements, but they agree on the principle of total evidence. This principle says that you should take account of everything you know. As stated, this idea is vague, but it gains precision when it is applied to concrete problems, as we shall see. It is a "pragmatic" principle in the philosophical sense of that term. This doesn't mean that it is something that cynics rather than idealists

---

[21] Although observed frequencies provide *evidence* concerning the values of probabilities, there are lots of contexts in which probabilities can't be *defined* in terms of (actual or hypothetical) frequencies; see Sober (1994, 2008b). For this reason, I prefer a "no-theory theory of probability," according to which probabilities are theoretical terms that cannot be defined in terms of observables.

embrace; rather, the point is that it gives advice about how probabilities should be *used* to solve problems. As there are many probability problems, the principle has many applications, and so the principle may be more plausible in some contexts than in others. I'll begin by describing a few settings in which the principle seems to make excellent sense. It will emerge in the next section that the principle of total evidence is controversial; it constitutes one of the fault lines that separate some central ideas in frequentism from both Bayesianism and likelihoodism.

Suppose two witnesses provide independent reports about what they saw at the scene of a crime. And suppose that each is at least minimally reliable in the sense described in §1.2, meaning that, for some relevant range of propositions:

$$Pr[W_i(P) \mid P] > Pr[W_i(P) \mid notP], \text{ for } i = 1, 2.$$

Here $W_i(P)$ means that witness i asserts that proposition P is true. The principle of total evidence says that you should take account of the testimony of *both* witnesses if that is the total evidence you possess. However, the principle is usually interpreted as saying that *more is better than less*; you should take account of both testimonies, rather than just one of them, even if there is more information available than what the two witnesses say.

Why are two witnesses better than one? If the witnesses agree that P is true, and the two witnesses go about their business independently,[22] the two pieces of testimony discriminate more powerfully between P and *notP* than either of them does by itself, in the sense that

$$\frac{Pr[W_1(P) \ \& \ W_2(P) \mid P]}{Pr[W_1(P) \ \& \ W_2(P) \mid notP]} > \frac{Pr[W_i(P) \mid P]}{Pr[W_i(P) \mid notP]} > 1, \text{ for each } i = 1, 2.$$

This is because

$$\frac{Pr[W_1(P) \ \& \ W_2(P) \mid P]}{Pr[W_1(P) \ \& \ W_2(P) \mid notP]} = \frac{Pr[W_1(P) \mid P]}{Pr[W_1(P) \mid notP]} \times \frac{Pr[W_2(P) \mid P]}{Pr[W_2(P) \mid notP]}$$

and each of the ratios on the right is greater than one. This just reflects the common sense fact that two independent and (at least minimally) reliable

---

[22] There can be (and will be!) a relation of *unconditional dependency* between what independent reliable witnesses say, in that $Pr[W_1(P) \mid W_2(P)] > Pr[W_1(P)]$. The relevant notion of independent witnesses is independence *conditional on the proposition reported*: $Pr[W_1(P) \ \& \ W_2(P) \mid P] = Pr[W_1(P) \mid P] \times Pr[W_2(P) \mid P]$.

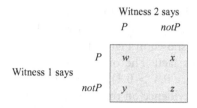

Figure 1.5 When two independent and reliable witnesses each report on whether proposition *P* is true, two yeses provide stronger evidence for *P* than one, and one yes provides stronger evidence than zero. Each cell represents the likelihood ratio $Pr(\text{testimony} \mid P)/Pr(\text{testimony} \mid notP)$ that goes with each of the four possible testimonies; $w > x, y > z$.

witnesses who agree that *P* is true provide stronger evidence in favor of *P* than either witness does alone.[23]

This example makes it look as if the principle of total evidence is justified by our hunger for strong evidence. But this can't be right. For suppose the two witnesses *disagree*. If you take both pieces of testimony into account, you may have no basis at all for discriminating between *P* and *notP*, whereas if you selectively focus on just one witness's testimony, you will. The principle of total evidence in this case tells you to resist the desire for telling evidence; if the total evidence says that you have little or no basis for discriminating between the two propositions, so be it.

When reliable witnesses reach their judgments independently of each other (conditional on *P*'s being true and conditional on *P*'s being false), this induces a kind of evidential *monotonicity*; if there are two witnesses, two votes for *P* provide stronger evidence that *P* is true than one vote would provide, and one vote provides stronger evidence for *P* than if neither witness had asserted that *P* is true. These comparisons are represented by the likelihood ratios depicted in Figure 1.5. As simple and familiar as this fact about multiple independent testimonies is, it is important to bear in mind that there is no rule written in Heaven that separate pieces of evidence must be independent. Suppose you are a cook in a restaurant. The waiter brings an order into the kitchen – someone in the dining room has ordered toast and eggs for breakfast. You wonder if this evidence discriminates between two hypotheses – that your friend Smith placed the order or that your friend Jones did so. You know the

---

[23] This point about multiple witnesses bears on Hume's analysis of the epistemology of reports about the alleged occurrence of miracles, on which see Earman's (2000) book and my review of it (Sober 2004d).

eating habits of each; the probabilities of different breakfast orders, conditional on Smith's placing the order, and conditional on Jones's placing the order, are shown in Figure 1.6. These probabilities give rise to the following curious fact: The order's being for *toast and eggs* favors Smith over Jones (since $0.4 > 0.1$); but the fact that the customer asked for *toast* provides no evidence on this question (since $0.5 = 0.5$); and the fact that the customer asked for *eggs* doesn't either (since, again, $0.5 = 0.5$). Here the whole of the evidence is more than the sum of its parts.

Figure 1.7 depicts the opposite pattern in which a new set of inclinations is attributed to your two friends. If Smith and Jones are disposed to behave as described, an order of *toast and eggs* fails to discriminate between the two hypotheses (since $0.4 = 0.4$). But the fact that the order included *toast* favors Smith over Jones (since $0.7 > 0.6$), and the same is true of the fact that the order included *eggs* (since $0.6 > 0.4$). Here the whole of the evidence is less than the sum of its parts.

Although the principle of total evidence says that you must use all the relevant evidence you have, it does not require the spilling of needless ink.

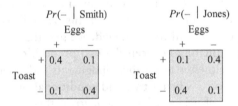

Figure 1.6 Smith and Jones differ in their inclinations to place different orders for breakfast. The breakfast order of toast and eggs provides evidence about which of them placed the order, although the fact that the order included toast does not, and neither does the fact that the order included eggs.

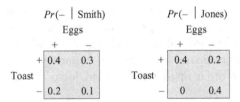

Figure 1.7 A new set of breakfast inclinations for Smith and Jones. Now the breakfast order of toast and eggs provides no evidence about which of them placed the order, though each part of the order favors Smith over Jones.

It does not require you to record irrelevant information. Consider the two hypotheses about coin tossing depicted in Figure 1.4. One of them says that $p = \frac{1}{4}$ while the other says that $p = \frac{3}{4}$, where p is the coin's probability of landing heads. I earlier described the data by saying that there were five heads in the twenty tosses of the coin. But why am I not obliged to describe the exact sequence of heads and tails that formed the data? There are many ways to get five heads in twenty tosses. A proposition that states just the sample frequency is *logically weaker* than a description of the exact sequence (in that the latter implies the former, but not conversely). Isn't it a violation of the principle of total evidence to use the sample frequency as a description of the data?

If we represent strength of evidence by the likelihood ratio, the answer is *no*. Consider each of the specific sequences in which there are five heads in twenty tosses. The two hypotheses we are considering ($p = \frac{1}{4}$ and $p = \frac{3}{4}$) agree that each of these exact sequences has a probability of $p^5(1-p)^{15}$ though they disagree about what the true value of p is. The likelihood ratio of $p = \frac{1}{4}$ to $p = \frac{3}{4}$, relative to a description of the exact sequence of heads and tails we observe, has the value:

$$\frac{Pr(\text{exact sequence} \mid p = \frac{1}{4})}{Pr(\text{exact sequence} \mid p = \frac{3}{4})} = \frac{(\frac{1}{4})^5 (\frac{3}{4})^{15}}{(\frac{3}{4})^5 (\frac{1}{4})^{15}} = 3^{10}.$$

If there are N exact sequences that can produce five heads in twenty tosses[24] the probability of obtaining *some sequence or other* in which there are five heads in twenty tosses has a value of $Np^5(1-p)^{15}$. Using this logically weaker description of the data, we obtain the following likelihood ratio:

$$\frac{Pr(5 \text{ heads} \mid p = \frac{1}{4})}{Pr(5 \text{ heads} \mid p = \frac{3}{4})} = \frac{N(\frac{1}{4})^5 (\frac{3}{4})^{15}}{N(\frac{3}{4})^5 (\frac{1}{4})^{15}} = \frac{(\frac{1}{4})^5 (\frac{3}{4})^{15}}{(\frac{3}{4})^5 (\frac{1}{4})^{15}} = 3^{10}.$$

Notice that the Ns have cancelled. There is no need to use the logically stronger description of the data that states the exact sequence of heads and tails, since it makes no difference to the likelihood ratio (Fisher 1922b; Hacking 1965: 80–1). In this sense, the sample frequency is a *sufficient statistic*. Notice the role played by the likelihood *ratio* in this argument; if you represented weight of evidence in some other way (e.g., via the

---

[24] N, the number of specific sequences in which there are m successes in n trials, is calculated by the formula for $\binom{n}{m}$, meaning *from n objects choose m*; $N = n!/m!(n-m)!$.

likelihood *difference*), maybe N would not disappear. Notice also how powerfully the data favor one hypothesis over the other, even though both say that the total data set was very improbable.

Whether the sample frequency is a sufficient statistic depends on the hypotheses being evaluated. In the example just described, the two hypotheses agree that tosses are independent of each other. But suppose this is something you want to test. And suppose further that the exact sequence of heads and tails is observed to be

$$\text{H T H T H T H T H T H T H T H T H T H T H T H T}$$

This sequence contains 50 percent heads, but it would be a mistake to think that this logically weakened description captures all the information in the data that is evidentially relevant. The *order* of heads and tails is evidentially relevant as well.

The logically weaker description of the data, the sample frequency, is a disjunction. One of the disjuncts describes the exact sequence that *did* occur; the other disjuncts describe exact sequences that *did not*. When $p = \frac{1}{4}$ and $p = \frac{3}{4}$ are the two hypotheses under test, there is nothing wrong with describing the data in this disjunctive form, saying that this sequence *or* that sequence *or* that other sequence was the one that occurred without saying which. The principle of total evidence is not a rule against disjunctions. Rather, the rule says that logically weakening your description of the data is not permitted when this changes your assessment of what the evidence indicates. Applying the principle requires a rule for interpreting what the evidence says about the hypotheses under test. At this point, likelihoodists appeal to the law of likelihood and use the likelihood ratio. Bayesians can agree with the above argument, since for them the likelihood ratio is *the* vehicle by which ratios of priors are transformed into ratios of posterior probabilities, as proposition (6) attests. Likelihoodists and Bayesians are on the same page when it comes to the principle of total evidence.[25]

The limits of likelihoodism

Likelihoodism addresses the first of Royall's three questions (§1.1) while remaining silent on the other two; it confines itself to the task of interpreting what the evidence says while giving no advice on what you should

---

[25] I will not try to address the deeper question of what the ultimate justification is of the principle of total evidence. I. J. Good (1967) provides a decision-theoretic justification.

believe or do. Even so, the question remains of whether likelihoodism accomplishes the relatively modest goal it sets for itself. The problem is that there are many scientific hypotheses of interest that are *composite*, rather than *simple*. These are technical terms. The two hypotheses about the coin (that $p = \frac{1}{4}$ and that $p = \frac{3}{4}$) depicted in Figure 1.4 are both simple in the sense that each says exactly how probable each possible outcome of the experiment is. Composite hypotheses are more ambiguous; they circumscribe a *family* of probabilities that an observation might have without singling out just one. An example would be the hypothesis that $p > \frac{1}{4}$; this hypothesis does not say what the probability is of observing exactly five heads in twenty tosses. There are many values that p might have if it exceeds $\frac{1}{4}$, and each specific value has its own likelihood relative to a given observation; composite hypotheses are disjunctions (sometimes infinite disjunctions) of simple hypotheses.

Hypotheses that look as if they are composite can in reality turn out to be statistically simple, if background information of a certain sort is available. Imagine that there are three kinds of coins that a factory manufactures – a third have $p = \frac{1}{4}$, a third have $p = \frac{1}{2}$, and a third have $p = 1.0$. If you chose a coin made at this factory at random, then if the coin before you has $p > \frac{1}{4}$, there are just two possibilities – that $p = \frac{1}{2}$ and $p = 1.0$ – and these are equiprobable. The average of these is $p = \frac{3}{4}$. Likelihoodists have no problem with assessing the hypothesis that $p > \frac{1}{4}$ in this kind of context. True to their antisubjectivist inclinations, they are happy to consider this hypothesis because there is an objective answer to the question of what observations we should expect to make if the hypothesis that $p > \frac{1}{4}$ is true. Absent this kind of information, they decline to assess the hypothesis at all. Rather, they relegate $p > \frac{1}{4}$ to the same epistemic limbo to which they consign *notGTR*, the catchall hypothesis that the GTR is false.

It is arguable that science often does not need to assess how the evidence bears on such catchall hypotheses. Eddington was able to compare the GTR with Newtonian theory, and maybe that is enough. However, other composite hypotheses seem to play a central role in the activity of science, so the likelihoodist denial that they can be handled should raise more eyebrows. For example, population geneticists often want to say whether the gene-sequence data gathered from a number of species favor the hypothesis of random genetic drift or the hypothesis of selection. The drift hypothesis is often statistically simple: For example, with respect to the two alleles A and a that might exist at a given genetic locus, the drift hypothesis says that they are identical in fitness. It says that $w_A = w_a$, which means that $w_A - w_a = 0$. In contrast, the hypothesis of selection is composite; it says that $w_A \neq w_a$; in other words, it says that $w_A - w_a = \theta$, where θ is a parameter whose value is not equal to zero. Notice that there are many different values that θ might have if it isn't equal to zero. Each of these specific values for θ entails its own probability for the data at hand. But what does the bare hypothesis of selection itself predict? As the previous example about the coin factory suggests, this question would be answerable if we had an objective basis for assigning probabilities to the different values θ might take if it were nonzero. But, alas, we often lack this type of information. For this reason, it is often impossible to compare drift with selection within the framework of likelihoodism. Although physicists may be content to compare the GTR with Newtonian theory and to feel no need to ponder the catchall hypothesis that the GTR is false, population geneticists have wanted to test drift against selection and have even claimed to have done so. We will examine the question of whether and how this is possible in Chapter 3. For now, the point is that we have isolated an issue that unites Bayesians and frequentists; these two old enemies maintain that likelihoodism is too austere. Frequentists think they have good methods for testing composite hypotheses and Bayesians deny that the hypotheses in question are really composite. Both rush in where likelihoodists fear to tread.

1.4. FREQUENTISM I: SIGNIFICANCE TESTS AND PROBABILISTIC *MODUS TOLLENS*

I began this chapter by painting with a broad brush. I said that Bayesians hold that science is in the business of determining which theories are probably true while frequentists hold that this is not at all what science is about. I then complicated the story by adding likelihoodists to the cast of characters. They often eschew the goal of assigning probabilities, but in many respects they are more like Bayesians than frequentists, as we now will see. The fact that there are three positions here, not two, complicates the problem of saying what frequentism amounts to. It is not enough to say that frequentists reject the goal of assigning probabilities to hypotheses, since that point, though correct, does not separate them from likelihoodists. What can be said that is distinctive of what frequentism is *for*? We will uncover some of its differences with the other two philosophies in due course. But we must bear in mind that frequentism is not a single unified theory. Rather, it is a motley of different techniques that are often only loosely connected with each other; sometimes they are even in conflict. In §1.2, I mentioned that Bayesianism gives *epistemological* advice about probability assignments; what probability statements *mean* (which "interpretation of probability" is correct) is a separate, semantic, question. A similar point applies to frequentism. Frequentism is not the thesis that probability statements are claims about actual or hypothetical frequencies, though this semantic thesis is something that many frequentists endorse. Rather, frequentism is a thesis about epistemology. Frequentists assess a rule of inference by examining the (expected) frequencies of good and bad outcomes when the rule is applied repeatedly.

The first frequentist method that I want to consider is R. A. Fisher's idea of *significance tests*. Fisher conceived of this procedure as a corrective to what he thought was wrong with the Neyman–Pearson theory of hypothesis testing, which I'll discuss in the next section. I take these two approaches in reverse chronological order because Fisher's theory is in some ways easier to grasp than the Neyman–Pearson approach and because its contrast with likelihoodism is more obvious.

To get started, let's consider a simple rule of deductive reasoning, *modus tollens*. This is a form of argument familiar to philosophers and scientists; it is the centerpiece of Karl Popper's views on falsifiability (which I'll discuss in §2.8):

(MT)
$$
\begin{array}{l}
\text{If } H, \text{ then } O \\
\underline{notO} \\
notH
\end{array}
$$

Modus tollens, like other rules of deductive logic, says what follows from what. It does not, in the first instance, give advice. Still, it is natural to interpret *modus tollens* as saying that if the hypothesis H entails the observation statement O, and O turns out to be false, then H should be *rejected*. I use a single line to separate premises from conclusion to indicate that *modus tollens* is deductively valid (meaning, recall, that if the premises are true, the conclusion must be). Since (MT) is valid, perhaps the following "probabilistic extension" of the rule constitutes a sensible principle of nondeductive reasoning:

(Prob-MT)
$$
\begin{array}{l}
Pr(O \mid H) \text{ is very high} \\
\underline{\underline{notO}} \\
notH
\end{array}
$$

According to *probabilistic modus tollens*, if the hypothesis H says that O will *very probably* be true, and O turns out to be false, then H should be rejected. Equivalently, the suggestion is that if H says that some observational outcome (*notO*) has a very low probability, and that outcome nonetheless occurs, then we should regard H as false. I draw a double line between premises and conclusion in (Prob-MT) to indicate that the argument form is not supposed to be deductively valid. But maybe it is a sensible form of inference nonetheless.

Before addressing whether probabilistic *modus tollens* is correct and how it is related to deductive *modus tollens*, I want to discuss a parallel question. Consider *modus ponens*:

(MP)
$$\frac{\text{If } O, \text{ then } H \\ O}{H}$$

Modus ponens is deductively valid, and this may suggest that the following probabilistic extension of the principle is also correct:

(Prob-MP)
$$\frac{Pr(H \mid O) \text{ is very high} \\ O}{H}$$

(Prob-MP) says that if O renders H very probable, and O is true, then we should accept H. My brief comments in §1.1 on the lottery paradox suggest that we should be wary of this rule of acceptance. But (Prob-MP) has a close cousin, which we have already examined:

(Update) $Pr_{\text{then}}(H \mid O)$ is very high
O

O is all the evidence we have gathered between then and now.

$$\overline{Pr_{\text{now}}(H) \text{ is very high}}$$

This is nothing other than the rule of updating by strict conditionalization. (Update) is a sensible rule, and it also has the property of being a generalization of deductive *modus ponens*. By parity of reasoning, should we conclude that probabilistic *modus tollens* is a good rule because it generalizes deductive *modus tollens*?

Friends of (Prob-MT) need to say where the probability cutoff for rejection is located. How low must $Pr(O \mid H)$ be for O to justify rejecting H? Richard Dawkins (1986: 144–6) addresses this question in the context of discussing how theories of the origin of life should be evaluated. He says that an acceptable theory can say that the origin of life on Earth was somewhat improbable, but it cannot go too far. If there are n planets in the universe that are "suitable" locales for life to originate, then an acceptable theory of the origin of life on Earth must say that that event had a probability of at least $\frac{1}{n}$. Theories that say that terrestrial life was less probable than this should be rejected. Creationists also have set cutoffs. For example, Henry Morris (1980) says that theories that assign to an event a probability less than $\frac{1}{10^{110}}$ should be rejected, and William Dembski (2004) says that a theory that assigns to a "specified event" (a technical term in Dembski's framework) a probability less than $\frac{1}{10^{150}}$ should be rejected.[26] Morris and Dembski obtain these numbers by attempting to calculate how many times elementary particles could have changed state since the universe began.

Dawkins, Dembski, and Morris have all made the same mistake. It isn't that they have glommed on to the wrong cutoff. The problem is deeper: *There is no such cutoff.* Probabilistic *modus tollens* is an incorrect form of inference (Hacking 1965; Edwards 1972; Royall 1997). Lots of perfectly reasonable hypotheses say that the observations are very improbable. As noted earlier, if H confers a very high probability on each of the observations $O_1, O_2, \ldots, O_{1,000}$ (but a probability that is short of unity), it will confer a very low probability on their conjunction, if the observations are independent of each other, conditional on H. A probability that is very large but less than one, when multiplied by itself a large number of times, will yield a very small probability. Adopting probabilistic *modus tollens* would have the effect of eliminating all probabilistic theories from science once they are repeatedly tested.

It may seem that the kernel of truth in (Prob-MT) can be rescued by modifying the argument's conclusion. If it is too much to conclude that H is false, perhaps we should conclude just that the observations constitute evidence against H:

(Evidential Prob-MT) $Pr(O \mid H)$ is very high.
 notO
 _____
 notO is evidence against H.

This principle is also unsatisfactory, as an example from Royall (1997: 67) nicely illustrates. Suppose I send my valet to bring me one of my urns.

---

[26] For discussion of Dembski's (1998) framework for inferring the existence of intelligent designers, see Fitelson et al. (1999).

I want to test the hypothesis (*H*) that the urn he returns with contains 0.2 percent white balls. I draw a ball from the urn and find that it is white. Is this evidence against *H*? It may not be. Suppose I have only two urns; one of them contains 0.2 percent white balls, while the other contains 0.01 percent white balls. In this instance, drawing a white ball is evidence *in favor* of *H*, not evidence *against* it.[27]

The use of genetic data in forensic identity tests provides a further illustration of Royall's point. Suppose that two individuals match at twenty independent loci; they are heterozygotes at each. At each locus, each individual has one copy of a rare allele (frequency $= 0.001$) and one copy of the alternative, common, allele (frequency $= 0.999$). The probability of this twenty-fold matching, if the two individuals are full sibs, is about $[(0.001)(0.5)]^{20}$. This is a very small number, but that hardly shows that the sib hypothesis should be rejected. In fact, the data *favor* the sib hypothesis over the hypothesis that the two individuals are unrelated. If they are unrelated, the probability of the observations is about $[(0.001)(0.001)]^{20}$. The two likelihoods are both very small, but the first is 500^{20} times larger than the second (Crow et al. 2000: 65–7).[28]

These examples reflect a central idea in the likelihoodist theory of evidence: judgments about evidential meaning are essentially *contrastive*. To decide whether an observation is evidence against *H*, you need to know what the alternative hypotheses are; *to test a hypothesis requires testing it against alternatives.*[29] In the story about the valet, observing a white ball is very improbable according to *H*, but in fact that outcome is evidence *in favor of H*, not evidence against it. This is because *O* is even more improbable according to the alternative hypothesis. Probabilistic *modus tollens*, in both its vanilla and evidential versions, needs to be replaced by the *law of likelihood*. The relevance of this point is not confined to urn problems and forensic DNA. It will play an important role in Chapter 4

---

[27] A third formulation of probabilistic *modus tollens* is no better than the other two. Can one conclude that *H* is *probably* false, given that *H* says that *O* is highly probable, and *O* fails to be true? The answer is *no*; inspection of Bayes' theorem shows that $Pr(notO \mid H)$ can be low without $Pr(H \mid notO)$ being low.

[28] Notice how the likelihood *ratio*, not the likelihood *difference*, figures in this argument.

[29] There are two exceptions to the thesis that testing is always contrastive. If a true observation statement entails *H*, there is no need to consider alternatives to *H*; you can conclude without further ado that *H* is true; this is just *modus ponens*. And if *H* entails *O* and *O* turns out to be false, you can conclude that *H* is false, again without needing to contemplate alternatives; this is just *modus tollens*. It is a separate question how often these forms of argument apply to testing in science. They rarely do. Observations almost never entail theories, and theories almost never entail observations. More on this later.

when we consider the question of why the similarities observed in two or more species is evidence for those species' having a common ancestor. Within the framework developed there, an observed similarity O provides *stronger* evidence in favor of the common ancestry (CA) hypothesis the *lower* the value is of $Pr(O \mid CA)$. The reason the evidence for CA is strengthened by lowering the value of this conditional probability is that lowering the value of $Pr(O \mid CA)$ leads the value of $Pr(O \mid SA)$ to plunge even more; here SA is the hypothesis of separate ancestry.

There is a reformulation of probabilistic *modus tollens* that makes sense, but it is Bayesian:

(Bayesian Prob-MT) $Pr_{\text{then}}(O \mid H)$ is very high.

$Pr_{\text{then}}(O \mid notH)$ is very low.

$Pr_{\text{then}}(H) \approx Pr(notH)$

not-O

$$\overline{Pr_{\text{now}}(H) \text{ is very low.}}$$

Although the conclusion of this argument follows *deductively* from the premises (given the rule of updating by strict conditionalization and that *notO* is all you learned between then and now), this is a form of argument that frequentists will not touch with a stick. The reason is not that it is invalid (it is not) but that it requires premises that frequentists regard as too subjective.[30]

Fisher's (1959) test of significance is a version of probabilistic *modus tollens* and that is bad enough. But it has the additional defect that it violates the principle of total evidence. In a significance test, the hypothesis you are testing is called the "null" hypothesis, and your question is whether the observations you have are sufficiently improbable according to the null hypothesis. However, you don't consider the observations in all their detail but rather the fact that they fall in a certain region. You use a logically weaker rather than a logically stronger description of the data. Here's an example (from Howson and Urbach 1993: 176) that illustrates the point. You want to test the hypothesis that a coin is fair (i.e., the hypothesis that the probability of heads is 0.5) by tossing the coin twenty times. Assume that the tosses are independent of each other. Suppose you obtain four heads. You then compute the

---

[30] Wagner (2004) shows that a bound on the value of $Pr(notH)$ can be derived from the values of $Pr(O \mid H)$ and $Pr(notO)$; he calls his result a probabilistic version of *modus tollens*. This is not the probabilistic *modus tollens* whose nonexistence I argue for above.

probability of a disjunction in which "four heads" is one of the disjuncts. You need to look at all the outcomes that the null hypothesis says are *at least as improbable* as the one you actually obtained:

Pr(0 or 1 or 2 or 3 or 4 or 16 or 17 or 18 or 19 or 20 heads |

the coin is fair and the coin is tossed 20 times) $= p$.

The probability of this disjunction, conditional on the null hypothesis, is called the *p*-value for the test outcome.

This *p*-value has two interpretations, corresponding to two different conceptions of what a significance test is supposed to accomplish. Sometimes significance testers draw a conclusion as to whether the null hypothesis should be rejected. To do this, they specify a value for α, the "level of significance"; the null hypothesis is rejected if the *p*-value is less than this cutoff. If $\alpha = 0.05$ is your level of significance, then four heads in twenty tosses will suffice to reject the null hypothesis, since the *p*-value of this outcome is 0.012; had you obtained six heads in twenty tosses, this outcome would not suffice to reject the null, since the *p*-value in this instance is 0.115. It is generally conceded that choosing a value for α is an arbitrary matter of convention. The other interpretation of significance tests is that they measure the strength of the evidence against the null hypothesis; the lower the *p*-value of the outcome, the stronger the evidence against. This comparative idea, by itself, does not say whether six heads in twenty tosses is (in an absolute sense) evidence against the hypothesis that the coin is fair, but it does say that four heads in twenty tosses would be stron*ger* evidence against it. If we stipulate that a *p*-value of 0.05 is the cutoff between "strong evidence against the null hypothesis" and not, then we know how to interpret six heads in twenty tosses, and also how to interpret four in twenty and two in twenty. The first of these is not strong evidence against the null while the second and third are. There is arbitrariness here as well.

Both interpretations of significance tests are vulnerable to the fact that there are many descriptions of the data that might be used, and changing these can lead to different conclusions about the null hypothesis. I mentioned that obtaining six heads in twenty tosses does not allow you to reject the null hypothesis (if you set $\alpha = 0.05$), since the probability of obtaining between zero and six or between fourteen and twenty heads is greater than 0.05. In this example, we thought of each possible number of heads that might occur in twenty tosses (0, 1, 2, ... 18, 19, 20) as an element in the outcome space and then gathered together the fourteen elements there that each has a probability of occurring under the null hypothesis that is less than or equal to the probability of obtaining exactly six heads. But the outcome space can be sliced up differently.[31] For example, instead of having twenty-one categories, you might decide to collapse some of these together. If you combine five heads and ten heads into one category, and fourteen heads and fifteen heads into another, you now have an outcome space with nineteen categories, not twenty-one. If you then construct a disjunction of the categories from this list that each has a probability that is less than or equal to the probability of getting exactly six heads, you'll discover that the probability of the relevant disjunction under the null hypothesis is 0.49, which will lead you to reject the null hypothesis (Howson and Urbach 1993: 182–3). Whether you reject the null depends on how you slice the cake.

It might be objected that collapsing the twenty-one categories into these nineteen is "unnatural," or that finer-grained taxonomies are preferable to ones that are coarser-grained. Defenders of significance tests have not attempted to develop an account of naturalness, and it is unclear how much help significance tests could extract from such an account. However, it is abundantly clear that insisting on logically stronger descriptions of the data does not help the significance tester. Instead of having twenty-one categories in the outcome space, why not treat each specific sequence of heads and tails as a separate element, with the result that our outcome space now has 2^{20} members, each with the same probability under the null hypothesis of $(\frac{1}{2})^{20}$? When we obtain a specific sequence of heads and tails (say, one containing two heads) and then collect the other elements in the outcome space that are no more probable according to the null hypothesis, the result is that we construct a disjunction that contains *all* 2^{20} elements; the probability of this disjunction, under the null hypothesis, is unity. With this fine-grained outcome space, we'll never reject the null, no matter what the outcome is.

Turning now to the evidential interpretation of significance tests, it is important to see how it conflicts with likelihoodism. According to the law of likelihood, whether the observations are evidence against the hypothesis that the coin is fair depends on which alternative hypothesis you consider. If the alternative to the null hypothesis says that the probability of heads is 0.8, then observing four heads in twenty tosses will

---

[31] Compare this point with considerations about cake slicing that arise in connection with the principle of indifference (§1.2).

be evidence *in favor* of the null hypothesis, not evidence *against* it. If the modest principle stated in §1.1 is correct, this point also bears on the idea that significance testing provides a rule of rejection. If an observation justifies you in rejecting *H*, and you were not justified in rejecting *H* before you obtained the observation, then the observation must be evidence against *H*. The fact that significance tests don't contrast the null hypothesis with alternatives suffices to show that they do not provide a good rule for rejection.

Another odd property of significance tests concerns the way in which they are sensitive to sample size. Howson and Urbach (1993: 208–9) explain this point by describing an example inspired by Lindley (1957). Suppose you wish to test the hypothesis (H_1) that 40 percent of the marbles in an urn are red. If you examine ten balls and choose $\alpha = 0.05$, you will reject H_1 if you see seven or more red balls. If you examine 100 balls and choose the same value for α, you will reject H_1 if you observed more than forty-eight red balls. And if you examine 1,000 balls, again with $\alpha = 0.05$, you will reject H_1 if you observe more than 423 red balls. As sample size increases, the observed frequency must be closer and closer to 40 percent for you to not reject H_1. With ten balls, you need to observe less than 70 percent; with 100 you need to observe less than 48 percent; and with 1,000, you need to observe less than 43 percent. This may not seem strange until you add the following detail. Suppose the alternative to H_1 is the hypothesis (H_2) that there are 60 percent red balls in the urn. The law of likelihood now entails that observing fewer than 50 percent red favors H_1 over H_2, that observing more than 50 percent red has the opposite evidential significance, and that *these interpretations of the observations are correct at all sample sizes*. If the law of likelihood is right, and if the modest principle stated in §1.1 correctly describes the connection between evidence and rejection, then we have here an objection to significance tests.

Although I have criticized the rejection and the evidential interpretations of significance tests, there is a more modest interpretation that is beyond reproach. Fisher (1956: 39, 43) put the point like this: If *H* says that *O* is very improbable, and *O* occurs, then we know that a disjunction is true – either *H* is false or something very improbable has occurred. This disjunction *does* follow. However, what does *not* follow is the first of Fisher's disjuncts; nor does it follow that we have obtained evidence against *H*. Another modest interpretation of significance tests is also appropriate: An observational outcome that a hypothesis says is very improbable may prompt you to search for a different hypothesis that says that the outcome was less surprising. This is how I understand the following remark that Gossett made in the 1930s:

[a significance test] doesn't in itself necessarily prove that the sample is not drawn randomly from the population even if the [*p*-value] is very small, say .00001; what it does is to show that if there is any alternative hypothesis which will explain the occurrence of the sample with a more reasonable probability, say 0.05 [...] you will be very much more inclined to consider that the original hypothesis is not true. (quoted in Hacking 1965: 83)

This gentle suggestion has good likelihoodist credentials.

If probabilistic *modus tollens* and significance tests have the flaws just described, can we abandon the *probabilistic* and simply rely on the *deductive* form? If H_1 entails O and O turns out to be false, it follows that H_1 is false. If H_2 is the only alternative to H_1, it further follows that H_2 is true. This is the pattern of reasoning that Sherlock Holmes endorses in *The Sign of Four* where Sir Arthur Conan Doyle has his hero say that "when you have eliminated the impossible, whatever remains, *however improbable*, must be the truth." The *correctness* of this pronouncement is not in dispute; rather, it is the *applicability* of Holmes's dictum that I contest. In science, it is rarely the case that the hypotheses under test deductively entail observational claims. This is obvious in the case of hypotheses that use the concept of probability (as in my running example of the hypothesis that a coin is fair). But the point often holds when hypotheses make no mention of probability. For example, when Eddington tested Newtonian theory against relativity theory, the competing hypotheses did not provide point predictions about what he should observe when he measured the bend in starlight during a solar eclipse. Because his measurements were imprecise, he could say only that the observations would *probably* fall in one value range if Newtonian theory were true and that they would *probably* fall in a second interval if relativity theory were true. The pervasive pattern in science is that hypotheses confer (nonextreme) probabilities on observations.[32]

It may seem not to matter much whether a hypothesis says that O *cannot* occur or says only that O *very probably* will not occur. In fact, the difference is profound. If you observe that O is true, the former allows you to reject H without your needing to consider an alternative hypothesis. In contrast, the latter does not license rejection, and there is

---

[32] The fact that scientific theories typically confer probabilities on observations only when auxiliary information is added will be explored in the next chapter in connection with Duhem's thesis.

no saying whether the observation is evidence against *H* unless an alternative hypothesis is specified.

1.5   FREQUENTISM II: NEYMAN–PEARSON HYPOTHESIS TESTING

The theory of hypothesis testing set forth by Neyman and Pearson (1933), and subsequently developed in detail by Neyman, gives advice about rejection, not, in the first instance, advice about the interpretation of evidence. As noted in §1.1, Neyman and Pearson state that they are not interested in interpreting evidence but only in stating general rules for guiding "behavior." This claim notwithstanding, the interpretation of evidence and the rational acceptance and rejection of hypotheses *are* related if the modest principle enunciated earlier is correct; if learning that *O* is true justifies rejecting *H*, where the rejection of *H* was not justified before that knowledge was gained, then *O* must be evidence *against H*. The Neyman–Pearson theory, as we will see, violates this principle.

If you are going to decide whether to accept or reject a hypothesis in the light of a set of observations, there are two kinds of error to which you are vulnerable. Consider the tuberculosis test discussed earlier, but this time let's frame the problem in terms of the task of acceptance and rejection, not as a question concerning the interpretation of evidence. You, the physician, receive the report of your patient's tuberculosis test result. The report is either positive or negative, and the patient either has tuberculosis or does not. You have two options: You can accept the hypothesis that your patient has tuberculosis or you can reject it. There are two kinds of error you might commit: You might reject the hypothesis that he has tuberculosis when it is true, or you might accept the hypothesis when it is false. These options are depicted in Figure 1.8, as are

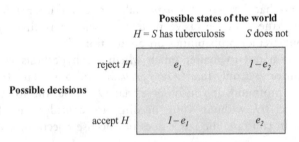

Figure 1.8   *S* either has tuberculosis or does not, and you, the physician, must decide whether to accept or reject the hypothesis *H* that *S* has tuberculosis. The four cells represent four possibilities; cell entries represent probabilities of the form *Pr*(decision | state of the world).

the probabilities of mistaken rejection (e_1) and of mistaken acceptance (e_2). If you were to ignore the report and merely toss a coin, your two error probabilities would then each have a value of 0.5. But you can do better if the test procedure you use is *reliable* in the sense described in §1.2; the more reliable the procedure, the smaller the error probabilities are. However, this does not mean that you will *probably* get the right answer if you use a reliable procedure. The error probabilities are of the form Pr(you accept hypothesis $H \mid H$ is false) and Pr(you reject hypothesis $H \mid H$ is true); they do not represent $Pr(H$ is false $\mid$ you accept $H)$ and $Pr(H$ is true $\mid$ you reject $H)$. Neyman–Pearson hypothesis testing is frequentist, not Bayesian.

Neyman–Pearson theory begins with the truism that it is better to have smaller error probabilities than larger ones. If you are going to base your decision about your patient's condition on what a test result says, you'll do better by using a more reliable testing procedure than one that is less. For example, suppose you can use a test kit that is made in Madison or one that is made in Middleton, where the two pairs of error probabilities are:

The Madison test kit: $Pr(-$ test result $\mid S$ has tuberculosis) $= 0.02$.

$Pr(+$ test result $\mid S$ does not have tuberculosis) $= 0.01$.

The Middleton test kit: $Pr(-$ test result $\mid S$ has tuberculosis) $= 0.04$.

$Pr(+$ test result $\mid S$ does not have tuberculosis) $= 0.03$.

Surely you'd want to use the Madison test kit, since both its error probabilities are lower. But how should you choose between the Madison kit and one made in Prairie du Chien? The error probabilities of this third test kit are:

The Prairie du Chien test kit: $Pr(-$ test result $\mid S$ has tuberculosis) $= 0.01$.

$Pr(+$ test result $\mid S$ does not have tuberculosis) $= 0.02$.

To choose between Madison and Prairie du Chien, you must decide which kind of error is worse to commit. Is it more important to avoid accepting that S has tuberculosis when he does not, or to avoid rejecting the hypothesis that S has tuberculosis when he does? One obvious way to decide this is to think about how your actions will be influenced by what you believe. Is it worse to treat someone for tuberculosis when he doesn't have the disease, or to fail to treat someone for tuberculosis when he does?

Notice how ethical considerations figure in this question. The issue is not strictly epistemological. In terms of Royall's three questions (§1.1), we are edging towards question (3) and away from questions (1) and (2).

The Neyman–Pearson theory recognizes that there are two types of error, but it does not treat them the same. First, you choose which of the two hypotheses under test you'll regard as the "null hypothesis." You then decide how large an error probability you will tolerate in connection with mistakenly rejecting the null:

$$Pr(\text{reject the null hypothesis} \mid \text{the null hypothesis is true}) < \alpha.$$

Scientists usually choose a value of $\alpha = 0.05$ while recognizing that this choice is arbitrary. The probability of rejecting the null hypothesis when it is true is called a Type-1 error. After putting an upper limit on how much Type-1 error you are prepared to tolerate, you then try to minimize the probability of the other kind of error:

$$Pr(\text{accept the null hypothesis} \mid \text{the null hypothesis is false}) = \beta.$$

The mistake of accepting the null hypothesis when it is false is a Type-2 error. So there are three steps in the Neyman–Pearson process: Decide which hypothesis is the null; set an upper limit on the probability of Type-1 error; and then minimize the probability of Type-2 error.

Suppose you decide that "*S* has tuberculosis" is your null hypothesis and you chose a value for α of 0.05; given these choices, all three test kits are acceptable so far. But now you want to minimize β. Madison does better on this score than either Middleton or Prairie du Chien. On the other hand, if you decide that "*S* does not have tuberculosis" is the null hypothesis while still hewing to the convention that $\alpha < 0.05$, you'll end up opting for the test kit from Prairie du Chien. Different decisions about what the null hypothesis is lead to different test procedures. Here is some more terminology: α (the probability of Type-1 error) is called the "size" of your test and $(1-\beta)$ is called its "power." Neyman–Pearson testing treats these asymmetrically: First get the size below some threshold, then maximize power.

To apply this framework in a way that brings out how it is related to likelihoodism, let's return to the coin-tossing problem discussed earlier. Suppose your plan is to toss the coin thirty times and that there are two hypotheses you want to consider. The first says that the probability of heads is $\frac{1}{4}$ on each toss while the second says that this probability is $\frac{3}{4}$. The probability that each hypothesis assigns to each possible outcome of your

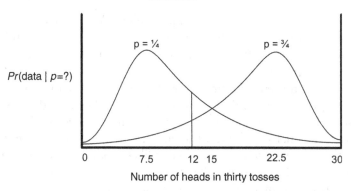

Figure 1.9 If $p = \frac{1}{4}$ is the null hypothesis and $p = \frac{3}{4}$ is the alternative to the null, and $\alpha = 0.05$ is chosen, the Neyman–Pearson theory says that the null hypothesis should be rejected if and only if twelve or more heads occur in thirty tosses of the coin.

experiment is depicted in Figure 1.9. Suppose you decide to regard the hypothesis that $p = \frac{1}{4}$ as your null hypothesis and you choose 0.05 as your value for α. You thereby require that the chance of rejecting this hypothesis, if it is true, must be less than or equal to 0.05. You now must use this stipulation to identify a "critical region." That is, you need to say what possible outcomes will suffice for rejecting the null hypothesis, given that you want to make sure that the probability of mistakenly rejecting the null hypothesis is no greater than 1 in 20. Many choices satisfy this requirement. For example, if you reject the null hypothesis precisely when there are *zero* heads in thirty tosses, the probability of rejecting the hypothesis that $p = \frac{1}{4}$ when the hypothesis is true is only $(0.75)^{30}$, which is tiny. The same can be said of the policy of rejecting the null hypothesis precisely when *all* thirty tosses land heads. With this policy, the chance of rejecting the null when it is true is only $(0.25)^{30}$, again a tiny number. Notice that neither of these judgments depends in any way on what the *alternative* to the null hypothesis happens to be. The fundamental difference between Neyman–Pearson testing and Fisher's test of significance is that the former is contrastive (pitting the null hypothesis *against a specified alternative*), while the latter is not. We now need to see what role the alternative to the null hypothesis plays in determining what the critical region will be. The critical region is determined by the joint fact that we want the chance of rejecting the hypothesis that $p = \frac{1}{4}$ if it is true to be no greater than 0.05 *and* we also want the chance of accepting this hypothesis if it is false (in which case $p = \frac{3}{4}$ is true) to be as small as possible. These two requirements result in a unique policy. We should reject the hypothesis that $p = \frac{1}{4}$ precisely when there are twelve or more heads in the thirty tosses. This cutoff is depicted in Figure 1.9 (the example is from Royall 1997: 16–17).

Notice that this cutoff differs from the one drawn by the law of likelihood, which says that a data set with fourteen or fewer heads favors the first hypothesis while a data set with sixteen or more heads has the opposite evidential significance. If there are exactly fifteen heads in thirty tosses, the two hypotheses have the same likelihood. As noted before, the law of likelihood answers Royall's first question (what does the evidence say?) while the Neyman–Pearson theory provides a policy for acceptance and rejection. However, the two come into contact (and are incompatible) if it is a mistake to reject a hypothesis because one has obtained a set of observations that, in fact, are evidence *for* the hypothesis, not evidence *against* it. This is precisely what happens if you observe twelve, thirteen, or fourteen heads in thirty tosses. If you obtain any of these outcomes in your experiment, the Neyman–Pearson theory says to *reject* $p = \frac{1}{4}$, while the law of likelihood interprets each of these outcomes as evidence *in favor of* $p = \frac{1}{4}$ (given that the alternative hypothesis is $p = \frac{3}{4}$). If the law of likelihood is right, the Neyman–Pearson theory is wrong.

What procedure would the Neyman–Pearson theory recommend if you were to decide that $p = \frac{3}{4}$ is your null hypothesis? You then would draw a different cutoff, but it, too, would fail to coincide with the boundary drawn by the law of likelihood. With the hypothesis that $p = \frac{3}{4}$ as your null, you will reject this hypothesis precisely when eighteen or fewer tosses land heads. This means that if you observe between twelve and eighteen heads, your decision about which of the two hypotheses you'll reject depends on which is the null and which is the alternative. If the hypothesis that $p = \frac{1}{4}$ is your null hypothesis, you reject it when any of these outcomes occurs; but if $p = \frac{1}{4}$ is the alternative to the null, you do not. Life is harder on a hypothesis if it is treated as the null. Notice that the law of likelihood does not depend on how you label the various hypotheses you wish to evaluate, and there is no need to choose a value for α, either. This is a good thing, since both choices are arbitrary.

As noted, Neyman–Pearson theory first fixes a value for α and then seeks to minimize the value of β. This is why the cutoff it draws differs from the one dictated by the law of likelihood. The history of statistics might have been different. If the two types of error had been treated as equally serious, the goal would have been to minimize the sum $(\alpha + \beta)$ of the two error probabilities. This would have provided no guidance in the choice between Madison and Prairie du Chien, but it would have resulted in a crossover point of fifteen heads in thirty tosses (Royall 1997: 17), thus bringing the Neyman–Pearson philosophy into accord with the law of likelihood. In fact, there are many policies that correspond to different ways of handling the disutilities that attach to Type-1 and Type-2 errors. Even if avoiding Type-1 error is more important than avoiding Type-2, why should this mean that we need to stipulate a value for α? For example, setting $\alpha = 0.05$ means that it doesn't matter to you whether the chance of Type-1 error is 0.04 or 0.004. If making α small matters more than making β small, why not require that the sum $(10\alpha + \beta)$ be minimized? This is why the behaviorist justification of the Neyman–Pearson philosophy does not work on its own terms. Even if "acceptance" and "rejection" are taken to be behaviors that need have no connection to an assessment of evidence, the desire to reduce the frequencies of errors in one's lifetime (or in the lifetime of the enterprise of science) does not automatically entail the policy of first choosing a value for α and then minimizing β.

In discussing the principle of total evidence (§1.3), I described a few examples in which logically strengthening or logically weakening one's description of the data affects which of two hypotheses has the higher likelihood. This principle is also relevant to thinking about how the Neyman–Pearson theory bears on the question of how evidence should be assessed. We have already seen, in connection with the coin-tossing example depicted in Figure 1.9, that observing twelve heads in thirty tosses leads the Neyman–Pearson theory to reject the null hypothesis that $p = \frac{1}{4}$ and to accept the hypothesis that $p = \frac{3}{4}$ (or to not reject it) even though the former has the higher likelihood. But now let us logically weaken the description of the observations. Instead of saying "we observed *exactly twelve*," let us say "we observed *twelve or more*." The law of likelihood judges that this logically weakened description of the data has a different evidential significance. Since α and β are both small, this weakened description of the data favors $p = \frac{3}{4}$ over $p = \frac{1}{4}$, and the likelihood ratio is $(1 - \beta)/\alpha$, a quantity substantially greater than unity. It *is* more probable that you'll get *twelve or more* heads in thirty tosses if $p = \frac{3}{4}$ than if $p = \frac{1}{4}$. Look at the areas under the two curves in Figure 1.9. The Neyman–Pearson theory and the law of likelihood are in accord with respect to how evidence should be interpreted *when information in the data set is thrown away*. However, this reconciliation has a price: We have violated the principle of total evidence. From the point of view of likelihoodism and Bayesianism as well, this is a serious defect in the Neyman–Pearson theory.

In addition to the difficulties already noted, which strike both likelihoodists and Bayesians as fatal, there is a further fact about the Neyman–Pearson theory that especially irks Bayesians. How can "acceptance" and "rejection" be based just on the evidence at hand? True, if your test procedure is very reliable, a positive test result provides evidence that strongly favors the hypothesis that S has tuberculosis over the hypothesis that he does not. However, this is consistent with its being very improbable, given the positive test result, that S has tuberculosis. The Neyman–Pearson policy sometimes recommends accepting a hypothesis in the light of evidence that renders the hypothesis very improbable. This is what can happen when acceptance and rejection are controlled by likelihoods and priors are ignored. This criticism of the Neyman–Pearson theory does not require that prior probabilities *always* make sense. All that is needed is that they *sometimes* do, and this is something that non-Bayesians should concede.

In order to bring out one last feature of the Neyman–Pearson approach, let us consider a fourth tuberculosis test kit; it is made in Mazomanie:

The Mazomanie test kit: $Pr(-$ test result $\mid S$ has tuberculosis$) = 0.902$

$Pr(+$ test result $\mid S$ does not have tuberculosis$) = 0.001$.

If you decide that "S has tuberculosis" is the null hypothesis and set $\alpha = 0.05$, you will decline to use this test kit. But suppose you did so anyway, perhaps by mistake, and you obtained a positive test result. How should you interpret this evidence? A likelihoodist will say that you have just obtained strong evidence favoring the hypothesis that S has tuberculosis since the relevant likelihood ratio is large:

$$\frac{Pr_{\text{Mazomanie}}(+ \text{ test result} \mid S \text{ has tuberculosis})}{Pr_{\text{Mazomanie}}(+ \text{ test result} \mid S \text{ does not have tuberculosis})} = \frac{0.098}{0.001} = 98.$$

In fact, this evidence is precisely as strong as the evidence that attaches to a positive result produced by the Madison test kit. Even though the two test kits have different values for α and β, a positive test result produced by using the Madison test kit also produces a likelihood ratio of $0.98/0.01 = 98$. Yet, the Neyman–Pearson methodology instructs you not to use the Mazomanie test kit and embraces the one from Madison. How can it do so, if the two are *evidentially equivalent* when a positive test result is produced? The answer is that the Neyman–Pearson theory addresses the question of how one should choose a *general policy*. If you, the doctor, have to choose between using the Madison test kit on all your patients and using the Mazomanie test kit on all of them, the plausible choice is to opt for the one from Madison. Notice that the previous sentence answers a question that falls under Royall's question (3): What should you do? That is, which test kit should you use in your medical practice? It is not an answer to question (1): What is the evidential meaning of S's positive test result? Nor does it address question (2): Should you believe that S has tuberculosis? Hacking (1965) makes this point by distinguishing the task of *before-trial betting* and *after-trial evaluation*. The first involves designing an experiment, the second the interpretation of the results you obtained on the experiment you actually ran. Likelihoodists and Bayesians hold that both tasks are important but maintain that they are distinct. The Neyman–Pearson philosophy does not distinguish these tasks; once a general procedure has been chosen, there is no additional question as to how the result obtained by applying the procedure on a single occasion should be interpreted. This difference between the philosophies becomes vivid when a less than optimal test procedure is used and one wishes to interpret the result. This was my point in introducing the Mazomanie test kit. If you use this procedure and obtain a positive result, Neyman–Pearson frequentists will say that you shouldn't have used that test kit and will refuse to interpret the outcome; Bayesians and likelihoodists will say that using that test kit rather than the one from Madison turned out not to matter and will be happy to interpret the test outcome. Philosophers will recognize that this difference between the two statistical frameworks parallels the distinction in ethics between rule and act utilitarianism.

I have described the rudiments of the Neyman–Pearson theory in the context of the simple example of coin tossing, and this has allowed me also to describe some standard criticisms of that approach. However, frequentists may want to object that it is silly to test the hypothesis that $p = \frac{1}{4}$ against the hypothesis that $p = \frac{3}{4}$. Instead, why not just *estimate* the value of p (and draw a confidence interval around that estimate)? For example, if there are twelve heads in the thirty tosses, you can simply say that the maximum likelihood (ML) estimate of p is 0.4; as already noted, this doesn't mean that p *probably* has a value of 0.4 or even that the true value is probably *close* to 0.4. However, in saying that this is the ML estimate, you can sweep aside the problem of deciding which of $p = \frac{1}{4}$ and

$p = \frac{3}{4}$ is the null hypothesis and what your value for α ought to be. ML estimation may sound like likelihoodism or even Bayesianism, but frequentists have their own special rationale for this procedure. Frequentists do not accept the law of likelihood. Rather, they see the method of ML estimation as justified, when it is, because it has certain virtues as a *general policy*; for frequentists, there is no additional question about the evaluation of an individual ML *estimate*. It is *estimators*, not *estimates*, that is their focus. A central concept in the frequentist theory of estimation is that an estimator (i.e., a procedure for making estimates) must be *admissible*. A method of estimation is *in*admissible if there is another estimator that has a smaller expected error for all possible values that the parameter being estimated might take. Whereas inadmissibility arguably suffices to not use an estimator, admissibility is not sufficient for a method to be used. The reason is that there can be multiple admissible estimators that give contradictory advice. In any event, it turns out that ML estimation is an admissible procedure when one or two parameters are being estimated but not when the estimation problem involves three or more. With more than two parameters, there is another procedure, involving shrinkage in accordance with a formula derived by James and Stein (1961) that has a lower expected error no matter what the true values are of the parameters being estimated (Efron and Morris 1977). This is not the place to pursue questions about estimation any further; suffice it to say that frequentists can decline to use the Neyman–Pearson theory to test the hypothesis that $p = \frac{1}{4}$ against the hypothesis that $p = \frac{3}{4}$ and insist that maximum likelihood estimation of the value of p is the way to go.[33]

Although estimation may make more sense than Neyman–Pearson hypothesis testing when the two hypotheses are statistically *simple*, this option is not available to the frequentist when both the hypotheses being tested are *composite*. In this case, the standard Neyman–Pearson approach is the *likelihood ratio test*. Don't let this terminology mislead you; this test is a frequentist construct even though the likelihood ratio also appears in the law of likelihood, which is the central concept of likelihoodism. Here's an example that illustrates what the likelihood ratio test involves. You conduct the following experiment in your kitchen: You heat a pressure cooker to a given temperature and then observe how much pressure there is in the container. You don't observe the temperature and

---

[33] It is worth emphasizing that this change in strategy does nothing to vindicate the Neyman–Pearson theory as it applies to simple hypotheses. The objections have not been *met*; rather an altogether different frequentist approach has been suggested.

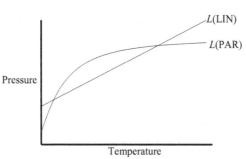

Figure 1.10 Each of the observations can be represented by a data point. L(LIN) is the straight line that fits the data best; L(PAR) is the parabola that fits best. The likelihood ratio test compares the models LIN and PAR by computing the likelihood ratio of L(LIN) and L(PAR).

pressure directly; rather, you observe the readings that a thermometer and a pressure meter provide. You know that these devices are reliable, but not perfectly reliable. You do this experiment multiple times, representing each observation by a point in the coordinate system depicted in Figure 1.10.

Suppose there are two models you want to test that both attempt to describe how temperature and pressure are related in this system. With the variables X and Y representing temperature and pressure, respectively, the two models are:

(LIN) $\qquad\qquad y = a + bx + e$

(PAR) $\qquad\qquad y = a + bx + cx^2 + e.$

LIN says that temperature and pressure are related linearly; PAR says that they are related parabolically. In these models, x and y are variables, while a, b, c, and e are parameters. Each model is an infinite disjunction; LIN is a disjunction of all straight lines in the X-Y plane; and PAR is a disjunction of all the parabolas. In other words, these models have existential quantifiers attached to their adjustable parameters; LIN, for example, says that *there exist* values for a, b, and e such that $y = a + bx + e$. The "e" in each model represents the fact that your observations are subject to error. Even if the true relationship between temperature and pressure is linear, you can't assume that the data you gather will fall exactly on a straight line. LIN postulates an error distribution around each of the straight lines it includes. Although a straight line is sometimes said to provide the "predicted" y-value for a given x-value, this is a bit misleading. What each straight line in LIN represents is the *average* (the

Evidence

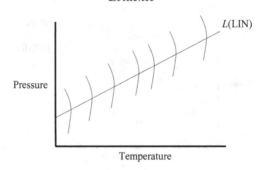

Figure 1.11 *L*(LIN) is the straight line that is closest to the data; the LIN model postulates an error distribution around this line. The observed pressure value for a given temperature need not coincide exactly with the average ("predicted") pressure value.

expected value; see §1.2) of the observed pressure-values that should be associated with a given value of temperature. This error distribution is depicted in Figure 1.11.

Here's how the likelihood ratio test applies to the comparison of LIN and PAR. First, you find the straight line that maximizes the probability of the data. This will be the straight line that is "closest" to the data; that is, the line that "fits" them best. Call this maximum likelihood straight line *L*(LIN). Then you do the same thing with PAR. There are many parabolas, some close to the data, others far away. You need to find the member of PAR that maximizes the probability of the data; this is *L*(PAR). These two "best cases" of LIN and PAR are depicted in Figure 1.10. In discussing how the Neyman–Pearson theory evaluates the two simple statistical hypotheses ($p = \frac{1}{4}$ and $p = \frac{3}{4}$) about coin tossing shown in Figure 1.9, I was able to discuss what each predicts about the data. But LIN and PAR are composite. Neither says how probable the data are that you generated in your kitchen (i.e., how probable the y values you observed are, given the x values you used). The Neyman–Pearson theory solves this problem by shifting from LIN to *L*(LIN) and from PAR to *L*(PAR). We test the two models by comparing the maximum likelihood members of each. It's as if LIN and PAR are two armies that compete by each sending forth its fittest champion. The armies stand idle and are evaluated by seeing which champion wins the *mano a mano*. The likelihood ratio test of LIN against PAR focuses on the likelihood ratio

$$\frac{Pr[\text{data} \mid L(\text{LIN})]}{Pr[\text{data} \mid L(\text{PAR})]}.$$

The question is whether this ratio is smaller than some arbitrarily chosen level of significance; if it is, you should reject LIN.

One interesting feature of the likelihood ratio test is that it avoids an arbitrariness that afflicts the Neyman–Pearson test of two simple hypotheses. In the coin-tossing example of testing $p = \frac{1}{4}$ against $p = \frac{3}{4}$, you need to decide which of these hypotheses is the null. As also was true in the example of the tuberculosis test, there is nothing inherent in these simple hypotheses that settles which is "really" the null. Considerations concerning which type of error you are more concerned to avoid are typically brought to bear, but this is a fact about *us*, not about the hypotheses themselves. Testing LIN against PAR is a different matter. Each of these models contains adjustable parameters, but it is LIN that says that $c = 0$ while PAR leaves open what value that parameter has. It is in this objective sense that LIN can be said to be the null hypothesis in this two-way competition. Frequentists sometimes describe the choice of null by talking about which of the hypotheses we want to nullify (i.e., reject), but there is no need for us and our desires to intrude into the story.

When I discussed the two simple hypotheses $p = \frac{1}{4}$ and $p = \frac{3}{4}$ about the coin and the problem of deciding which of them is the null hypothesis and what level of α to use, I considered the possibility that frequentists might decline to apply the Neyman–Pearson theory to this problem and instead would insist that the problem to address is how best to estimate the value of the parameter p, where $p = Pr$(the coin lands heads | the coin is tossed). Estimation is an issue that arises *after* you have settled on a given model of the experiment. You have already decided that each toss of the coin has the same probability of landing heads as every other and you have decided that the tosses are independent of each other. Given this framework, you can estimate p. Testing the composite hypotheses LIN and PAR is different. The problem of choosing a level of significance can't be set aside and an estimation problem considered in its stead. The reason is that the competition between LIN and PAR is a competition *between* models, while estimation is a task that is carried out *within* the confines of a single model. True, if you assume that LIN is true, you can estimate the values of the parameters in it; the same goes for PAR. But that hardly suffices to test LIN against PAR. In fact, you know in advance that L(LIN) can't have a higher likelihood than L(PAR). This is because LIN is *nested* in PAR. LIN is a special case of PAR; the equation for LIN can be obtained from the equation for PAR by setting the parameter $c = 0$. Given that the ratio on which the likelihood ratio test focuses can't have a value that is greater than unity, the frequentist's question is whether the ratio is *significantly* less than unity; you have to look at the data to see whether this is so.

It is interesting to reflect on what the frequentist advice to "accept" or "reject" means in the context of these two composite models. LIN is *nested* in PAR, meaning that LIN logically entails PAR. If so, what would it mean to accept LIN and reject PAR? You can't regard LIN as true and PAR as false if the former entails the latter. It also makes no sense to decline to reject LIN and to reject PAR; if PAR is false, so is LIN. It might be replied that the frequentist can eliminate this problem by stipulating that the models worth talking about are *not* nested; this can be achieved by requiring that all the parameters in the two models have nonzero values. Now the models are incompatible. The problem with this reply is that the mathematics that underlies the likelihood ratio test requires that models be nested (Burnham and Anderson 2002).

Bayesians have an additional criticism of the Neyman–Pearson treatment of composite hypotheses, one that does not apply when only simple hypotheses are considered. The Neyman–Pearson theory compares LIN and PAR by comparing the members of each that have maximum likelihood, namely $L(\text{LIN})$ and $L(\text{PAR})$. But the likelihoods of LIN and PAR, the Bayesian will observe, are not these *maxima* but rather are their *average* likelihoods. Since LIN is a disjunction of straight lines (L_1, L_2, ...), it has a likelihood of the following form:

$$Pr(\text{data} \mid \text{LIN}) = \sum_i Pr(\text{data} \mid L_i) Pr(L_i \mid \text{LIN}).^{34}$$

Frequentists don't want to discuss these average likelihoods because it often is impossible to empirically justify an assignment of values to the weighting terms that have the form $Pr(L_i \mid \text{LIN})$. If the temperature and pressure in your pressure cooker are linearly related, what is the probability of the different specific straight-line relations that might obtain (and please answer this question without looking at the data you drew from your pressure cooker)? This is one motive that frequentists have for shifting from the *average* likelihood of the infinitely many straight lines that belong to LIN to the *unique* likelihood value that attaches to just one of them, namely to $L(\text{LIN})$. This *is* a motive for shifting, but not a justification for the likelihood ratio test. The justification offered is that if you follow the Neyman–Pearson procedure again and again, the expected value of your

---

[34] This should be an integral, not a discrete summation, but I prefer to use the latter to make this material accessible to a wider readership. Aficionados know how to correct this crudity.

Type-1 errors will be no more than α, and the expected value of your Type-2 errors will be β. It's the *general policy* that has this property, but the question may be asked of why this property of the general policy shows, in the concrete situation of evaluating LIN and PAR with the data you have from your kitchen experiment, that you should evaluate the two models by examining the maximum likelihood special cases of each. Frequentists regard this question as irrelevant, while Bayesians regard it as central.

Even if there is nothing arbitrary about saying that LIN is the null hypothesis when LIN is compared with PAR in a likelihood ratio test, there is another detail of this procedure that introduces a kind of arbitrariness that did not appear in the example of testing the two simple hypotheses $p = \frac{1}{4}$ and $p = \frac{3}{4}$ about coin tossing. To see what this new arbitrary element is, we need to consider a hierarchy of nested models, not just two. LIN and PAR are both polynomials; each has the form:

$$y = b_0 + b_1x + b_2x^2 + \cdots + b_{n-1}x^{n-1} + b_nx^n.$$

LIN is a first-degree polynomial and PAR is second-degree. Let's consider a hierarchy of five polynomials by adding to our list three more – a third, fourth, and fifth degree. For simplicity, let's call these five models A, B, C, D, and E. We need to fit each of these five models to the data from our stovetop experiment and then figure out the likelihood ratios for adjacent pairs of fitted models. Suppose we obtain the following left-to-right likelihood ratios:

$$L(A) \leftarrow (0.1) \rightarrow L(B) \leftarrow (0.3) \rightarrow L(C) \leftarrow (0.05) \rightarrow L(D) \leftarrow (0.5) \rightarrow L(E).$$

There are two ways to apply the likelihood ratio test to this hierarchy: *step-up* and *step-down*. In each case, a level of significance needs to be chosen; suppose you select α = 0.15. In step-up testing, you begin with the simplest model A and ask whether the likelihood ratio of $L(A)$ to L(B) is less than 0.15. If it is, you reject A and then compare B and C and ask the same question. You continue to step-up until you can't anymore. The result of step-up testing on this sequence of models is to reject A in favor of B but then to fail to reject B in favor of C. The process terminates with B. In step-down testing, you begin with the most complex model, E, and compare it with the model that is one step down, namely D. The question is whether the likelihood ratio of $L(D)$ to $L(E)$ is less than 0.15. If it is, you stay with E. If it is not, you move from E to D. Given the numbers shown above, this step-down process terminates with D. The choice between step-up and step-down testing is arbitrary and yet it can influence which models you accept and reject (Burnham and Anderson 1998).

1.6 A TEST CASE: STOPPING RULES

There is a classic puzzle that illustrates the clash between Bayesianism and likelihoodism on the one hand and significance tests and the Neyman–Pearson theory on the other. It concerns the "stopping rule" used when observations are gathered. This rule determines when the inquiry is over. In the example about coin tossing that I used to explain significance tests in §1.4, the stopping rule was to stop after the coin is tossed twenty times; it then turned out that six heads had occurred. The same outcome can occur if a different stopping rule is used. For example, you might decide to toss the coin until you obtain six heads, and it then turns out that the sixth head occurs on the twentieth toss. Here's the question: if you obtain the sixth head on your twentieth toss, should your interpretation of this result depend on which of the two stopping rules you used? Likelihoodism and Bayesianism say *no*, whereas the two versions of frequentism examined so far say *yes*.[35]

Let's begin with the likelihood analysis, which the Bayesian accepts; the issue about prior probabilities plays no role here. Although this problem is sometimes described as if it is supposed to be obvious that Bayesianism entails that the choice of stopping rule is irrelevant, the reason for this is worth tracing carefully. For the sake of a simpler example, let's shift for a moment to comparing a fixed-length experiment that involves tossing a coin three times with a flexible-length experiment in which you toss the coin until it lands heads. The possible outcomes of each of these experiments are depicted in Figure 1.12. If the coin is fair ($p = 0.5$), each specific sequence that can occur in the fixed length experiment has a probability of $\frac{1}{8}$; in the flexible length experiment, the probabilities of the different outcomes (reading from left to right) are $\frac{1}{2}$, $\frac{1}{4}$, $\frac{1}{8}$, and so on. Suppose you obtain tails on the first two tosses and heads on the third but don't know what the value of p is. The probability of obtaining the sequence TTH is the same regardless of which experiment was performed:

$$Pr(\text{TTH} \mid \text{there will be 3 tosses}) = Pr(\text{TTH} \mid \text{there will be exactly one H})$$

$$= p(1-p)^2.$$

---

[35] This example is from Howson and Urbach (1993: 210–12); it is similar to an example given by Lindley and Phillips (1976).

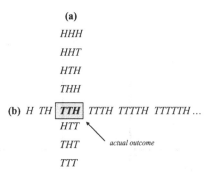

Figure 1.12 If a coin lands tails on the first two tosses and heads on the third, this outcome might be the result of two different experiments: (a) toss three times; (b)toss until heads occurs once (from Goodman 1999: 1000). The possible outcomes of both experiments are shown.

This means that if you are testing the hypothesis that $p = 0.5$ against the hypothesis that $p = 0.9$, the following equality obtains

$$\frac{Pr(\text{TTH} \mid p = 0.5 \ \& \text{ there will be 3 tosses})}{Pr(\text{TTH} \mid p = 0.9 \ \& \text{ there will be 3 tosses})}$$

$$= \frac{Pr(\text{TTH} \mid p = 0.5 \ \& \text{ there will be one heads})}{Pr(\text{TTH} \mid p = 0.9 \ \& \text{ there will be one heads})}.$$

This equality indicates why Bayesians say that the choice of stopping rule is not relevant to the interpretation of the observations; the weight of evidence (as measured by the likelihood ratio) is the same, regardless of which experiment you performed. Returning to our initial example, I hope it is clear why it doesn't matter to the likelihoodist whether you obtained six heads in a fixed length experiment of twenty tosses or if it took you twenty tosses to obtain six heads in a flexible length experiment – the meaning of the evidence is the same.

Why does the difference between the two experimental designs matter to the significance tester? The answer begins with the fact that significance tests require you to consider the probability under the null hypothesis of a logically weaker description of the data – that you obtained the test result *or ones that are at least as improbable*. If the null hypothesis says that $p = 0.5$, the probabilities you need to think about to perform a test of significance for the fixed and the flexible length experiments are, respectively,

(Fixed) $Pr(0-6 \text{ or } 14-20 \text{ heads} \mid p = 0.5 \ \& \text{ there will be 20 tosses})$

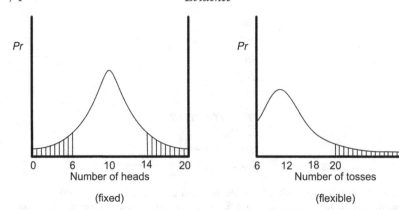

Figure 1.13 Suppose that a fixed-length experiment in which a coin is tossed twenty times and a flexible-length experiment in which a coin is tossed until six heads occur both result in six heads in twenty tosses. In each case, a significance test of the null hypothesis that the coin is fair focuses on the probability of obtaining that result or ones that are at least as improbable.

and

(Flexible) $Pr(20 \text{ or more tosses} \mid p = 0.5 \text{ \& there will be 6 heads})$.

The relevant regions of the two outcome spaces that the two significance tests consider are shown in Figure 1.13. It turns out that (Fixed) has a value of 0.115 and (Flexible) has a value of 0.0319. If you set your level of significance at $\alpha = 0.05$, you should not reject the null hypothesis if you performed the fixed experiment, but you should reject the null if you performed the flexible. Which experiment you performed to obtain your six heads in twenty tosses makes all the difference.[36]

It is not a unique feature of significance tests that the probability the null hypothesis confers on a logically weakened description of the data depends on which experiment was performed. Consider the simpler example depicted in Figure 1.12. You tossed the coin three times and obtained the exact sequence TTH. As already noted, this description of the data has a probability of $\frac{1}{8}$ under the null hypothesis that $p = 0.5$ regardless of which experiment was performed. However, with a logically

[36] The same result can arise in Neyman–Pearson hypothesis testing, for example, if the null hypothesis is tested against the composite alternative that $p \neq 0.5$ (Howson and Urbach 1993: 211).

weaker description of the data (in which you describe the mix of heads and tails but omit to mention their order), this agreement dissolves:

$$Pr(2T \text{ and } 1H \mid \text{Null \& there will be 3 tosses}) = \frac{3}{8}.$$

$$Pr(2T \text{ and } 1H \mid \text{Null \& there will be just } 1 H) = \frac{1}{8}.$$

As shown in Figure 1.12, there is just one way to get two tails and one heads in the flexible experiment, but there are three in the fixed-length experiment. The point is that likelihoodists don't care about the values of these single conditional probabilities but only about the values of various *ratios*, whereas significance testers think that what matters is the value of a single conditional probability – (Fixed) or (Flexible) as the case may be.

Given the importance that significance testers assign to the choice of stopping rule, what should they say about experiments in which it is unclear which stopping rule was actually used? Howson and Urbach (1993: 212) describe the following example. Suppose two scientists collaborate to perform a coin-tossing experiment; they obtain six heads in twenty tosses (with the sixth head occurring on the last toss) and then sit down to write an article in which they report their results, thinking that nothing is amiss. It then emerges that they had different plans in mind; the first scientist thought the plan was to toss twenty times; the second thought the plan was to toss until six heads occur. Of course, they should have talked things through beforehand, but what are they now to do? According to the logic of significance tests, they need to figure out what they would have done if other results had emerged. If they had obtained the sixth head on the nineteenth toss, would they have continued the experiment? If they had obtained only five heads by the twentieth toss, would they have persevered? Answering these questions requires information about the power relations between the two experimenters. Perhaps you are inclined to say that it doesn't matter what they would have done if the results had been otherwise; what matters is the results they in fact obtained and this result can be interpreted without psychoanalyzing the two scientists. If so, you are thinking like a likelihoodist.

Defenders of significance tests often suggest that Bayesians are hopelessly uncritical of how experiments are designed but that frequentists, in this respect, have their heads screwed on right. Suppose I decide to continue tossing a coin until I obtain results that go against the null hypothesis. If so, I apparently know in advance what conclusion I'll draw.

But if I cannot fail to reject the null hypothesis, regardless of whether that hypothesis is true, how can the experiment be said to test that hypothesis? And if the experiment doesn't test the null, why bother to run it in the first place? Frequentism explains why it is pointless to do this experiment, but frequentists often claim that Bayesians have a blind spot here; Bayesianism, they say, holds that there is nothing wrong with running this type of "try-and-try-again" experiment (Mayo 1996). What is even more galling to frequentists is that Bayesians have the temerity to proclaim this a *virtue* of their position, rather than acknowledging it to be the embarrassment to Bayesianism that it truly is.

This criticism of Bayesianism is sometimes stated as a very general claim: That Bayesianism *never* accords any epistemic import to the design of experiments and can offer no rationale for declining to perform experiments whose outcomes are known in advance. This criticism is vastly overstated, as a simple example from Eddington (1939) illustrates. You throw a net in a lake and wait until fifty fish have been caught. You pull the net out and see that all fifty fish are more than 10 inches long. How does this observation bear on the following two hypotheses? H_1 says that all the fish in the lake are more than 10 inches long; H_2 says that 50 percent of the fish are more than 10 inches long. Your first impulse is to think that the observations favor H_1 over H_2, but then you realize that this interpretation depends on what the net was like. If the net has 1 inch holes, the interpretation makes sense, but if the holes are 10 inches across, the observation fails to discriminate between the two hypotheses. The general point is that the bearing of observations on hypotheses often depends on the methods used to obtain the observations. When the outcome of an experiment is knowable beforehand and does not depend on which hypothesis is true, there is no point in performing this experiment; the law of likelihood provides a perfectly straightforward explanation of why this is so.[37]

Setting this hyperbolic criticism of Bayesianism to one side, let us look in more detail at fixed- and flexible-length experiments of the kind described in Figures 1.12 and 1.13. Let's begin by getting the facts straight in connection with frequentism. Consider an experiment that ends precisely when a significance test takes the data to indicate that the null hypothesis should be rejected. It is a certainty that this experiment will end if one uses the "nominal" value for the level of significance (Anscombe 1954). Using the nominal value means that at each stage one pretends that

---

[37] I discuss Eddington's example of an *observation selection effect* in connection with the fine-tuning version of the design argument in Sober (2004b).

the data were the result of an experiment designed to have that number of observations. Since this experiment's outcome is known in advance and does not depend on whether the null hypothesis is true, frequentists think there is an excellent reason not to run it. However, they are not opposed to "sequential trials." Armitage (1975) has described a protocol for such experiments in which one uses the "overall," rather than the "nominal," value for the level of significance. This new concept has the consequence that it is no longer a certainty that the experiment will end, and so it is no longer crazy, from a frequentist point of view, to run it. Armitage also describes how sequential trials can be structured so that accepting the null hypothesis as well as rejecting it is a possible outcome.

To understand what Bayesianism and likelihoodism say about this problem, we must be careful not to saddle these frameworks with ideas that are alien to them. Neither uses significance tests, and their experiments don't end with the "acceptance" or "rejection" of the null hypothesis. Both interpret experimental results by using the law of likelihood, so we need to be explicit about the alternative to the null hypothesis that is in contention. To this end, let's suppose that the null hypothesis (H_0) says that $p = 0.5$, that the alternative hypothesis (H_1) says that $p = 0.9$, and that the experiment you undertake will stop precisely when the frequency of heads engenders a likelihood ratio of H_0 to H_1 that is less than or equal to $1/k$ (where $k \geq 1$). If H_0 is true, is this experiment bound to end, thus resulting in misleading evidence that favors H_1? Robbins (1970) has shown that the probability of this experiment's ending when H_0 is true is less than or equal to $1/k$. If you define "strong evidence against the null" to mean a ratio that is less than $\frac{1}{8}$, then the probability of this misleading result is less than $\frac{1}{8}$. Commenting on this point, Royall (1997: 7) says that "if an unscrupulous researcher sets out deliberately to find evidence supporting his favorite but erroneous hypothesis [...] over his rival's [...] which happens to be correct, by a factor of at least k, then the chances are good that he will be eternally frustrated." Notice that this point has nothing to do with the prior or posterior probabilities of the hypotheses; it falls strictly within the likelihood framework.[38]

---

[38] Kadane et al. (1996) obtain similar results but within a fuller Bayesian framework and using the strong assumption of countable additivity. Suppose you decide to end the experiment precisely when the posterior probability assigned to H_1 exceeds some value v. If your prior for H_1 is r, the probability that the experiment will end, if H_0 is true, is no more than $r(1-v)/(1-r)v$. So if H_0 and H_1 each have priors of $\frac{1}{2}$, and you don't stop the experiment until H_1 has a posterior probability of at least 0.9, the probability of the experiment's ending is no more than 0.11. Notice

Thus, the try-and-try-again design in which you end the experiment only when you've obtained strong evidence against H_0 is *not* bound to end, if the criterion for its ending is formulated in terms of the likelihood ratio. If there is something wrong with this experimental design, it is not that you know in advance what will happen. One defect, noted by Teddy Seidenfeld, is that if the null hypothesis is true, this experiment has a serious chance of going on forever; if experiments cost money to run, Bayesians with finite funds have a good reason not to use this experimental design (Backe 1999: S360).[39] Fortunately, there are other designs that are far more sensible; for example, you could continue drawing evidence until strong evidence favoring H_0 over H_1, or strong evidence favoring H_1 to H_0, is obtained. The probability that this even-handed experiment will end, sooner or later, is unity (Wald 1947: 37–40; Backe 1999: S359); of course, it is not a foregone conclusion which result you'll obtain.

Where do these points leave the optional stopping problem? Significance testers abhor the try-and-try-again experimental arrangement when carried out with "nominal" p-values. However, with "overall" p-values, sequential experiments are not beyond the frequentist pale. And if you organize your test along Bayesian or likelihoodist lines, it is not true that try-and-try-again must result in the experiment's ending (where ending means attaining a likelihood ratio that represents strong evidence against the null). This shows that if the experiment *does* end, you really do have evidence (as defined by the likelihood ratio). Bayesians think that both the design of experiments and the interpretation of the results obtained are important topics; this is Hacking's (1965) distinction between before-trial betting and after-trial evaluation (§1.5). It is frequentists who often do not see the second as a problem separate from the first.

1.7   FREQUENTISM III: MODEL-SELECTION THEORY

The keys and the lamppost

When I raised the objection that Bayesianism often has no objective basis for assigning values to prior probabilities or to the likelihoods of catchall hypotheses, I did not describe a *different* theory for assigning such values

---

the relationship to the likelihood ratio in this example; given these values for r and v, the experiment ends precisely when the likelihood ratio of H_0 to H_1 is $\frac{1}{9}$ or less.

[39] Compare Jeffrey's (1983: 154) response to the St. Petersburg paradox: The bargain you are offered must be fraudulent, since no one has an infinite amount of money.

and then argue that this different theory is *better* than Bayesianism. No, what I did was change the subject. I retreated to likelihoodism, which addresses a different question – the question of how evidence ought to be interpreted. This pattern of shifting questions is not unique to the foundations of statistics nor is it unique to philosophy. Though politics is often called "the art of the possible," science deserves to be described in this way as well. If one problem cannot be solved, there is no reason why another should not be taken up that can. The only sin is to give the false impression that a new theory solves the same problem that an old one was unable to address. Science is sometimes like the man searching under the lamppost for the keys that he misplaced. When asked why he is searching there, he replies that that is where the light is. He does not reply that that is where his keys probably are.

In the previous two sections, I explained the rudiments of significance tests and of the Neyman–Pearson theory of hypothesis testing. I described some serious (and standard) objections to each. However, as mentioned at the start of the discussion of frequentism, this statistical philosophy is not a unified theory; rather, it is a loose confederation of ideas. The criticisms I've made of significance tests and of hypothesis testing don't necessarily attach to other frequentist ideas. The part of statistics called *model-selection theory* may have its problems, but it avoids the problems we so far have identified. There is no need to decide which hypothesis to call the null, and there is no need to choose a value for α. Indeed, there is no such thing as acceptance and rejection in model-selection theory. The name of this part of statistics is misleading; the problem addressed is one of model *comparison*, not model *selection*. Before we consider some of the solutions that have been proposed to the problem of model comparison, we need to understand what the problem is. An important element in this field has been the articulation of a new question: How should we estimate how accurate a theory's predictions will be?

Model building in science: Two pervasive patterns

In many areas of scientific research, a great deal of effort goes into the construction and evaluation of "models." This term has a technical meaning in statistics and a somewhat different nonmathematical meaning in the sciences themselves. As noted in the discussion of LIN and PAR in the previous section, models in the statistical sense of that term contain adjustable parameters; the statement that X and Y are related linearly is a model, while the statement that $y = 3 + 4x$ is not. This specific straight-line equation has been obtained from the linear model by substituting point values for adjustable parameters. When scientists use the term "model," they often have a different idea in mind. For them, a model is a simplified hypothesis; it purports to explain or predict a set of observations without trying to represent all the factors that are relevant. Models are not fully *realistic*; rather, they contain *idealizations*. Physicists work with models that assume that planets are spherically symmetrical, that particles collide with perfect elasticity, and that balls roll down inclined planes that are perfectly frictionless; evolutionary biologists consider models that assume that populations are infinitely large, that mating is perfectly random, and that a trait has a single unchanging fitness value in each of the many generations of the population in which the trait evolves. These models are known to be false, but they are not dismissed out of hand. The hope is that there may be truth in these falsehoods. If the idealizations are *harmless*, their departures from the truth won't matter much; these idealized models will yield accurate predictions even though they are false (McMullin 1985; Hausman 1992). If your goal is to predict how much time a ball will take to roll down a ramp, assuming that the ramp is *perfectly* frictionless may be fine if the ramp is *nearly* frictionless and your measurements are somewhat imprecise.[40]

There are two pervasive facts about the use of models in science that are of considerable philosophical significance. The first is that scientists often test models that they know are false. This is especially clear for many of the hypotheses that are labeled *null models*. This term is often applied to hypotheses that say that there is no difference between two quantities. The hypothesis that two fields of corn plants have the same mean height is a null hypothesis in this sense; the same is true of the hypothesis that a coin is fair (since it says that there is no difference between the chance of heads and the chance of tails). It is interesting that we often know, with as much certainty as we can ever have in science, that these so-called null hypotheses are false. Consider the coin. Do you really think that the coin is *exactly* symmetrical, that the chance of heads (p) is *exactly* equal to the chance of tails ($1 - p$) on each toss and that this precise symmetry remains in place each and every time the coin is tossed? I, personally, do not. My expectation is that there are modest asymmetries in the shape and balance

---

[40] There is a third use of the term "model" found in the part of logic called model theory. Here a model is a set of objects, properties, and relations that make a set of sentences true. In this usage, models are not propositions. For historical and philosophical reflections on the use of models in science, see Hesse (1966), Morgan and Morrison (1999), Da Costa and French (2003), and Frigg and Hartmann (2006).

of the coin; I am virtually certain that $p \neq \frac{1}{2}$. I also feel pretty sure that the coin changes its shape, if only slightly, during its lifetime. So why do scientists bother to test the simple hypothesis that $p = \frac{1}{2}$ against the composite hypothesis that $p \neq \frac{1}{2}$? Or consider the two fields of corn. The null hypothesis says that there is no difference in their average heights. Again, I find myself as certain as I am about almost anything that this null hypothesis is not true. The falsity of the null hypothesis, of course, is not an *a priori* matter; however, I suggest that our empirical experience of the world assures us that the two means are not *exactly* the same (to 1 million decimal places and more). Yet, scientists test the null hypothesis that the difference is zero against one or another alternative hypothesis.

Given that null hypotheses are often known to be false before any statistical test is run, it is not surprising that statisticians sometimes argue that these null hypotheses are not worth testing (see, for example, Yoccoz 1991 and Johnson 1995). I do not draw this conclusion. If the goal of scientific inference were just to find out which theories are true, dismissing such null hypotheses without testing them would make sense. But if the goal is to discover which theories will make accurate predictions, there may be a point in testing null hypotheses. Maybe hypotheses known to be false will make accurate predictions. And if *all* the hypotheses under test are known to be false (since all contain idealizations), it may still be worthwhile to determine which of them can be expected to make the most accurate predictions. If idealized (and therefore false) models are proper objects of scientific testing, we need to change our conception of what the goal of scientific reasoning is. Bayesianism is usually understood as a theory for deciding which hypotheses are probably *true*; the Neyman–Pearson theory concerns which hypotheses we should accept as *true* and reject as *false*; and likelihoodism tells us whether our evidence favors the hypothesis that H_1 is *true* over the hypothesis that H_2 is *true*. Truth enters into each of these theories of inference. This obsession needs to be overcome.

A second fact about model building in science also is pregnant with philosophical meaning. It concerns an experience that scientists often have when they use models that are very complex. When scientists consider a body of data that they suspect was produced by multiple causes that interacted in complex ways, they may be tempted to invent a complex model as an explanation. Doesn't a complex reality need a complex theory to do it justice? However, when such models are fitted to the data by finding the maximum likelihood estimates of their adjustable parameters (as we did in the example in §1.5 about the pressure cooker), those fitted models often do a terrible job of predicting new data drawn from the same system. Here's an example that illustrates the kind of pattern I have in mind. Suppose you made *n* observations of <*xy*> pairs during your experiment with the pressure cooker. It is a mathematical fact that a polynomial of degree *n* − *1* can be found that fits those *n* data points *perfectly*. If you made two observations, there is a straight line (a first-degree polynomial) that passes exactly through them; if you made three observations, there is a parabola (a second-degree polynomial) that does the same thing. And so on. Sadly, the mathematical assurance that a sufficiently complex polynomial will fit the *old* data perfectly is no guarantee that the fitted polynomial will do a good job predicting *new* data. In fact, scientists often find that complex models do very poorly in predicting new data when fitted to old. Simpler models often do better. Here the complexity of a model corresponds to the number of adjustable parameters it contains.

Given this common experience that model-builders have, it may seem that the only lesson is the following vague rule of thumb: Don't make your models too complicated or too simple, either. This advice is sensible, but it isn't very helpful. How complicated is too complicated? What is remarkable is that this advice can be made more precise. Work in model-selection theory has shown that, in a variety of circumstances, it is possible to *estimate* how accurately a model will predict new data when it is fitted to old. There is much that remains to be learned about the mathematical underpinnings of this area, but what is striking is that there are mathematical structures here to be investigated. The fact that models that are very complex are often not good at predicting new data when fitted to old is not a brute fact. Rather, there is a body of mathematics that *explains* why complex models are often poor predictors and allows scientists to take measures to avoid using models that are too complex.

Akaike's framework, theorem, and criterion

Model-selection theory began as a subject in statistics with Hirotugu Akaike's 1973 paper. Akaike identified a problem, and he proposed a solution to it. It is important to keep separate these two parts of what he accomplished, since the problem he singled out for study has an importance that transcends the solution to the problem that he proposed. This is because the subject he founded led to the discovery of different solutions that are appropriate in different settings. There now are multiple model-selection criteria on the market, and it is widely recognized that different criteria should be used for different model-selection tasks.

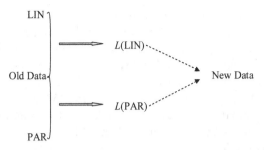

Figure 1.14 The prediction problem that Akaike considered. From the old data and a model you can deduce what the likeliest member of that model is. That likeliest model then makes probabilistic predictions about new data. Which model, LIN or PAR, will do better in predicting the new data when fitted to the old? Deductive relations are indicated by solid arrows, probabilistic by broken.

A simple version of the kind of problem that Akaike discussed is depicted in Figure 1.14. LIN and PAR are the two models we examined in §1.5 of how temperature and pressure are related in the pressure cooker. You use the available data to find the maximum likelihood estimates of the parameters that each model contains; that is, you use the data to find $L(LIN)$ and $L(PAR)$; $L(LIN)$ is the member of LIN that has the highest likelihood, and $L(PAR)$ is likewise the maximum likelihood member of PAR. You then ask the following question: If you were to draw new data from the pressure cooker, would $L(LIN)$ or $L(PAR)$ do a better job predicting this new data? I hope the reader finds it puzzling how this question could be answered. The only data you can consult is the old data you already drew from the pressure cooker. Since LIN is nested inside PAR, you know in advance that $Pr[\text{old data} \mid L(PAR)] \geq Pr[\text{old data} \mid L(LIN)]$, no matter what the old data are. The only way the two fitted models can have exactly the same likelihoods is if the data fall exactly on a straight line; otherwise $L(PAR)$ will have the higher likelihood. As already noted, more complex models inevitably fit the data at hand better than simpler models. But we know from experience that more complex models often do worse at predicting, not better. What else is there to consider here besides the likelihoods of $L(LIN)$ and $L(PAR)$?

Bayesians may feel inclined at this point to appeal to the prior probabilities of LIN and PAR. But here we run into a wall. Since LIN entails PAR, $Pr(LIN) \leq Pr(PAR)$. The simpler model cannot have the higher prior probability – a point that Popper (1959) emphasized. This problem can be circumvented if we stipulate that the only models we are willing to consider must be incompatible with each other. For example, if we require that LIN and PAR be set aside, and LIN* and PAR* considered instead (where all the parameters in this latter pair of models have non-zero values), the axioms of probability theory do not settle in advance which of the two has the higher prior probability. But even though it is logically consistent to say that $Pr(\text{LIN*}) > Pr(\text{PAR*})$, it is hard to see how this can be anything more than a stipulation. Consider the parameter c that attaches to the squared term in PAR*. The claim that $Pr(\text{LIN*}) > Pr(\text{PAR*})$ is equivalent to the claim that $Pr(c = 0) > Pr(c \neq 0)$. What objective reason could there be for thinking that this inequality is true?

Let us make the prediction problem more precise. After you draw the old data and use them to identify $L(\text{LIN})$ and $L(\text{PAR})$, you want to know how well these two curves will predict new data drawn from the same pressure cooker. But there is no one form that the new data set must take. Different data sets will differ from each other, though all are produced by the same underlying mechanism. The reason for expecting variability among data sets is that observations are subject to error. This means that when we ask how well a given fitted model will do in predicting new data, what we want to ascertain is how well it will do *on average* in this prediction task. $L(\text{LIN})$ may accurately predict one new data set but do less well in predicting another. By the *predictive accuracy* of a model M we mean how well *on average* M will do when it is fitted to old data and the fitted model is then used to predict new (Forster and Sober 1994). Imagine carrying out the task described in Figure 1.14 again and again. The *expected* performance of a model is what we want to know about.

There is a second refinement that needs to be added to this definition of predictive accuracy. What does it mean to talk about how accurately $L(\text{LIN})$ or $L(\text{PAR})$ predicts a single new observation in the pressure-cooker experiment? This new observation takes the form of a pair of temperature and pressure values $<x, y>$. When the temperature value x is fed into $L(\text{LIN})$ or into $L(\text{PAR})$, the output is a predicted value for the pressure y. We then can determine how close or far away the predicted value for y is from the observed value. We might do this by taking the difference between the two values and squaring it. Greater accuracy then means a smaller squared distance. Or we might compute the value of $Pr(\text{observed pressure value } y \mid \text{fitted model \& temperature value } x)$, with a larger likelihood indicating a higher degree of accuracy. These two approaches are related, in that squared distance is inversely related to likelihood, given some standard assumptions. The next question is how we should measure accuracy of prediction when there is more than one data point in the new data set. We could sum the squared distances or we could compute the likelihoods relative to all the new data. But notice that as the data set we are trying to predict gets larger, the sum of squares increases and the likelihood must decline as more and more terms are multiplied together. The problem is that we want to define the notion of predictive accuracy so that it does not turn out that a model is automatically less predictively accurate with respect to a larger data set than it is with respect to a smaller one. This is a general point about how we want to conceptualize measurement devices. When we ask about the accuracy of a bathroom scale or a thermometer or a tuberculosis test, the answer should not depend on how many times the device is used. A natural solution is to think of the predictive accuracy of a model as its average accuracy *per datum* (Forster and Sober 1994; Forster 2001). This point about the concept of predictive accuracy is not in Akaike (1973); he was thinking about model comparison where all the models are asked to predict the same new data set, which contains the same number of observations that the old data set contained. In this context, the difference between *per datum* predictive accuracy and total accuracy over the entire data set does not matter. For the moment, I'll follow Akaike's lead and omit mention of the fact that predictive accuracy is a *per datum* quantity. Later on, I'll return to the question of how larger and larger data sets should be brought into the picture.

After isolating the prediction problem of estimating how accurately a model will predict new data when fitted to old, Akaike (1973) derived a result that bears on it:

Akaike's Theorem: An unbiased estimate of the predictive accuracy

of model $M = \log\{Pr[\text{data} \,|\, L(M)]\} - k$.

We use the old data to find the likeliest member of model M and then take the natural logarithm (= base e) of its likelihood. We then subtract k, which is the number of adjustable parameters that the model contains. Notice that Akaike's estimate pays attention to both the model's fit to data and its simplicity. Akaike's theorem led to the formulation of the following model-selection criterion:

The Akaike information criterion (AIC): The AIC score of a model M,

$\text{AIC}(M) =_{\text{def}} \log\{Pr[\text{data} \,|\, L(M)]\} - k$.

The absolute value of a model's AIC score is not what is interesting about this criterion. What matters is how the scores of different models *compare* when the models are fitted to the same data set. AIC is a proposal that addresses the task of model *comparison*, not the task of model *acceptance* and *rejection* (Sakamoto et al. 1986: 84).[41]

How will the AIC scores for LIN and PAR compare? PAR will have a higher value for the first addend than LIN; L(PAR) will have the higher likelihood, and therefore the higher log-likelihood. But PAR contains one more adjustable parameter than LIN; in this respect, PAR is worse than LIN. Each model has one piece of good news and one piece of bad. The AIC scores of the two models depend on the character of the data. With some possible data sets, LIN will score better; with others, PAR will. The question is whether the data at hand depart sufficiently from linearity to justify the loss in simplicity that comes from shifting from LIN to PAR. AIC provides a principled basis for deciding how fit-to-data should be *traded off* against simplicity.

Three questions need to be answered about Akaike's theorem. What does "unbiased" mean? How is Akaike's theorem related to AIC? And what are the assumptions from which the theorem was derived?

A bathroom scale provides an unbiased estimate of your weight if the average of its values over many weighings is your true weight. In this hypothetical run of tests, we assume that your true weight remains the same. An unbiased estimator is *centered* on the true value; if your true weight is x, the *expected value* of the scale's readings is x. However, an unbiased scale may on any given occasion provide an estimate that is way too high or an estimate that is way too low. How much (squared) variation there will be, on average, among different estimates is called the *variance* of the estimator. A reading produced by a scale that has a high variance may have a very small probability of being close to the true value.[42] The best bathroom scale would be unbiased and have very small variance, but suppose you had to choose which of these virtues you prize more. Suppose one scale is perfectly unbiased and has large variance while a second has a small bias and a small variance. It is easy to imagine preferring the second to the first; this scale tends to read too high or tends

---

[41] I have presented AIC as a measure of predictive *accuracy*, so models with bigger scores are better than models with smaller ones. The reader should realize that AIC is usually presented as an expected *distance* from new data, in which case models are better the smaller their scores.

[42] Consider Figure 1.2 and suppose that the curves shown there represent the readings that three different unbiased scales might produce when an object that weighs 78 pounds is placed upon them.

to read too low (you don't know which) but it rarely is off by more than an ounce. This scale might be better than one that is centered on your true weight but tends to swing from 10 pounds too light to 10 pounds too heavy. These considerations indicate that it would be desirable to show, not just that AIC is an unbiased estimator but also that it is an estimator of minimum variance, or that it has a lower variance than other estimators that one might use. Akaike's theorem does not address this question; Sakamoto et al. (1986: 76–80) describe the variance of AIC estimates, but there is more to be learned about this subject. In any event, recall the frequentist setting of these questions about unbiasedness and variance. We are discussing the "operating characteristics" of a general policy, that of using AIC to estimate the predictive accuracy of models. Even if AIC is unbiased and has a low variance, that does not entail that when LIN scores better than PAR with respect to the data drawn from your pressure cooker, that LIN will *probably* be more predictively accurate. Posterior probabilities require priors, and this is something that frequentists disdain.

I turn next to the assumptions that Akaike used to prove his theorem. Akaike's proof uses the "normality assumptions" that are frequently exploited in mathematical statistics. This means, roughly, that each of the parameters in the model whose value you need to estimate will be such that repeated estimates form a normal distribution. There is, second, the assumption that old and new data sets are drawn from the same underlying reality. When you accumulate a data set on your pressure cooker and decide which of LIN and PAR will be more predictively accurate with respect to a new data set you have yet to see, you need to assume that the true but unknown law governing the pressure cooker won't change between your old observations and your new ones. Temperature and pressure must be related to each other in the same way across all data sets. There is a third assumption that goes into the proof. Your old data were accumulated by looking at various temperature values. These were chosen in accordance with some sampling procedure; perhaps these temperature values were drawn at random from a range of values. The theorem assumes that your new data will be drawn from the same distribution of temperature values. So, the relationship of X and Y is the same across data sets and so is the distribution of X values. Together these constitute a Humean *uniformity of nature assumption* (Forster and Sober 1994).

This last assumption means that Akaike's theorem and his criterion do not apply to inference problems in which you are trying to *extrapolate* – situations in which your sample is constrained to come from one range of temperature values and you want to make a prediction concerning what is true outside that range (Forster 2000b, 2001). It is useful to think about this point in relation to the fact that AIC is asymptotically equivalent with a different model-selection method that is called take-one-out cross-validation (Stone 1977). The cross-validation criterion makes no mention of simplicity. Rather, to test a model like LIN, you set aside one of the n data points in the sample, fit LIN to the $n - 1$ data points that remain, and then see how well $L(LIN)$ predicts the data point that was set aside. The procedure is repeated for each of the other data points; then you compute the average performance of the model across these n trials. That's the cross-validation score of LIN. The same procedure is carried out for other models and then the scores of different models are compared. Cross-validation is a general kind of procedure in which one gauges a model's performance by dividing one's data into a training (or calibration) set and a prediction (or test) set. In the application just described, the n data points are divided into $n - 1$ for training and 1 for prediction. This is called take-one-out cross-validation. The cross-validation framework also allows you to consider take-two-out, take-three-out, and so on. It is possible that one model scores better than another in terms of take-one-out, while the reverse is true for take-ten-out. The fact that AIC is equivalent with the former rather than the latter is telling. AIC is a solution to *one* prediction problem, but there are others.

It is interesting how AIC's comparison of LIN and PAR changes as the size of the data set increases. Consider the fact that a model's AIC score is influenced by two quantities, but only one of them changes as more data accumulate. The log-likelihoods of $L(LIN)$ and $L(PAR)$ both decline as more data roll in, but the number of adjustable parameters in each model of course stays the same. We know from the definition of AIC that AIC(LIN) > AIC(PAR) precisely when

$$\log\{Pr[\text{data} \mid L(\text{LIN})]\} - \log\{Pr[\text{data} \mid L(\text{PAR})]\} > -1.$$

The reason "−1" is on the right-hand side of this inequality is that PAR has one more adjustable parameter than LIN. Thus, what we want to know is whether

$$\log\left\{ \frac{Pr[\text{data} \mid L(\text{LIN})]}{Pr[\text{data} \mid L(\text{PAR})]} \right\} > -1$$

and (since the logarithm is base $e \approx 2.72$) this is true precisely when

(A) $$\frac{Pr[\text{data} \mid L(\text{LIN})]}{Pr[\text{data} \mid L(\text{PAR})]} > \frac{1}{2.72} \approx 0.37.$$

This inequality describes what it takes for LIN to have the higher of the two AIC scores. It may be true for small and middling sized data sets, but, with a sufficiently large data set, the inequality must be false; PAR must score better.[43] This is a sensible feature of AIC; the greater simplicity of LIN over PAR can compensate for $L(\text{LIN})$'s lower likelihood for some sample sizes, but eventually it cannot. If there is a slight parabolic bend in the data, you might want to ignore this when sample sizes are small, but if the bend is still there when you have lots of data, you'd be foolish to ignore it. The impact of simplicity on model evaluation *should* depend on sample size. The prediction problem that AIC is meant to address involves using an old data set to predict a new one *of approximately the same size*. If LIN scores better than PAR, given the data you have at hand, it does not follow that LIN would score better on a data set that is vastly larger.

The likelihood ratio test (§1.6) also applies to models like LIN and PAR, so it is useful to review some differences between it and AIC. First, there is the fact that the likelihood ratio test gives advice about whether the null hypothesis should be rejected; it therefore requires an arbitrary decision about how small the likelihood ratio of the two fitted models must be for one to reject the null. In contrast, AIC gives advice about model comparison, not model acceptance and rejection, and what it compares are estimates of predictive accuracy, not truth. Second, the mathematical underpinnings of the likelihood ratio test sanction its use only on nested models, but what could it mean to accept LIN and reject PAR, given that LIN entails PAR? The mathematics behind AIC justify its use on nested and non-nested models alike. Third, the likelihood ratio test violates the principle of total evidence; one doesn't look at the point

---

[43] Recall the point in §1.2 about the two witnesses whose testimonies agree. The fact that the testimonies are *independent* of each other, conditional on the proposition reported, was important in that discussion; similarly, in the present context, each datum is independent of every other, conditional on $L(\text{LIN})$ and conditional on $L(\text{PAR})$. When a small number of independent and reliable witnesses all say that proposition P is true, it is an open question whether the likelihood ratio of P to $notP$ will exceed some threshold; but for any threshold you specify, the likelihood ratio must exceed that threshold if there are sufficiently many unanimous witnesses. Similarly, if the data set is large enough, the log-likelihood of $L(\text{PAR})$ will exceed that of $L(\text{LIN})$ by any threshold you name, including the one described in proposition (A).

values in the data but only at a logically weaker description that says whether or not those data points fall in a given region. AIC abides by the principle of total evidence.

Identifiability

AIC penalizes models for being complex, but there are some models that are so complex that AIC does not even apply. It isn't that models that have more than 23,453,450,965 parameters are in principle beyond the pale. Rather, the limitation I have in mind comes from the number of data points in the observations one has at hand. A model with more parameters than there are data points will (typically) not be *identifiable*. A failure of identifiability means that there is no such thing as *the* maximum likelihood estimate of the parameters that the model contains. This failure of uniqueness can occur even in simple models, provided that the data set is sufficiently small. Consider our old friend LIN, a simple model if ever there was one. Suppose your data set consists of a single data point. There are infinitely many straight lines that pass exactly through this point; each has a likelihood that cannot be bettered. What would it mean to talk about LIN's predictive accuracy in this case? One would have to envision fitting the model to this single datum and then using "the" fitted model to predict other data sets that contain a single new data point. However, there is no such thing as "the" fitted model in this case. AIC does not even apply.

A data set that contains a single observation may seem like a joke, but the point about identifiability applies to the larger data sets that scientists actually use. AIC cannot be applied to models that are not identifiable. This means that our data limit the kinds of theories we can evaluate. In contrast, Bayesianism does not prohibit the assignment of prior probabilities and likelihoods to such models; for subjective Bayesians, such quantities are always well defined. Wittgenstein says in the last line of the *Tractatus* that *whereof one cannot speak, one must remain silent*. AIC embodies a kind of Wittgensteinian circumspection; Bayesianism is bolder.

Is AIC statistically inconsistent?

I mentioned earlier that estimators can be assessed for their unbiasedness and for their variance. I now want to consider a third property of estimators that one might value or even demand. This is the property of *statistical consistency.* Don't confuse this with the property of *logical* consistency. An estimator is statistically consistent when it converges on the true value of the parameter being estimated as more and more data are added. For example, suppose you want to infer the probability a coin has of landing heads when it is tossed. The policy of using the frequency of heads in a sample of tosses as your estimate is statistically consistent (a point that arose in connection with Reichenbach's straight rule in §1.2). This is the method of maximum likelihood estimation; by using this procedure, the estimate will converge on the true probability of heads as the number of tosses is increased. Does AIC converge on the true value of a model's predictive accuracy when the size of the data set is increased? That is, if one model is in fact more predictively accurate than another, can AIC be relied upon to award the first model the higher score as the size of the data set is increased without limit?

The question of AIC's consistency has often been misunderstood. The question is not whether AIC converges on the *true* model. AIC is not a device for assessing which model is true but provides an estimate of a model's predictive accuracy (Forster 2001); as already noted, it is perfectly legitimate to use AIC to evaluate a set of models all of which are known to contain idealizations and so all are known at the outset to be false. Also, when models are nested, you know in advance that the most complex model is true if any of them are. There is no need to use data or a model-selection criterion to ascertain this fact. Sometimes the question of consistency has been taken to be whether AIC converges on the true model that has the smallest number of adjustable parameters. So, if LIN and PAR are both true, the task assigned to AIC is to converge on LIN when the data are made large without limit. I pointed out before that this is not something that AIC will do. As a data set is made larger and larger, eventually the most complex model will have the best AIC score if the models considered are nested. This is not a defect in AIC. This most complex model *is* the model of greatest predictive accuracy for data sets that are large enough; AIC has succeeded in converging on the best model in that sense. However, the point of AIC is not to ascertain which models will be most predictively accurate for enormous or infinite data sets; the problem is to cope with the finite data sets one has at hand. If you make 200 observations of pressure and temperature on your pressure cooker, the problem is to figure out which model will do best in predicting what you'll observe if you draw another 200; it is a different problem to figure out which model will do best if old and new data sets contain 2,000,000,000,000,000 observations (Burnham and Anderson 2002: 298).

Demanding that AIC converge on the most predictively accurate of the models considered as data sets are made larger and larger is a bit like demanding that a bathroom scale converge on your true weight as you get heavier and heavier. The scale will fail to converge on a single value because the target is moving, not stationary. It makes more sense to demand that the scale's readings be centered on your true weight. If you weigh a single object of fixed weight again and again, will the average of these weighings converge on the object's true weight as the number of weighings increases? This is what the scale will do if it is unbiased. Repeatedly "weighing" a set of models using AIC will do the same thing, since AIC is an unbiased estimator.

Bayesian model selection

The criticism that AIC is statistically inconsistent is often voiced in the context of claiming that the Bayesian information criterion (BIC) derived by Schwarz (1978) is better. BIC will converge on the smallest true model, if the set of models you are considering includes one that is true. However, it is questionable why consistency in this sense should be thought a virtue if the competing models considered are not exhaustive; in this case, there is no guarantee that any of them is true. Also, if the models are nested, you know in advance that the largest model is true if any of them are. Why is it important to converge on the *smallest* true model, rather than on *a* true model? The latter task is easily achieved (if one of the models is true) and no model-selection criterion is needed to do this; the fact that the former task is harder does not explain why it is worthwhile.

Logically prior to this question about consistency is a more fundamental point of difference that separates BIC and AIC. As noted, the goal of AIC is to compare different models for their expected predictive accuracies. The goal of BIC, however, has nothing to do with predictive accuracy. This model-selection criterion has a Bayesian goal: to estimate the average likelihoods of composite models. LIN, for example, is an infinite disjunction of different straight lines, each of which confers its own probability on the data at hand. We saw earlier that the likelihood of LIN must be a weighted average over the likelihoods of these different straight lines, where the weighting terms have the form $Pr(L_i \mid LIN)$. Since BIC aims to estimate $Pr(\text{data} \mid LIN)$, the method must make assumptions as to what values these weighting terms have. Those not sold on Bayesianism despair of grounding these weighting terms in anything objective, and for that reason will be skeptical of BIC. Although a commitment to the values of these weighting terms must figure in any valid derivation of BIC, the weighting terms do not appear in the final product, which is the criterion that Schwarz (1978) derived for the average likelihood:

The Bayesian information criterion: The likelihood of model $M \approx \log\{Pr[\text{data} \mid L(M)]\} - k[\log(n)]/2$.

Here, k is the number of adjustable parameters in the model, and n is the number of data. BIC imposes a bigger penalty for complexity than AIC does; notice also that the second addend in BIC increases as the sample size increases, which is not true of the second addend in AIC. Schwarz (1978) derives BIC by assuming that the models under consideration have the same priors. Given this assumption, the criterion not only estimates average likelihoods; it also estimates posterior probabilities.

BIC is often applied to nested models, the idea being that BIC identifies the model in the set of competitors that has the highest posterior probability. But, as already noted, no matter what the data say, LIN cannot be more probable than PAR if LIN entails PAR. When models are nested, one can tell *a priori* which model has the highest prior and the highest posterior probability; there is no need to consult the data to figure this out and no need to consult a model-selection criterion. If the data lead BIC to say that LIN has a higher posterior probability than PAR, the Bayesian criterion has simply made a mistake and its testimony should be set aside. This problem can be avoided by restricting the application of BIC to non-nested models.

Although BIC was derived as a device for estimating average likelihoods and posterior probabilities, we still may ask how well it performs as an estimator of predictive accuracy. We know from Akaike's theorem that AIC is unbiased; since BIC differs from AIC by a constant, BIC must therefore be a biased estimator of predictive accuracy. A further defect in BIC also follows: BIC's estimates of predictive accuracy have a larger expected squared error than the ones generated by AIC (Forster and Sober, in preparation).

The debate over AIC and BIC needs to be understood, in the first instance, as a debate over choice of goals – estimating predictive accuracy versus estimating average likelihood. Only after a goal has been chosen can the question be raised as to which criterion does better in achieving that goal.

The subfamily problem

A curve, since it contains no adjustable parameters, is a member of many models. For example, "$y = 3 + 4x$" is a member of LIN, but it also is a member of PAR and of lots of other models besides. Given this, how is a curve's AIC score to be computed? Its log-likelihood is univocal, but what penalty should we impose on it for its degree of complexity? If we view the curve as a member of one model, we'll apply one penalty term, but if we view it as a member of a different model, we'll apply another. This is the subfamily problem (so called by Forster and Sober 1994).

One step towards solving this problem is to recognize that AIC applies to *models* and that there is no need for AIC to say which model is the one to which a curve "really" belongs. The predictive accuracy of a model is its average performance as it is fitted to old data sets and then makes predictions about new ones. There is no paradox in saying that LIN and PAR may differ in their predictive accuracies even if L(LIN) and L(PAR) happen to be identical curves in virtue of the (collinear) data set one has at hand. AIC also applies to *curves*, but this is because curves are a limit case; they are models that contain zero adjustable parameters. A curve's AIC score is just its log-likelihood (since its complexity penalty is zero). Thus, it can turn out that "$y = 3 + 4x$" has a lower likelihood than "$y = 3 + 4x + 0.001 \, x^2$," and so the former has the lower AIC score, and yet LIN has a higher AIC score than PAR, where the two curves happen to be the best-fitting members of the two models, respectively. The two curves have their own AIC scores, LIN has a third, and PAR has a fourth.

Although this point shows that AIC is not guilty of contradicting itself (or of arbitrarily deciding which model a curve "really" belongs to), it does leave another question unanswered. How should we *use* AIC to make predictions? This is a *pragmatic* question in the sense of that term discussed earlier in connection with the principle of total evidence (§1.3). Should we apply AIC to the two curves L(LIN) and L(PAR) and therefore use the latter to make our predictions? Or should we apply AIC to LIN and PAR and allow the data to help us decide which model is better? Focusing exclusively on curves has the result that we always choose the curve that comes from the largest model. The motivation for using AIC is to find models that make accurate predictions; applying AIC only to fitted models prevents the criterion from helping us to achieve that end. But there is another reason to decline to use AIC in this way. AIC provides unbiased estimates of predictive accuracy, regardless of whether it is applied to LIN and PAR, or to L(LIN) and L(PAR), or to all four. One reason to score LIN and PAR, rather than L(LIN) and L(PAR), is that AIC has greater variance when it is applied to smaller models (Escoto 2004); applying AIC to fitted models is more apt to produce inaccurate estimates of predictive accuracy.

There is another dimension to this pragmatic problem. The fact that AIC is a comparative principle, not a criterion for acceptance, shows that it would be a mistake to make a prediction by using the model that has the best AIC score while ignoring all the other models that were considered. After all, AIC is an estimator that is subject to error. This suggests that predictions should be made by *model averaging* (Burnham and Anderson 2002). If you want to predict the pressure that will result when you set your pressure cooker to a given temperature, you should consider the prediction made by the model with the best AIC score, the prediction made by the second best, and so on. You can average these different predictions by using *AIC weights* – giving more weight to predictions that come from models that have better AIC scores.

The scope of AIC

I have used the models LIN and PAR to explain what AIC amounts to, but this should not be taken to mean that AIC is relevant only to "curve-fitting problems." Philosophers sometimes disparage curve fitting as a kind of naïve inductive inference in which the hypotheses considered seek merely to identify patterns that hold among observational quantities. Model-selection criteria, including AIC, are not limited to such problems. They also apply to *causal models* that say that an effect term is influenced by the values of any number of input variables. In Chapters 3 and 4, we will see how model-selection ideas apply to problems in evolutionary biology.

Although I have argued that the dispute over AIC versus BIC is based on a failure to realize that they are estimators of different quantities, the fact remains that there are different model-selection criteria that all focus on the goal of estimating predictive accuracy. For example, there is a version of AIC derived by Sugiura (1978) that is better to use when some of the models under evaluation have a large number of parameters relative to the number of observations available; it is called AIC_c and imposes a larger penalty for complexity than AIC does.[44] There is also a criterion (TIC) derived by Takeuchi (1976). These criteria all compute the likelihood of the best fitting member of a model and then impose a penalty for complexity; they differ over what that penalty term is. I mentioned earlier

---

[44] Burnham and Anderson (2002: 50) recommend using AIC_c precisely when $n/k < 40$, where n is the number of observations and k is the number of parameters in the largest model under evaluation.

that AIC is equivalent to take-one-out cross-validation; this raises the question of what the statistical properties are of cross-validation methods that take more than one out, and of what use such methods are in different inference problems (Forster 2006, 2007). And there is also the question of what model-selection criteria are best when the goal is extrapolation, not interpolation. What I find striking in this diversity of problems and solutions is what they have in common. This is the Akaike framework, within which all these approaches are to be understood. We want to know how accurately a model will predict new data when it is fitted to old. How well the model fits the old data is relevant to this question, but so is the model's complexity (the number of adjustable parameters it contains). This framework helps explain why scientists should bother to test models that they know are false. If the goal were to decide which models are true, there would be little point in testing idealizations. But predictive accuracy is a different story, and it has its own epistemology. Bayesianism, likelihoodism, and the Neyman–Pearson framework each have their different drawbacks when applied to this kind of problem. The subject that Akaike initiated throws new light on these issues, and there is the promise of more light to come.

Realism and instrumentalism

Virtually everyone who follows professional basketball believes that players sometimes have "hot hands." When players are hot, their chance of scoring improves, and teammates try to feed the ball to them. Gilovich et al. (1985) tested this widespread belief by doing a statistical analysis of scoring patterns in the National Basketball Association. Their conclusion was that one cannot reject the null hypothesis that each player has a constant probability of scoring throughout the season; belief in hot hands, they say, is a "cognitive illusion."[45] Basketball mavens reacted to this statistical pronouncement with total incredulity. Placing this dispute in the Akaike framework allows it to make more sense. Scientists should not feel shy about admitting that the null hypothesis is false. The idea that each player never wavers in his probability of scoring *is* preposterous. But

---

[45] See Wardrop (1999) for a skeptical assessment of Gilovich et al.'s analysis. Wardrop argues that Gilovich et al. tested hypotheses about *correlation* (whether a player's probability of scoring on a given shot if he scored on earlier shots is greater than his probability of scoring if he missed previously), but did not assess the issue of *stationarity* (maybe a player's probability of scoring suddenly shifts from one value to another).

even if this silly hypothesis is false, there still may be a point to seeing how accurately it predicts new data. Perhaps the truth about basketball players is very complex; their scoring probabilities change as subtle responses to a large number of interacting causes. If so, players and coaches may make better predictions by relying on simplified models. Even if hot hands are a reality, trying to predict when players have hot hands may be a fool's errand.

The problem of evaluating how accurately models predict new data when fitted to old has a philosophically interesting property: a model known to be false will sometimes be more predictively accurate than a model known to be true. What is perhaps more surprising is that we can sometimes *estimate* which of them we should expect to be more predictively accurate and the methods available for assessing this sometimes favor false models over true ones. The Akaike framework thus breathes new life into an old philosophy. *Instrumentalism* is the view that the goal of scientific inference is to find theories that make accurate predictions, not to find theories that are true.[46] It stands opposed to *scientific realism*, which holds that the goal is to find true theories.

The debate between realism and instrumentalism can't be resolved by polling scientists as to what their goals are. Some scientists say that they want to find out what is true while others say that their object is to find theories that make accurate predictions; all may be sincerely reporting their personal goals, but that is not what is at issue. The philosophical debate concerns what *scientific inference* is able to attain, not what *scientists* yearn for. If the inference procedures used in science are able to discover which theories are true, or which are probably true, then realism is correct. If those procedures are capable only of discovering which theories will make the most accurate predictions, then instrumentalism is. Both philosophies need to be tempered by the fact that scientists rarely are able to examine a set of hypotheses that exhaust the possibilities (Stanford 2005). Scientists deal with the theories that have been developed thus far, and no one can foresee the novel theories that future innovators may put on the table. This sobering fact about the limitations that scientists perpetually face means that the best that scientists can do at any time is to render comparative judgments. Realism should be understood as the

---

[46] Instrumentalism is sometimes also formulated as a semantic thesis – that scientific theories are neither true nor false, but are merely instruments for making predictions. The proper response is that there is no reason to think that theories lack truth values, and no reason to burden an epistemological thesis with an outmoded philosophy of language (Sober 2002).

claim that scientific modes of inference indicate which of a set of competing hypotheses is the best candidate for being true; instrumentalists think that science is in a position only to say which of the competitors can be expected to make the most accurate predictions.

Instrumentalism and realism are usually formulated as *global* theses. They are claims about *all* the hypotheses that scientists investigate. It doesn't matter whether the hypotheses in question are models or fitted models, any more than it matters whether they are part of the subject matter of one science or another. The Akaike framework shows that this global formulation of the problem needs to be recast. The framework makes room for an instrumentalist philosophy of *models*. The fact that one model (M_1) has a better AIC score than another (M_2) is grounds to think that the first will be more predictively accurate; it is not grounds for thinking that M_1 is true, or more probably true, or better supported as a candidate for being true. However, this difference in the scores of the two models has another implication concerning the truth of the *fitted models* – Akaike's theorem can also be formulated as the thesis that the AIC score of a model M is an unbiased estimate of the closeness to the truth of the fitted model $L(M)$, where closeness is measured by the Kullback–Leibler distance.[47] With respect to the pressure cooker in your kitchen, there is a true but unknown curve that describes how temperature and pressure are related. Specific curves have different Kullback–Leibler distances to that true curve. Models are instruments for finding curves that are close to the truth and models are compared with each other to determine how well they advance that goal.[48] The Akaike framework therefore makes plausible a mixed philosophy: instrumentalism for models, realism for fitted models (Sober 2002b). When a false model F and a true model T are both fitted to the data, $L(F)$ will sometimes be closer to the truth than $L(T)$. AIC and other model-selection criteria seek to provide guidance as to when this is so.

---

[47] Suppose t is the true distribution $(p_1, p_2, \ldots, p_n)$ of a discrete random variable and c is a candidate distribution $(\pi_1, \pi_2, \ldots, \pi_n)$. The KL distance from the candidate c to the truth t is $I(t,c) = \sum p_i \log(p_i/\pi_j)$. Notice that the true distribution provides the weighting on the log of the ratio. KL is a "directed distance;" the distance from c to t (where t is true) doesn't have to be the same as the distance from t to c (where c is true). See Burnham and Anderson (1998) for further discussion.

[48] The relation of AIC to Kullback–Leibler distances provides an easy answer to the question of why one should care about AIC estimates if one has no interest in using fitted models to predict *new* data. One still might care about finding fitted models that are close to the truth when Kullback–Leibler distance is used to measure closeness.

One challenge to this limited form of instrumentalism begins with the idea that instrumentalism and realism should be thought of as claims about the *ultimate* goals of science. Maybe finding models that make accurate predictions is a mere tactic that science deploys in the larger campaign. A realist can grant that it is useful to find idealized models that make accurate predictions if such models are worth having because they help one get to the truth, and truth is the ultimate goal. A defense of this response requires more than the psychological fact that scientists often would *like* to find true theories. What is needed is an account of how scientific inference makes it possible to turn assessments of the predictive accuracy of models into claims about which theories are true. I've already mentioned that fitted models may be nearer or farther away from the truth, and that there is an intimate connection between M_1's being a better predictor than M_2 and $L(M_1)$'s being closer to the truth than $L(M_2)$. Perhaps the objection can then be put by saying that the real goal of science is to discover which fitted models are true and that models themselves are mere means to that end. Again, this may or may not be true as a psychological claim about what interests various scientists (though, in fact, scientists are often more interested in models than in fitted models). But how can it be justified as a claim about scientific inference, not about the psychology of scientists? If finding models that are accurate predictors and fitted models that are close to the truth go hand in hand, then it is hard to see that one is logically prior to the other. Given this, the mixed thesis of "instrumentalism for models, realism for fitted models" may be more satisfactory than either global realism or global instrumentalism.

What is a parameter?

AIC says that the complexity of a model is relevant to estimating its predictive accuracy; BIC says that a model's complexity is relevant to estimating its average likelihood. Both measure complexity by counting parameters. This raises an important question. A model is a *proposition*, distinct from the sentence in some language in which it happens to be expressed; the proposition that temperature is linearly related to pressure is no more a part of English than it is part of Chinese. Yet, the number of parameters in a model seems to be a syntactic feature of how the model happens to be described; by changing the language used, you seemingly can change the number of parameters the model contains. If so, how could the number of parameters be relevant to ascertaining these epistemically relevant properties of the model itself – its predictive accuracy or its average likelihood?

This question can be fleshed out by way of our running example, the comparison of LIN and PAR. I've said that LIN has two parameters and PAR has three (ignoring, for the moment, the error term that each deploys). Any straight line of the form $y = mx + b$ can be represented as a point in a two-dimensional parameter space in which one axis is its slope (m) and the other is its y-intercept (b). A straight line in the x-y plane is just an ordered pair of numbers $<m,b>$ in this parameter space. In the nineteenth century, Georg Cantor discovered that the number of points in a plane is the same as the number of points on a line. This means that there is a one-to-one (injective) mapping from ordered pairs to single numbers. An example of this kind of mapping is provided by *interleaving*. Consider a plane whose possible m values run from 0 to 1 and whose b values do the same. Each point in this unit square can be expressed as an ordered pair, each of whose members is a decimal expansion of the form

$$m = 0.m_1 m_2 m_3 \cdots \quad b = 0.b_1 b_2 b_3 \cdots$$

By interleaving we can represent this pair of numbers as a single number

$$i = 0.m_1 b_1 m_2 b_2 m_3 b_3 \cdots$$

Notice that there is a function from each $<m,b>$ pair to a single number i, and another function from each possible value of i back to that single $<m,b>$ pair. So, in what sense are there *two* parameters (m and b) in LIN? Why not say, instead, that there is just *one* (namely i)? And if LIN has just one parameter, so does PAR (since you can interleave triplets just as well as pairs). The difference in complexity of the two models seems to be an artifact of the notation we arbitrarily choose.

This question was important in nineteenth-century mathematics where the problem was to describe what *dimension* means. Is there a rigorous and linguistically invariant way to express the thought that a plane has two dimensions while a line has just one? The problem was solved in the twentieth century by Brouwer, who isolated a concept of dimension that is *topologically invariant* (Courant and Robbins 1959: 249–51; Dauben 1994). The idea of interleaving can be used to convey the intuitive idea.

Consider three straight lines (one of which is true); each is defined by its coordinates in the $<m,b>$ parameter space:

$$\text{Truth} = <1,1> \qquad L_1 = <2,1> \qquad L_2 = <1,3>.$$

Notice that L_1 is closer to Truth than L_2 is. If we interleave each of the ordered pairs, we obtain:

$$\text{I(Truth)} = 11 \qquad \text{I}(L_1) = 21 \qquad \text{I}(L_2) = 13.$$

Notice that $\text{I}(L_2)$ is closer to I(Truth) than $\text{I}(L_1)$ is. Although the mapping achieved via interleaving is injective, it is not *distance preserving*. The mapping does not have the property that points that are close together in the $<m,b>$ plane have images in the line that are always close together. There is more to the idea of topological invariance than that of a mapping that is distance-preserving, but the example of interleaving helps elucidate what a parameter is in model-selection theory. If a space has n dimensions, then there is no one-to-one, continuous, and distance-preserving mapping from that space to another space that has m dimensions, if $n \neq m$. Dimensionality is in this sense an invariant quantity.

What does this imply about the dimensionality of LIN? Is it two, or one, or some other number? By definition, it must be unique, the possibility of interleaving notwithstanding. To answer this question would lead us too far afield. But I hope the following two comments are helpful. First, consider the relationship of *LIN* to *PAR*. *LIN* is nested in *PAR*. This is a fact about the two propositions and has nothing to do with the language in which they happen to be expressed. It is a consequence of this nesting relationship that *LIN* cannot have a higher dimensionality than *PAR*. And since the fact about the nesting relationship is invariant, the same holds for the fact about dimensionality (Forster 1999). The second comment returns us to the content of Akaike's theorem. As noted, the theorem identifies an unbiased estimate of the predictive accuracy of a model M, or, equivalently, an unbiased estimate of the Kullback–Leibler distance from $L(M)$ to the true but unknown probability distribution T. Expressed in this second way, Akaike's theorem states that:

$$E[\textit{KL–Closeness of } L(M) \text{ to } T] = [\text{Log-likelihood of } L(M)] - k.$$

The left-hand side describes a language-independent quantity, and the same is true of the first addend on the right. It follows that k must be language independent as well. Again, this does not tell you how to determine what value of k a model has. But it does assure you that, whatever it is, it is not an artifact of notation.

Is AIC frequentist?

I have classified AIC as a type of frequentism; I now want to consider briefly whether this classification makes sense. I have emphasized that AIC isn't a criterion for acceptance and rejection and that it does not violate the principle of total evidence. What is more, the AIC score of a model does not depend on the stopping rule used. These properties of AIC separate it from significance tests and the Neyman–Pearson theory. If AIC is frequentist, it is a different kind of frequentism.

Akaike (1973) refers to his result as "an extension of the maximum likelihood principle," but this phrase should not lead us to conclude that AIC is a form of likelihoodism. AIC does not say that the best model is the one that has the highest average likelihood, nor does it say that model M_1 is better than model M_2 precisely when L(M_1) has a higher likelihood than L(M_2). It is even clearer that AIC is not Bayesian. In using AIC, you are not estimating the probability that a model is true, nor are you estimating the probability that one model will be more predictively accurate than another. To reach conclusions about such posterior probabilities, you would need prior probabilities, and these play no role in AIC.

The main reason that AIC is viewed as a frequentist construct is the character of Akaike's theorem, which establishes that this estimation procedure has the long-run operating characteristic of being unbiased. This is just the sort of property that frequentists care about. Of course, they recognize that other operating characteristics are relevant as well. Is a procedure statistically consistent? What is its variance? Is it admissible? As noted in §1.5, Bayesians and likelihoodists do not object to the evaluation of procedures; they find nothing amiss in comparing the Madison tuberculosis test with the one manufactured in Prairie du Chien. However, they insist that there is a further question that needs to be asked: How should one evaluate a given *estimate* (never mind what method of estimation was used to construct it)? Likelihoodists want to know how well supported the estimate is, where support is understood in terms of the law of likelihood. Bayesians want to know how probable it is that the estimate is true (or close to the truth). Frequentists deny that this second question makes any sense; they hold that *estimators* have long-run operating characteristics, but there is nothing further to be said about the individual *estimates* that those estimators generate.

The fact that Akaike's *theorem* addresses a kind of question that frequentists think is important does not show that *AIC scores* are meaningless from a Bayesian or likelihoodist point of view. Of course it is possible for M_1 to have a better AIC score than M_2 even though M_1 has the lower average likelihood and even though $L(M_1)$ is less likely than $L(M_2)$. But the law of likelihood and AIC still could join hands in friendship if AIC scores provided evidence concerning the predictive accuracies of different models, where evidence is understood in terms of the law of likelihood. Think of AIC as a measurement device, like a thermometer; perhaps AIC scores are to predictive accuracy as thermometer readings are to temperature. If a thermometer assigns a higher number to one object than it does to another, we take that to be evidence that the first object has a higher temperature than the second. Perhaps the same is true of AIC scores. The relevant property of thermometers can be described as follows. Suppose the thermometer readings on objects O_1 and O_2, $R(O_1)$ and $R(O_2)$, are such that $R(O_1) - R(O_2) = x > 0$. This observation indicates that the best point estimate of the temperature difference is positive when

There exists a $y > 0$ such that for all $z < 0$,

$$Pr[R(O_1) - R(O_1) = x \mid \text{Temp}(O_1) - \text{Temp}(O_2) = y]$$
$$> Pr[R(O_1) - R(O_2) = x \mid \text{Temp}(O_1) - \text{Temp}(O_2) = z].$$

What would it take for the same thesis to hold for AIC scores and their relationship to the predictive accuracies of different models? What would be true is that, when we observe that model M_1 has an AIC score that is x units larger than the AIC score of model M_2, that the best point estimate of the difference in predictive accuracies is positive. That is,

There exists a $y > 0$ such that for all $z < 0$,

$$Pr[\text{AIC}(M_1) - \text{AIC}(M_2) = x \mid \text{PA}(M_1) - \text{PA}(M_2) = y]$$
$$> Pr[\text{AIC}(M_1) - \text{AIC}(M_2) = x \mid \text{PA}(M_1) - \text{PA}(M_2) = z].$$

This inequality does not follow from Akaike's theorem. And it may not hold for *all* values of x – e.g., when x is very close to zero (Forster and Sober, in preparation) – but when it *does* hold, Bayesians and likelihoodists should have no qualms about viewing AIC scores as evidence. AIC began life with a frequentist pedigree, with Akaike's theorem. But

AIC scores may be essentially tied to frequentism no more than thermometer readings are.

1.8 A SECOND TEST CASE: REASONING ABOUT COINCIDENCES

When Evelyn Marie Adams twice won the New Jersey lottery, the *New York Times* said that the odds of this happening by chance are 1 in 17 trillion; this is the probability that Adams would win both lotteries if she had purchased a single ticket for each and the drawings had been at random. In fact, the newspaper made a small mistake. If the goal is to calculate the probability of Adams' winning those two lotteries, the reporter should have taken into account the fact that Adams purchased multiple tickets; the newspaper's very low figure should therefore have been somewhat higher. However, the typical response of statistical sophisticates is that this modest correction misses the point. For sophisticates, the relevant event to consider is not that Adams won those two lotteries, but the fact that someone won two state lotteries at some time or other. Given the many millions of people who have purchased lottery tickets, this is "practically a sure thing" (Diaconis and Mosteller 1989: 859).

Was Adams' double win a mere coincidence? Or were these two lotteries rigged in her favor? Diaconis and Mosteller say that the relevant principle to use when reasoning about coincidences is the *law of truly large numbers*. This says that, "with a large enough sample, any outrageous thing is likely to happen." They cite Littlewood (1953) as having the same thought; with tongue in cheek, Littlewood defined a miracle as an event whose probability is less than 1 in 1 million. Using as an example the US population of 250 million people, Diaconis and Mosteller observe that if a miracle "happens to one person in a million each day, then we expect 250 occurrences a day and close to 100,000 such occurrences a year" (1989: 859). If the human population of the earth is used as the reference class, miracles can be expected to be even more plentiful.

How should the law of truly large numbers be applied to Adams' double win? One possibility is to change our description of the observations from

(1) Evelyn Marie Adams, having bought four tickets in each of two New Jersey lotteries, wins both.

to the logically weaker statement that

(2) Someone at sometime, having bought some number of tickets in two or more lotteries in one or more states, wins at least two lotteries in a single state.

If you are using probabilistic *modus tollens* (§1.4) to think about this problem, and if you believe that Adams' double win does not warrant rejecting the hypothesis that the lotteries were fair, then weakening the data description from (1) to (2) may be appealing. It provides a simple strategy for neutralizing the appeal of conspiracy theories. But even if this strategy leads to the conclusion about Adams' good fortune that you find intuitive, it raises the question of when and how much a description of the data should be weakened. Without some guidance on this issue, you run the risk of weakening the data whenever they go against your pet theories. This allows you to be complacent about what you already believe and skeptical about the hobbyhorses that others have chosen to ride – a satisfying state of mind perhaps, but one that cannot stand up to rational scrutiny.

A second approach, which abides by the principle of total evidence (§1.4), is Bayesian. It concedes that the hypothesis that the lotteries were fair has a much lower likelihood than the hypothesis that the two lotteries that Adams won were rigged in her favor, but then invokes prior probabilities to show that Adams' double win does not make it *probable* that the two lotteries were rigged. My objection to invoking priors here is not that they are subjective. After all, we may have evidence that lotteries are usually fair, though developing this point would require us to consider the fact that people who rig lotteries have a powerful incentive to insure that their chicanery remains secret. Rather, my reservation about this Bayesian reply is that it concedes that the observations favor the hypothesis that the two lotteries were rigged in Adams' favor. The law of likelihood, which is central to Bayesianism, obliges Bayesians to make this concession. I suggest that it is possible to show that the observations do not have this evidential significance. The model-selection framework allows this kind of argument to be developed, although it must be recognized that the goal has been changed; we no longer are trying to figure out which hypothesis is probably true or which has the highest likelihood; rather, we are aiming to discover which will be most predictively accurate.

The model-selection approach agrees with Bayesianism that data cannot be discarded. Rather, the right approach is to *add* observations. Instead of weakening the observations by discarding (1) and focusing on (2), we should include additional observations about the people who won and lost other lotteries and how many tickets they purchased. Once the data set is augmented, we can consider multiple models. One of them says that each lottery is fair:

(Fair)   For each ticket i purchased in New Jersey lottery j,

Pr(ticket i wins | ticket i was purchased in lottery j) $= \frac{1}{n_j}$

(where n_j is the number of tickets purchased in lottery j).

This model has one parameter for each lottery. It is far simpler than the following model:

(Rigged)   For any ticket i purchased in New Jersey lottery j by person k, Pr(ticket i wins | ticket i was purchased in lottery j by person k) = p_{jk}.

The (Rigged) model has a separate parameter for each person buying a ticket in each lottery. If the data on lottery winners and losers favors (Fair) over (Rigged), they do so not by showing that (Fair) is more probable than (Rigged), nor by showing that (Fair) has the higher likelihood, but by showing that (Fair) can be expected to be more predictively accurate than (Rigged).

(Fair) is a model that *unifies* the data far more than (Rigged) does. (Fair) says that all the tickets sold in a given lottery are subject to the same probabilistic process, whereas (Rigged) says that each person buying tickets in a given lottery is a law unto herself. Because AIC and other model-selection criteria value paucity of parameters, they offer an explanation of why a model that applies k parameters to an entire data set often has a leg up on a disunified model that subdivides the data into parts, supplying a different set of k parameters to each.

It is important to realize that whether a more unified model has a better AIC score than a less unified model depends on the data. There is no categorical imperative that says that unified models are always better. For example, it is not inevitable that Fair is superior to the following even simpler model:

(One)   For each ticket i purchased in any New Jersey lottery, Pr(ticket i wins | ticket i was purchased in any New Jersey lottery) = p.

The (One) model lumps together all New Jersey lotteries; tickets purchased in different lotteries are said to have the same chance of winning. This model is even more unified than (Fair), but that does not guarantee that its estimated predictive accuracy will be greater.

Although the models just considered exhibit a virtue of the model-selection framework, there is a model not yet mentioned that exhibits one of its limitations. The conspiracy model (Rigged) gets lower marks than the (Fair) model, but what about the following (Mixed) model?

(Mixed)   For each ticket k purchased by Evelyn Marie Adams, Pr(ticket k wins | ticket k was purchased by Evelyn Marie Adams) = p. For each other ticket i purchased in New Jersey lottery j, Pr(ticket i wins | ticket i was purchased in lottery j) = $\frac{1}{n_j}$ (where n_j is the number of tickets purchased by people other than Adams in lottery j).

Suppose, to make things simple, that Evelyn Marie Adams bought tickets only on the two lotteries that she ended up winning and bought a few tickets on each. This means that L(Mixed) fits the data far better than L(Fair). And (Mixed) has just one more parameter than (Fair). This means that (Mixed) may have a better AIC score than (Fair). If so what's wrong with this mixed model? The Bayesian has an answer: It has a lower prior probability. It is not obvious what the model selectionist can say here.

This question aside, there is a point here on which defenders of different statistical frameworks can agree. The human mind often imposes patterns where none exist. Repeatedly tossing a fair coin will inevitably produce runs of heads; it is tempting to think that the coin has suddenly become biased ("hot"). Part of what facilitates this kind of over-interpretation is that we tend to focus on observations that are vivid. We narrow the data set. We focus on the run of heads, and not on all the tosses. It is Adams' double win that excites our curiosity, not a boring compilation of all the winners and losers in all New Jersey lotteries. In all these cases, we need to embed what we find vivid in a more inclusive data set; we then need to formulate models that apply not just to what is vivid but to what is quotidian as well.

1.9 CONCLUDING COMMENTS

The claim that science aims to discover which theories are probably true may sound like a truism, but there are two reasons to pause over this formula. The first is that one must be wary of an equivocation. In ordinary English, to say that a theory is "probably true" just means that it is plausible or reasonable, given the evidence at hand; praising a theory in this way leaves open what relevance the mathematical theory of probability might have to such judgments. Bayesianism is a substantive epistemology, not a truism. The second reason for pausing is that scientists often work with idealized models that are known to be false. How can a model known to be false probably be true? There needs to be a place in our epistemology for comparisons of such theories.

Royall's three questions (§1.1) are different; questions about *evidence* must be separated from questions about *acceptance* and from questions about *action*. This threefold distinction will be important in what follows when we consider evidential questions such as the following:

- Are the imperfect adaptations that organisms exhibit evidence that they were not produced by an intelligent designer?

- Is the fact that bears in cold climates have longer fur than bears in warm climates evidence that fur length evolved by natural selection as an adaptive response to ambient temperature?
- Are the similarities that species exhibit evidence that they stem from a common ancestor?

Perhaps you find it obvious that the answer in all three cases is *yes*. If so, what's the point of taking on the job of figuring out why? The answer is that the book you are reading is a work of philosophy, not biology, and so the exploration of what seems obvious is of central importance. Even when a proposition strikes us as obvious, it is often not so obvious why the proposition is *true*. This is the occasion for philosophical exploration. One possible result is that what seems obvious turns out *not* to be true unrestrictedly, but is true only in a restricted set of circumstances. Another is a deeper grasp of the assumptions we tacitly make that underlie our convictions.

The law of likelihood is common ground for Bayesians and like-lihoodists. It will provide the starting point for several of the questions about evidence and evolution that I will examine. Putting the law to work in the next chapter will require us to consider a new complication. The hypotheses we wish to test often do not have likelihoods when considered all by themselves; they need to be supplemented by additional information if they are to confer probabilities on the observations. An important question will be how this "additional information" should be obtained. There also will be a place in what follows for ideas about evidence that derive from a model-selection framework. Just as the readings of an unbiased scale can provide evidence as to which of two people is heavier, so AIC scores can provide evidence as to which of two models is apt to be more predictively accurate. The law of likelihood is central to understanding what evidence is, but it is not the only idea we will use. The law applies to simple statistical hypotheses and produces a verdict about whether the observations favor the hypothesis that H_1 is true over the hypothesis that H_2 is true; AIC and other model-selection criteria apply to composite statistical models and help us discern which models will be more predictively accurate. The law of likelihood and AIC are not in conflict, given their different goals and their different realms of applicability.

CHAPTER 2

Intelligent design

2.1 DARWIN AND INTELLIGENT DESIGN

The first edition of Darwin's *On the Origin of Species by Means of Natural Selection, or the Preservation of Favoured Races in the Struggle for Life* (1859) begins with quotations from two philosophers:

But with regard to the material world, we can at least go so far as this – we can perceive that events are brought about not by insulated interpositions of Divine power, exerted in each particular case, but by the establishment of general laws. (W. Whewell, *Bridgewater Treatise*)

To conclude, therefore, let no man out of a weak conceit of sobriety, or an ill-spirited moderation, think or maintain, that a man can search too far or be too well studied in the book of God's word, or in the book of God's works; divinity or philosophy; but rather let men endeavour an endless progress or proficience in both. (F. Bacon, *Advancement of Learning*)

William Whewell was Darwin's contemporary and rejected his theory of evolution, a result that Darwin probably anticipated when he wrote *The Origin of Species*.[1] Francis Bacon wrote more than 200 years earlier. The two quotations are interesting because of what they reveal about Darwin's views on the relationship of belief in God and belief in evolution.

Bacon's remark harks back to an old distinction between the Bible (God's word) and nature (God's work). Sacred texts and natural phenomena provide separate pathways for learning about God. This two-pathway picture was important in the formation of the Royal Society in

---

[1] The *Bridgewater Treatises* were a series of books that developed the argument for the existence of God that we will consider in detail in this chapter – the argument from design. In the 1833 book from which Darwin drew this quotation, Whewell embraced the view that the origin of species and the origin of languages are beyond the reach of present-day science and are likely to remain so; he argued that both require divine intervention. Darwin's quoting from Whewell does not mean that he expected Whewell to like how he used this passage. See Ruse (1979), Hodge (1991), Brooke (2003), and Snyder (2006) for different views of Darwin's relation to Whewell.

109

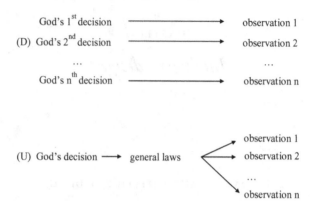

Figure 2.1 Two theistic hypotheses. (D) says that each of our observations traces back to a separate decision made by God; (U) says that God creates a single set of general laws that produces all the details we observe. (D) is a disunified model; (U) is unified.

London and to the philosophy within which the scientific revolution of the seventeenth century developed. The founders of the Royal Society included many clerics who saw "the new science as itself a witness to the Deity's handiwork and therefore to his existence" (Hacking 1975: 169). Darwin quotes Bacon to make a more specific point: that there is no conflict between theism and the theory of evolution. Darwin aimed to describe the processes in nature that account for the features of organisms that we observe; it is logically consistent to add to this biological claim the theological thesis that the evolutionary process occurs because God put it in place. This is the idea that evolution is God's way of making organisms. It now goes by the name *theistic evolutionism*.

The passage from Whewell expresses a different idea. Whewell is discussing the two hypotheses depicted in Figure 2.1. We make a vast number of observations. Should we view each of these observations as the direct result of God's separate decree? Or should we view those observations as knitted together, as flowing from a single cohesive set of laws that God ordained? Whewell's view is that the unified hypothesis is superior to the disunified hypothesis. Darwin used this idea to draw a conclusion that Whewell did not anticipate. The hypothesis that each kind of animal and plant was separately created by an intelligent designer is inferior to the hypothesis that each evolved according to a single set of laws that God created. Darwin's theory of evolution by natural selection was intended to specify the laws that unify the enormous variety of observations we have made and continue to make of the living world (Kitcher 2003).

Darwin went beyond the thesis that the unified hypothesis is *superior* to the hypothesis of disunity; he additionally thought that the disunified hypothesis is *empty*. The "theory" it embodies is easy to state: Whenever you observe something, you simply declare that this is what the designer wanted. In *The Origin of Species*, Darwin ([1859] 1964: 435) puts the point with a touch of irony: "On the ordinary view of the independent creation of each being, we can only say that so it is – that it has so pleased the Creator to construct each animal and plant." If this simple formula were enough to explain the observations in question, there would be no need for science. Not only would Darwin's own theory be unnecessary; there would be no need for theories in any other area of science, either. This does not mean that "God did it" is false, only that it is no substitute for science. Theists regard scientific theories as describing how God brought about various observations, while atheists and agnostics decline to interpret them in this way. Whether this theistic gloss is added or withheld, the practice of science should be the same.[2]

Darwin sometimes failed to live up to his own principles. Consider, for example, the famous sentence that ends *The Origin of Species*:

There is grandeur in this view of life, with its several powers, having been originally breathed into a few forms or into one; and that, whilst this planet has gone cycling on according to the fixed law of gravity, from so simple a beginning endless forms most beautiful and most wonderful have been, and are being, evolved. (Darwin [1859] 1964: 490)

In saying that life was "breathed into a few forms or into one," Darwin seems to concede that the origin of life is to be understood as the act of a creator. In a letter to J. D. Hooker written four years after *The Origin of Species* appeared, Darwin says what he thinks of his earlier choice of words:

I have long regretted that I truckled to public opinion, and used the Pentateuchal term of creation, by which I really meant "appeared" by some wholly unknown process. It is mere rubbish, thinking at present of the origin of life; one might as well think of the origin of matter. (Darwin 1887: II, 202–3)

Darwin's considered view was that the origin of life, being an event that occurred *in* nature, needs to be understood in terms of natural processes,

---

[2] If natural science seeks to answer questions about what happens *in* nature and has nothing to say about *super*natural beings, then "methodological naturalism" is an appropriate scientific research strategy; for discussion of the history of this idea, see Numbers (2003).

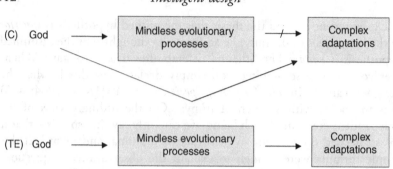

Figure 2.2 Creationism (C) holds that mindless evolutionary processes are incapable of producing the complex adaptations we observe in nature and that God directly produced what we observe. Theistic evolutionism (TE) holds that God indirectly produced complex adaptations in nature by setting the evolutionary process in motion.

not by the facile declaration that it was God's will (Brown 1986). Notice, by the way, how Darwin concludes his book by putting the process of evolution along side "the fixed law of gravity." Newton was a devout theist, but his theism was no substitute for the *Principia*.

It is important not to lose sight of the possibilities that Darwin saw so clearly. Many creationists describe theism and evolutionary theory as if they are incompatible: that the proposition that God exists entails that evolutionary theory is untrue. Some defenders of evolutionary theory agree with creationists on this point (Dawkins 1986; Dennett 1995, 2006; Provine 1989), only they reason from the truth of evolutionary theory to the falsehood of theism. Both parties are invoking a false dichotomy (Ruse 2000). Theistic evolutionism is logically consistent; it also happens to be the viewpoint that many religious people have embraced. My point here is not that theistic evolutionism is *true* or *plausible* (a question to which I'll return in §2.21), but just that it isn't *contradictory*.

Creationism isn't simply the claim that organisms exist and have the features we observe because of a plan that God decreed. This mischaracterization of creationism blurs its difference with theistic evolutionism. The real difference between creationism and theistic evolutionism is depicted in Figure 2.2. Creationists hold that the evolutionary process is fundamentally incapable of producing the complex adaptations we observe; these features require God's *direct* intervention. For theistic evolutionists, God produces complex adaptations *indirectly*, by way of the natural processes he put in place. In addition to creationism and theistic evolutionism, there are other possibilities, such as atheistic evolutionism and agnostic evolutionism.[3] All these positions are internally consistent; biology does not address the question of whether the universe was created by an intelligent designer. The theory of evolution, which is a theory about *living* things, not about *the origin of the entire universe*, is silent on the question of whether there is a God.

2.2 DESIGN ARGUMENTS AND THE BIRTH OF PROBABILITY THEORY

Bacon's distinction of God's word from God's works means that there are two types of theology: revealed and natural. Revealed theology bases its claims on sacred texts; natural theology seeks to ground its claims on natural observations. When creationism is developed within the context of natural theology, its central organizing concept – both before and after 1859 – is the design argument for the existence of God. Creationists *after* Darwin reject evolutionary theory and claim that the only plausible explanation of the complex adaptations we observe in nature is the hypothesis of intelligent design. Creationists *before* Darwin of course could not have considered Darwin's theory, but they too argued that the complex adaptations we observe in nature provide compelling evidence for the existence of an intelligent designer.

The design argument has evolved, so it might be better to regard it as a family of arguments. I don't propose to give anything like a complete account of the history of this family; rather, in this section I want to mention a few landmarks. My goal in this chapter is to arrive at the strongest, most defensible, version of the argument, and then to say why I think the argument is defective.

The design argument differs from the cosmological argument. The latter argues that the universe as a whole is the result of a first cause (i.e., God). In contrast, the design argument is a claim about what we find *in* nature, not about the existence *of* nature as a whole. The argument from design usually focuses on the complex adaptive features that organisms possess, but other versions of the design argument have been stated. Kepler, for example, believed that the face we see when we look at the moon requires explanation in terms of intelligent design; Newton had the same thought concerning the planets' revolving around the sun in the same direction and in the same plane. And, more recently, the fine-tuning

---

[3] And, of course, the *rejection* of evolutionary theory is logically consistent with theism, atheism, and agnosticism.

argument maintains that the values of the fundamental physical constants show that the universe was created by an intelligent designer (see Sober 2004b for discussion). These extrabiological examples will not concern us in what follows; our subject will be the *organismic* design argument.

Here is a classic formulation of the design argument developed in the thirteenth century by Thomas Aquinas in his *Summa Theologica* (Part I, Question 2, Article 3):

We see that things which lack intelligence, such as natural bodies, act for an end, and this is evident from their acting always, or nearly always, in the same way, so as to obtain the best result. Hence it is plain that not fortuitously, but designedly, do they achieve their end. Now whatever lacks intelligence cannot move towards an end, unless it be directed by some being endowed with knowledge and intelligence; as the arrow is shot to its mark by the archer. Therefore some intelligent being exists by whom all natural things are directed to their end; and this being we call God.

To which objects in nature is this argument intended to apply? Aquinas thinks that the adaptive features of plants and animals that lack minds must be explained in terms of intelligent design. Yet, the human mind does not fall within this argument's purview. If the arrow's trajectory requires the archer's mind, doesn't the archer's mind also require an intelligent creator? As we shall see shortly, more recent versions of the design argument usually point to the complexity and functionality of an organ as evidence that it was produced by an intelligent designer, thus providing an opening for the argument to offer an explanation of the human mind. It also is worth noting that Aquinas takes his argument to extend beyond the realm of biology. Being an Aristotelian, Aquinas thinks of lifeless physical objects as goal-directed systems. When we drop an object, it falls towards the center of the earth; for Aristotelians, this is the goal the object has, and it falls in order to achieve that end. The mechanical philosophy that developed within the scientific revolution of the seventeenth century discarded this teleological conception of the behavior of physical objects. Falling objects, planets, and projectiles obey laws, but the idea that they have goals or purposes gradually lapsed from scientific discourse. It might be thought that Darwin did for biology what Newton and others did for physics two centuries earlier – that is, that Darwin demonstrated that it is a mistake to regard organisms as goal-directed systems. I do not agree. Darwin's biology and the evolutionary biology that he inspired seek to understand how features of organisms contribute to their survival and reproduction. These are the *functions* those features subserve. For example, the function of the heart is to pump blood; its function is not to make noise. Darwinism does not reject these claims; rather, it provides a framework within which they can be understood. The heart evolved because there was selection for pumping blood; it did not evolve because there was selection for making noise. The concept of function does not require that the heart be capable of conscious striving.[4]

Aquinas's formulation of the design argument concludes that there is one intelligent designer responsible for the goal-directed behavior of all mindless objects. However, what follows is something more modest: that for *each* such object there must be an intelligent designer. It does not follow that *all* such objects trace back to a single intelligent designer.[5] In addition, it is a further step in the argument, requiring further defense, to conclude that this single designer is God. However, the main point to which I want to draw the reader's attention is the connection Aquinas sees between goal-directed behavior and the existence of an intelligent designer. Consider the following two interpretations of his argument:

- If a mindless system exhibits goal-directed behavior, it *must* have been made by an intelligent designer.
- If a mindless system exhibits goal-directed behavior, it *probably* was made by an intelligent designer.

I won't address which of these is the interpretation that Aquinas intended, nor even whether he was aware of the distinction involved here. The point of importance is logical, not biographical: *the distinction between necessity and high probability makes a huge difference for the argument's defensibility.*

The modern mathematical theory of probability began to develop in the seventeenth century. This theory led the design argument to evolve. It became clear to many defenders of the argument that the first of the two versions of the argument just described, which we might

---

[4] See Wright (1976) for a definition of function according to which function claims are claims about why a feature or organ is present. The present point, that Darwinism does not entail that function talk be discarded, does not require that this definition is correct.

[5] Aquinas's argument commits the *birthday fallacy* (Sober 1990). In the following argument, the premise does not entail the conclusion:

> Everyone has a birthday.
> ---
> There is a single day on which everyone was born.

summarize with the slogan "no design without a designer," is a mistake. The reason is very simple: a mindless random process *can* produce complex and useful devices. It is possible, as we now would say, for monkeys pounding at random on typewriters to eventually produce the works of Shakespeare.[6] The problem is that this outcome, given some fixed number of monkeys and typewriters and a limited amount of time, is very improbable. What is true is that monkeys pounding at random on typewriters *probably* will not produce the works of Shakespeare. For just this reason, it is a mistake for the design argument to claim that complex adaptations *cannot* arise by a mindless random process. The probabilistic formulation of the argument is more defensible.

The birth of probability theory not only transformed the design argument into a probabilistic argument. It also led defenders of the argument who absorbed the point about monkeys and typewriters to think contrastively. Instead of simply declaring that design requires a designer, they were led to consider possible alternatives to intelligent design. Until 1859, the main alternative they considered was *Epicureanism*; here I don't mean the philosophy of eat, drink, and be merry, but the hypothesis due to Epicurus and his followers that physical particles whirling at random in the void eventually combine to produce orderly, stable, and functional arrangements. Design theorists repeatedly held this alternative up to ridicule. For example, Jonathan Swift satirizes Epicureanism in Book 3 of *Gulliver's Travels* (published in 1726) by describing a distinguished professor at the Grand Academy of Lagado who sought to "improve speculative knowledge by practical and mechanical operations"; his innovation was to produce random arrangements of words by twiddling the handles of a device that resembles a foosball game (illustrated in Plate 5 of *Gulliver*). The probability of successfully generating a well-formed sentence of the language – and one that is a new and useful contribution to speculative knowledge as well – is not zero; rather, it is exceedingly tiny. It is not impossible that Chance should produce this result, just very improbable that it should do so.

Swift's satire of Epicureanism may have been inspired by an argument that Richard Bentley, an important figure in the Royal Society, made from "linguistic combinatorics." In his inaugural Boyle lectures of 1692, Bentley asks what the probability would be that a male and

---

[6] The earliest source I have been able to find for the metaphor of monkeys and typewriters is Borel (1913); Eddington (1928: 72) says that "if an army of monkeys were strumming on typewriters they *might* write all the books in the British Museum."

a female of the same species should each arise by chance. He answers by proposing an analogy, derived from Cicero's *De natura deorum*, between the gigantic number of sequences that can be constructed from the Latin alphabet of twenty-four letters and the still greater number of arrangements there can be of the 1,000 or more parts that comprise the human body (Shoesmith 1987: 136). For both the complex adaptive features of organisms and the orderly pattern of letters in a book, it is absurd to claim that they are due to chance. However, this is not because a random process *cannot* yield the results we observe; rather, the reason is that the *probability* of these results is tiny if a mindless random process is doing the work.

One landmark in the development of probability theory during this period was a version of the design argument published by John Arbuthnot, who was physician to Queen Anne and inventor of the satirical character John Bull. Arbuthnot's "Argument for Divine Providence, Taken from the Constant Regularity Observ'd in the Births of Both Sexes" appeared in the *Philosophical Transactions of the Royal Society* for 1710. The paper provides a tabulation of eighty-two years of London christening records; more boys than girls are listed for each year. Arbuthnot takes this difference at face value; he must have realized that not every birth was recorded, but he nonetheless assumes that the records reflect a real difference in the frequencies of male and female births. The main part of the paper is given over to the task of calculating the probability that this pattern would obtain if the sex ratio were due to chance. By "chance" Arbuthnot means that each birth has a probability of $\frac{1}{2}$ of being a boy and $\frac{1}{2}$ of being a girl. According to this hypothesis, there being more boys than girls in a given year has the same probability as there being more girls than boys in that year; the chance hypothesis also allows for a third possibility, namely, there being exactly as many girls as boys:

Pr(more boys than girls born in a given year | Chance)

$= Pr$(more girls than boys born in a given year | Chance)

$\gg Pr$(exactly as many boys as girls born in a given year | Chance) $= e$.

Although Arbuthnot goes to the trouble of explaining how e might be calculated, the details of his calculation don't matter to the argument; the point is just that for each of the years surveyed, e is tiny. Arbuthnot concludes that the probability of there being more boys than girls in a given year, according to the chance hypothesis, is just under $\frac{1}{2}$, and so the probability of there being more boys than girls in each of eighty-two years is less than $\left(\frac{1}{2}\right)^{82}$. He further asserts that if we were to tabulate births in other years and other cities, we would find the same male bias. So, the probability of all these data – both the data that Arbuthnot presents and the data that he does not have but speculates about – is "near an infinitely small quantity, at least less than any assignable fraction." The conclusion is obvious: "it is Art, not Chance, that governs."

Arbuthnot also notes that males have a higher mortality rate than females, so that the male bias at birth gradually gives way to an even sex ratio at the age of marriage. "We must observe," he says,

that the external accidents to which males are subject (who must seek their food with danger) do make a great havock of them, and that this loss exceeds far that of the other sex, occasioned by diseases incident to it, as experience convinces us. To repair that loss, provident Nature, by the disposal of its wise creator, brings forth more males than females.

At the end of the paper, Arbuthnot adds, as a *scholium*, that

polygamy is contrary to the law of nature and justice, and to the propagation of the human race. For where males and females are in equal number, if one man takes twenty wives, nineteen men must live in celibacy, which is repugnant to the design of nature, nor is it probable that twenty women will be so well impregnated by one man as by twenty.

In Arbuthnot's hands, the design argument begins as an explanation of what *is*, but ends as an argument concerning what *ought to be*.[7]

2.3 WILLIAM PALEY: THE STONE, THE WATCH, AND THE EYE

William Paley published his book *Natural Theology, or, Evidences of the Existence and Attributes of the Deity, Collected from the Appearances of Nature* in 1802; it appeared after more than a century of defenses of intelligent design and attacks on Epicureanism. Paley's was neither the first nor the last, but the way he puts the argument is famous:

In crossing a heath, suppose I pitched my foot against a *stone* and were asked how the stone came to be there, I might possibly answer that for anything I knew to the contrary it had lain there forever; nor would it, perhaps, be very easy to show the absurdity of this answer. But suppose I had found a *watch*

---

[7] For discussion of the eighteenth-century reaction to Arbuthnot's argument, and of Darwinian theorizing about sex ratio evolution, see Sober (2007b).

upon the ground, and it should be inquired how the watch happened to be in that place, I should hardly think of the answer which I had before given, that for anything I knew the watch might have always been there. Yet why should not this answer serve for the watch as well as for the stone? Why is it not as admissible in the second case as in the first? For this reason, and for no other, namely, that when we come to inspect the watch, we perceive – what we could not discover in the stone – that its several parts are framed and put together for a purpose, e.g., that they are so formed and adjusted as to produce motion, and that motion so regulated as to point out the hour of the day; that if the different parts had been differently shaped from what they are, of a different size from what they are, or placed after any other manner or in any other order than that in which they are placed, either no motion at all would have been carried on in the machine, or none which would have answered the use that is now served by it. To reckon up a few of the plainest of these parts and of their offices, all tending to one result; we see a cylindrical box containing a coiled elastic spring, which, by its endeavor to relax itself, turns round the box. We next observe a flexible chain – artificially wrought for the sake of flexure – communicating the action of the spring from the box to the fusee. We then find a series of wheels, the teeth of which catch in and apply to each other, conducting the motion from the fusee to the balance and from the balance to the pointer, and at the same time, by the size and shape of those wheels, so regulating that motion as to terminate in causing an index, by an equable and measured progression, to pass over a given space in a given time. We take notice that the wheels are made of brass, in order to keep them from rust; the springs of steel, no other metal being so elastic; that over the face of the watch there is placed a glass, a material employed in no other part of the work, but in the room of which, if there had been any other than a transparent substance, the hour could not be seen without opening the case. This mechanism being observed – it requires indeed an examination of the instrument, and perhaps some previous knowledge of the subject, to perceive and understand it; but being once, as we have said, observed and understood – the inference we think is inevitable, that the watch must have had a maker-that there must have existed, at some time and at some place or other, an artificer or artificers who formed it for the purpose which we find it actually to answer, who comprehended its construction and designed its use. (Paley 1809: 1–3)

Four chapters later, Paley connects his discussion of the watch[8] with a claim about the complex adaptations that organisms have. One of his many examples is the eye: "Every observation which was made in our first chapter concerning the watch may be repeated with strict propriety

---

[8] Paley did not invent the analogy with a clock. William Derham, an expert on the mechanics of time pieces, gave the Third Boyle lectures, published as *Physico-Theology* in 1711, developing the analogy between watches and watchmakers on the one hand and the universe and God on the other (Hacking 1975: 169–70).

concerning the eye, concerning animals, concerning plants, concerning, indeed, all the organized parts of the works of nature" (Paley 1809: 11).

How did Paley understand the logic of his argument? He often writes as if complex adaptations *must* be the result of intelligent design. For example, in Chapter 2 he says that

There cannot be design without a designer; contrivance without a contriver; order without choice; arrangement without anything capable of arranging; subserviency and relation to a purpose without that which could intend a purpose [...] Arrangement, disposition of parts, subserviency of means to an end, relation of instruments to a use imply the presence of intelligence and mind.   (Paley 1809: 268–9)

Paley's repetitions make his point more than clear. Yet, in other passages, Paley seems well aware of the relevant fact about monkeys and type-writers. For example, in Chapter 15 he considers the fact that "the eyes are so placed as to look in the direction in which the legs move and the hands work" (an example he may have drawn from Plato's *Timaeus* 44D–45B). The obvious explanation, Paley says, is intelligent design. This is because the alternative explanation is chance; if the direction in which our eyes point were "left to chance [...] there were at least three-quarters of the compass out of four to have erred in" (Paley 1809: 269). Paley here grants that it *is* possible for the adaptive arrangement to arise by chance.

2.4   FROM PROBABILITIES TO LIKELIHOODS

The simple point about monkeys and typewriters shows that it is a mistake to claim that complex adaptive features *cannot* be brought into existence by a mindless random process. In response, I suggested that Paley's argument should be formulated as a claim about probability: That the complex adaptive features we observe were *probably* put in place by an intelligent designer. However, a question arises when we look at the details of how this probabilistic argument should be articulated. It concerns the important distinction drawn in Chapter 1 between the probability a hypothesis has in the light of evidence and the probability that the hypothesis confers on the evidence. This is the distinction between $Pr(H \mid O)$ and $Pr(O \mid H)$; the former, recall, is the *posterior probability* of the hypothesis H, whereas the latter is called (unhelpfully) the *likelihood* of H. When Paley talks about the fact that our eyes point in the direction in which we walk, he considers Pr(Observations | Chance) and says that this has a value of $\frac{1}{4}$. And when Arbuthnot talks about his sex ratio data, he points out that Pr(Observations | Chance) $< \left(\frac{1}{2}\right)^{82}$. These assessments say nothing about the value of Pr(Chance | Observations). The question we now need to address is whether a suitably probabilistic version of the design argument should describe the probability of intelligent design, or only its likelihood.

We know from Bayes' theorem (§1.2) that the prior and posterior probabilities of the two hypotheses, intelligent design (*ID*) and chance, are related to their likelihoods as follows:

$$\frac{Pr(ID \mid O)}{Pr(\text{Chance} \mid O)} = \frac{Pr(O \mid ID)}{Pr(O \mid \text{Chance})} \times \frac{Pr(ID)}{Pr(\text{Chance})}.$$

The ratio of the posterior probabilities (the "odds") equals the ratio of the likelihoods times the ratio of the priors. Paley's and Arbuthnot's assessments of intelligent design and chance involve a comparison of their likelihoods:

$$Pr(O \mid ID) \gg Pr(O \mid \text{Chance}).$$

However, this does not entail that

$$Pr(ID \mid O) \gg Pr(\text{Chance} \mid O).$$

To reach that further conclusion, we need further assumptions about the prior probabilities (Keynes 1921: 298; Himma 2005).

As I explained in Chapter 1, I am disinclined to appeal to prior probabilities when they reflect only a subjective degree of certainty; however, I have nothing against priors when they can be justified by sampling data or by an empirically well-established theory. It is for this reason that an assessment of Newton's theory of gravitation or of Darwin's theory of evolution should not be formulated so as to depend on assigning them prior probabilities. The same holds, I suggest, for intelligent design. What is the prior probability that the vertebrate eye was the result of intelligent design? I see no way to answer this question in a way that allows that probability to be objective. For this reason, I don't want to formulate the design argument as an argument that seeks to establish that the hypothesis of intelligent design has high probability. Better to think of the argument from design as a likelihood argument. The law of likelihood (§1.3), applied to the alternatives that Paley and Arbuthnot considered, says that

Observation *O* favors *ID* over Chance if and only if
$Pr(O \mid ID) > Pr(O \mid \text{Chance})$.

Understood in this way, the design argument does not seek to establish that an intelligent designer *must* exist, nor even that such a being *probably* exists. The likelihood argument is more modest than these alternatives, and therein lies its strength. I'll consider two nonlikelihood formulations of the design argument in §2.18 and §2.19. But for now, let's go with likelihoods.

2.5   EPICUREANISM AND DARWIN'S THEORY

Arbuthnot, Paley, and the many other defenders of the design argument who wrote before 1859 naturally did not know about the Darwinian theory of evolution. The alternative to intelligent design that they knew about was Epicureanism. Post-Darwinian creationists often write as if Darwin's theory is nothing new – that evolution by natural selection is just like monkeys and typewriters. This is what they intend to convey when they claim that natural selection has the same chance of producing complex adaptations that a hurricane blowing through a junkyard has of assembling scattered pieces of metal into a functioning airplane.[9]

This analogy is fundamentally misleading. In colloquial usage, a random process is one in which all outcomes have the same (or nearly the same) probability. Gambling devices are the paradigm. A fair coin and an unrigged roulette wheel are randomizing devices.[10] However, the essence of the process of natural selection is that some outcomes are far more probable than others. Traits that help an organism survive and reproduce have a higher probability of evolving than traits that hurt. Natural selection is a *biased* process, not a *random* process. When biologists talk about the "random" element in Darwinian evolution, they usually have in mind the origination of novel variants by mutation. The idea is not that a mutation has a probability of $\frac{1}{2}$ of being advantageous and a probability of $\frac{1}{2}$ of being deleterious; the conventional wisdom is that most mutations are

---

[9] The analogy is due to the astronomer Fred Hoyle.
[10] Probabilists use "random" in a wider sense. They would apply the term to the sequence of heads and tails produced by repeatedly tossing a highly biased coin (probability of heads = 0.9999 on each toss). Ordinary usage is closer to what probabilists call a "uniform" or "flat" distribution.

deleterious. This is because a random change in a complex functioning machine is very unlikely to improve its performance; it is mutation, not natural selection, that resembles the hurricane blowing through the junkyard. The two-part process of mutation + selection contains a random element and a nonrandom element. Do not confuse the part for the whole.

One important difference between a purely random process and a two-step process in which variation is randomly generated and then there is nonrandom retention of favorable variants concerns *time*; the first process will take longer than the second for advantageous features to evolve. Dawkins (1986) provides a nice illustration of this point by using Simon's example (1981) of a combination lock. Suppose that the lock opens when its nineteen windows spell out *METHINKSITISAWEASEL*. Each window has twenty-six alternative states, one for each letter of the alphabet. If the nineteen tumblers are simultaneously spun at random, the chance that this exact sequence of letters will appear on a given spin is $\left(\frac{1}{26}\right)^{19}$. Imagine doing a very large number of experiments with this combination lock; in each experiment, you repeatedly spin all the wheels at random until all of them fall into place and spell the target sentence. Some experiments will hit the target sooner and others will take much longer. But, *on average*, it will take 26^{19} spins to hit the target; this is the *expected value* (§1.2). Now consider a second kind of experiment. You spin the first wheel at random until it hits *M*; after it hits its target, that letter is frozen in place and the second wheel is spun until it hits *E*, and so on. Imagine doing this kind of sequential experiment a large number of times. Some experiments will achieve the target sentence sooner and some will take much longer. But, on average, this sequential process will take $26 \times 19 = 494$ spins to attain the target sentence. The purely random all-at-once process takes much longer on average than the sequential and partly random process; the nonrandom element in this second process involves the selective retention of letters that match the target. There is a third experiment that will probably hit the target even faster. In this experiment, all the wheels are spun simultaneously until one or more target letter is attained, after which those wheels are frozen and the remaining wheels are spun at random. In this third experiment you don't postpone spinning the second wheel until the first wheel hits the target. In all three of these experiments, the target sentence for the combination lock is set by an intelligent being (the designer of the lock), but that isn't relevant to the present point. The point is that a *purely* random process takes longer to evolve adaptive configurations than the *partly* random *partly* nonrandom process of mutation plus selection. This is why analogizing the process of natural selection with a hurricane blowing through a junkyard is fundamentally misleading.[11]

Another difference between Epicureanism and Darwinism consists in the fact that the evolutionary process involves branching (as lineages split) and the sequential accumulation of different modifications (both those that are adaptive and those that are not) in different lineages. Darwinism includes the idea of common ancestry as well as the idea of natural selection, and the former introduces an element that was not standard in the Epicurean picture. According to Epicureanism, there was a time of random mixing at the end of which all the stable configurations we now observe had come into existence; it is not intrinsic to this picture that the stable configurations that now exist share common ancestors. To see how this marks an important difference between Epicureanism and Darwinism, consider what Paley says about Epicureanism in Chapter 5 of *Natural Theology* (1809: 49–51). He argues that Epicureanism makes the false prediction that we should see unicorns and mermaids. He also says that it mistakenly predicts that organisms should fail to form a nested taxonomic hierarchy (e.g., *Mammalia* within *Vertebrata* within *Animalia*). There is an irony in Paley's second objection, in that Darwin later claimed, correctly, that his own theory *predicts* hierarchy.[12] Nor is it surprising, on Darwin's theory, that some conceivable organisms do not exist. Evolution by natural selection is path-dependent; the traits that evolve earlier in a lineage constrain the traits that will probably evolve later. One lineage leads to fish and another to human beings; there is no reason to expect this process to

---

[11] Because he had long admired the philosophical writings of John Herschel, Darwin must have been disappointed to read the following footnote in the 1867 edition of Herschel's *Physical Geography of the Globe* (quoted in Hull 2000: 59):

> We can no more accept the principle of arbitrary and casual variation and natural selection as a sufficient account, *per se*, of the past and present organic world, than we can receive the Laputan method of composing books (pushed *a l'outrance*) as a sufficient one of Shakespeare and the *Principia*. Equally, in either case, an intelligence, guided by a purpose, must be continually in action [...] We do not believe that Mr. Darwin means to deny the necessity of such intelligent direction. But it does not, so far as we can see, enter into the formula of his law.

[12] In *The Origin of Species*, Darwin ([1859] 1964: 128–9) says that

> it is a truly wonderful fact ... that all animals and all plants throughout all time and space should be related to each other in group subordinate to group [...] On the view that each species has been independently created, I can see no explanation of this great fact in the classification of all organic beings; but to the best of my judgment, it is explained through inheritance and the complex action of natural selection.

produce organisms that are half-fish and half-human. Perhaps Paley was right about Epicureanism, but Darwinism is different. The misleading analogy between natural selection and a hurricane blowing through a junkyard should be junked. Darwin suggests something better in his book *The Variation of Animals and Plants Under Domestication*:

> Let an architect be compelled to build an edifice with uncut stones, fallen from a precipice. The shape of each fragment may be called accidental; yet the shape of each has been determined by the force of gravity, the nature of the rock, and the slope of the precipice, – events and circumstances all of which depend on natural laws; but there is no relation between these laws and the purpose for which each fragment is used by the builder. In the same manner the variations of each creature are determined by fixed and immutable laws; but these bear no relation to the living structure which is slowly built up through the power of natural selection, whether this be natural or artificial selection. (Darwin 1876: 236)

Natural selection is no more a random process than intelligent design is. As for the "randomness" of variation, the point is that novel variants do not arise because they would be useful; this does not mean that they are uncaused.

2.6 THREE REACTIONS TO PALEY'S DESIGN ARGUMENT

There are three possible reactions one might have to the design argument, once it is formulated as an argument about likelihoods and we acknowledge that Paley was comparing intelligent design with chance and did not consider Darwin's theory of evolution.

One reaction, now common among biologists, is that Paley reasoned correctly given the alternatives he was considering but that the dialectical landscape shifted profoundly when a third hypothesis was formulated. Translated into the language of likelihoods, this reaction consists in the thought that Paley was right in his claim that

(L) Pr(Observations | Intelligent design) $\gg Pr$(Observations | Chance).

What he could not have anticipated is that

$$Pr(\text{Observations} \mid \text{Darwinian evolution})$$
$$\gg Pr(\text{Observations} \mid \text{Intelligent design}).$$

The pattern here is familiar. A better theory displaces an inferior one, but then a new theory comes along that is better still.

The second possible reaction is that Paley's argument is flawed and that it doesn't take the development of Darwin's theory to see what is wrong with the theory of intelligent design. I associate this reaction with David Hume's *Dialogues Concerning Natural Religion*, which appeared posthumously in 1779. Hume, of course, didn't know about Darwin any more than Paley did. The point is that the *Dialogues* present a number of serious criticisms of the design argument (some of which I'll consider in §2.11). If any of these criticisms are correct, they show that there are flaws in Paley's argument that we can recognize without knowing anything about Darwin's theory. And even if Hume's criticisms miss the mark, perhaps there are other criticisms of Paley's reasoning that do not depend on the theory of evolution by natural selection.

The third possible reaction to Paley's argument is the one that post-Darwinian creationists have. Whether or not they acknowledge that chance and Darwinian evolution are different hypotheses, their view, expressed in a likelihood framework, is that Paley was not only correct in asserting the inequality (L); essentially the same argument also shows that

$$Pr(\text{Observations} \mid \text{intelligent design})$$
$$\gg Pr(\text{Observations} \mid \text{Darwinian evolution}).$$

This is the thought that intelligent design is better supported by what we observe than both the hypothesis of chance and the hypothesis of Darwinian evolution.

So as to leave no doubt in the reader's mind as to which of these reactions I favor, let me say this: *I stand with Hume*. Although I think that some of Hume's criticisms of the design argument are off the mark, I do think there is a devastating objection to Paley's argument that does not depend in any way on Darwin's theory.

2.7   THE NO-DESIGNER-WORTH-HIS-SALT OBJECTION TO THE HYPOTHESIS OF INTELLIGENT DESIGN

Paley's analogy between the watch and the eye is seductive. Surely Paley reasoned correctly when he considered the watch. And since his reasoning about the eye apparently follows the same pattern, it seems irresistible to conclude that he also reasoned well about the eye. Evolutionists who grant Paley this much usually hasten to point out that Paley did not know about Darwin's theory (much less about the evolutionary biology that developed after 1859); their point is that even though Paley demonstrated that intelligent design is more plausible than chance, it also is true that Darwinian evolution is more plausible than intelligent design. This is the first possible reaction to Paley's argument described in the previous section. But why think that Darwin's theory is better supported than the hypothesis of intelligent design? A standard way to defend this assessment is to point to the many imperfect adaptations that are found in nature. This style of argument has a long history. Darwin gives voice to it in Chapter 14 of *The Origin of Species* when he says "on the view of each organic being and each separate organ having been specially created, how utterly inexplicable it is that parts [...] should so frequently bear the plain stamp of inutility!" More recently, Stephen Jay Gould (1980) used the example of the panda's "thumb" to make the same point.

The name for this feature is misleading since pandas don't have opposable digits. Rather, they have a spur of bone that sticks out from their wrists. The thumb and the paw together form a V through which the panda repeatedly runs branches of bamboo, laboriously stripping the stalks to get them ready to eat. Pandas spend a large portion of their waking lives at this task. The thumb is extremely inefficient. Gould's point is that no designer worth his salt (the phrase is due to Raddick 2005) would have given the panda this device for preparing its food. A truly intelligent designer would have done better. On the other hand, Darwin's theory of evolution by natural selection says that inefficient devices of this kind are not at all surprising. Darwin thought of natural selection as a gradual process that improves adaptedness; natural selection does not necessarily lead to perfect adaptation, whatever that might mean. Selection modifies the traits found in ancestors by small changes; the result is not that the *best* of all *conceivable* adaptations evolves; rather, natural selection causes traits to evolve that do a *better* job than the alternatives that are *actually* present in the evolving lineage.[13]

I hope it is clear that Gould's argument is a likelihood argument. He claims that the hypothesis of intelligent design makes the panda's thumb very improbable, whereas the hypothesis of evolution by natural selection makes the result much more probable. Creationists have a serious objection to Gould's argument. It can be expressed by a rhetorical question: How does Gould know what God (or some unspecified designer) would have wanted to achieve in building the panda? Gould is assuming that an

---

[13] Darwin was inspired by Lyell's geology, according to which huge changes in the Earth were brought about by a long series of small alterations. See, for example, Darwin ([1859] 1964: 95) and also Darwin's comment about "absolute perfection" (1964: 202).

intelligent designer would have wanted to supply pandas with a super-efficient device (like a stainless-steel can opener) for preparing bamboo and would have had the ability to achieve this objective. But why is it so clear that God would have wanted to do this? Perhaps God realized that if pandas had better tools, they would eat all the bamboo, which would cause the extinction of the bamboo forest and of pandas as well. And maybe these two extinctions would have triggered a cascade of others. Perhaps God realized that these bad consequences would follow if pandas had better tools, and so he decided to slow them down. Creationists don't need to assert that *they* know what God would have had in mind if he had built the panda. All they need to say is that *Gould* does not know this. Gould adopts assumptions about the designer's goals and abilities that help him reach the conclusion he wants – that intelligent design is implausible and Darwinian evolution is plausible as an explanation of the panda's thumb. But it is no good simply *inventing assumptions* that help one defend one's pet theory. Rather, what is needed is *independent evidence* concerning what God (or some other intelligent designer) would have wanted to achieve if he had built the panda. And this is something that Gould does not have. I think creationists are right to object in this way to Gould's argument.[14] We will see in §2.12 that this good point comes back to haunt the theory of intelligent design.

Paley anticipates the no-designer-worth-his-salt objection. After describing the watch found on the heath, he responds to various objections that might be made against the hypothesis of intelligent design. Here is what he says about one of them:

Neither [. . .] would it invalidate our conclusion, that the watch sometimes went wrong, or that it seldom went exactly right. The purpose of the machinery, the design, and the designer, might be evident, and in the case supposed would be evident, in whatever way we accounted for the irregularity of the movement, or whether we could account for it or not. It is not necessary that a machine be perfect, in order to show with what design it was made: still less necessary, where the only question is, whether it were made with any design at all.

Paley then says that what is true of imperfect adaptations is also true for traits of unknown function. The watch manifestly has the function of telling time and it is a complex machine. If it is imperfect and if it has parts whose functions are unknown, that does not matter.

---

[14] Behe (1996: 223) says that "another problem with the argument from imperfection is that it critically depends on a psychoanalysis of the unidentified designer." See also Nelson (1996).

2.8 POPPER'S CRITERION OF FALSIFIABILITY

If the hypothesis of intelligent design is not refuted by the fact that organisms have imperfect adaptations, maybe the problem with the hypothesis is that it can't be refuted by any observation at all. Here we come to a second standard criticism that has been made of creationism: that it is *untestable*. But what does testability mean? Scientists often answer by using the philosopher Karl Popper's (1959) concept of falsifiability, thinking that this provides the needed clarification. Unfortunately, there are serious problems with his account.

Popper's idea is that a statement is falsifiable precisely when it rules out a possible observational outcome. Popper understood "ruling out" in terms of deductive logic; the hypothesis must deductively entail that some observation statement is false. According to Popper, a hypothesis that is logically consistent with all possible observations is unfalsifiable. Falsifiable statements need not be false; rather, they must have the following property: *If* they are false, a finite set of observations can prove that they are. For Popper, *modus tollens* (MT) (§1.4) is a good representation of the logic of testing a hypothesis (*H*) by seeing if its observational implications (*O*) hold true:

(MT)
$$\frac{\begin{array}{l} \text{If } H \text{ then } O \\ not O \end{array}}{not H}$$

If the hypothesis's prediction fails to be true, the hypothesis is refuted. But what are we to say if the prediction comes true? Does this mean that the theory has been proven true? Popper rejected this, since the *fallacy of affirming the consequent* (FAC) is deductively *in*valid:

(FAC)
$$\frac{\begin{array}{l} \text{If } H \text{ then } O \\ O \end{array}}{H}$$

Because *modus tollens* is deductively valid and FAC is deductively invalid, Popper held that scientific theories can be proved false but can never be proved true. He further suggested that falsifiability provides a *demarcation criterion*, separating science from nonscience.

Popper's criterion entails that some creationist claims are falsifiable and, hence, are scientific. A blatant example is provided by the hypothesis that an omnipotent supernatural being wanted everything to be purple and had this as his top priority. Of course, no creationist has endorsed this proposition. However, it is inconsistent with what we observe, so purple ID is falsifiable, its invocation of the supernatural notwithstanding. For Popper, purple ID is a scientific theory. The same is true of more familiar formulations of intelligent design; for example, the claim that an intelligent designer gave vertebrates their eyes entails that vertebrates have eyes, so it too is falsifiable.

In addition to judging that many intelligent design claims pass the test of falsifiability, Popper's criterion also entails that many apparently scientific statements are unfalsifiable and hence are not scientific at all. This is the situation for all the *probabilistic theories* that have been developed in different scientific disciplines. Consider a simple example: the statement that a coin has a probability of 0.5 of landing heads each time it is tossed. This statement is logically consistent with all possible sequences of heads and tails in any finite run of tosses. The probability statement is testable, but it does not satisfy Popper's criterion. Popper recognized this problem and proposed to remedy it by expanding the concept of falsification. Rather than saying that H is falsified only when an observation occurs that is logically inconsistent with H, Popper suggested that we regard H as false when an observation occurs that H says is very improbable. But how improbable is improbable enough for rejecting H to be warranted? Popper (1959: 191) thought that there is no objectively correct answer to this question and suggested that we solve the problem by adopting a convention – a "methodological decision." Popper's idea has much in common with R. A. Fisher's test of significance (1956) (§1.4). *Modus tollens*, though deductively valid, is not a good model of how probabilistic theories are tested, and I argued in §1.4 that probabilistic *modus tollens* is not a principle of inference that we should endorse.

Although Popper's falsifiability criterion should not be used to define what testability is, or to criticize creationism,[15] there are two lessons we can learn from it. The first is that a testable statement makes predictions, either by deductively entailing that an observation will occur or by conferring a probability on an observational outcome. The hypothesis that the coin is fair satisfies this requirement. The second point is that we must

---

[15] Here is a curious bit of history: Popper (1976: 151, 170) claimed that "the theory of natural selection is not a testable scientific theory, but a metaphysical research program," though he allowed that the theory is "invaluable" because "it sheds much light upon very concrete and very practical researches" (1976: 171). Two years later, Popper (1978) rejected this position.

think of testing contrastively (§1.3–1.4); to test an intelligent-design claim is to test it against some alternative. It is true that "an intelligent designer gave eyes to vertebrates" entails that vertebrates have eyes, but if this intelligent-design hypothesis is to be tested against the Epicurean hypothesis that a mindless chance process gave vertebrates their eyes, one needs to find predictions over which the two hypotheses *disagree*. We will return to the task of constructing a more adequate criterion of testability along these lines in §2.14.

2.9 SHARPENING THE LIKELIHOOD ARGUMENT

Within a likelihood framework, there is no beating a hypothesis that *entails* the observations. If we are trying to explain the observation

O = human beings have eyes that have features $F_1, F_2, \ldots, F_n$

the hypothesis

$ID++$ = an intelligent designer gave human beings eyes that
have features $F_1, F_2, \ldots, F_n$

entails O, so

$$Pr(O \mid ID++) = 1.0.$$

When we use observation O to test $ID++$ against an alternative hypothesis, it is *impossible* for O to favor the alternative hypothesis, no matter what that alternative is. This may sound like a point in favor of creationism. However, this fact about the relation of $ID++$ to O does not embody a *victory* for the design hypothesis; rather, the result is a *stalemate*, since it is easy to construct numerous other hypotheses that also entail the observation O. Since the point is to find the strongest representation of Paley's argument, and Paley's alternative to intelligent design is Epicureanism, not Darwinism, consider the following:

Chance++ = a purely random process caused human beings to
have eyes that have features $F_1, F_2, \ldots, F_n$.

The law of likelihood says that O does not discriminate between $ID++$ and *Chance*++. Something has gone wrong here. Surely Paley's argument is not undermined by this fact about the two hypotheses.

The problem we are considering has nothing special to do with the argument from design. Consider an example that came up briefly in the previous chapter: Arthur Stanley Eddington's testing the general theory of relativity (GTR) against Newtonian theory by observing an eclipse. Relativity theory predicted that the bend in starlight would probably fall in one interval of values, while Newtonian theory predicted that the bend in the light would probably fall in another. A natural way to represent Eddington's observation and its impact on the two theories is in terms of a likelihood inequality:

$$Pr(O \mid GTR) \gg Pr(O \mid \text{Newtonian theory}).$$

The point of importance here is that the evidential significance of Eddington's observation will be thoroughly obscured *if we build the observational outcome into the theories we wish to test.* This maneuver has the result that the likelihoods are both equal to unity:

$$Pr(O \mid \text{a general relativistic process produced outcome } O)$$
$$= Pr(O \mid \text{a Newtonian process produced outcome } O) = 1.0.$$

The defect in this equality is not that it is false (it isn't) but that it fails to represent the import of Eddington's test.

It is a mistake to represent Paley's design argument as a competition between *ID++* and *Chance++*. These formulations of the hypotheses under test are *inflated* (hence my use of "++" to label them). Taking account of this point leads to the following improved representation of the argument. We stick with the previous description of the observations (O), but now we represent the intelligent-design and chance hypotheses as follows:

(*ID*)   An intelligent designer designed and produced human beings.
(*Chance*)   A mindless chance process produced human beings.

This reformulation does not complete the task of obtaining a good representation of the problem at hand, but it is a beginning. Paley's claim is that

$$Pr(O \mid ID) > Pr(O \mid Chance).$$

He may or may not be right in asserting this inequality, but at least we will not be distracted by the true but irrelevant point that

$$Pr(O \mid ID++) = Pr(O \mid Chance++).$$

There is a big gap between *ID++* and *ID*, just as there is between *Chance++* and *Chance*. Why must the bubble be burst so completely? What would be wrong with a more modest deflation of the two hypotheses to the following?

(*ID+*)   An intelligent designer gave human beings eyes.
(*Chance+*) A purely chance process gave human beings eyes.

The problem here is that we have built the presence of eyes into both hypotheses. Surely Paley would want to insist that the observation that human beings have eyes favors the hypothesis of design over the hypothesis of chance. If so, we must omit this fact from the competing hypotheses. Of course, a similar point applies to *ID* and *Chance*; since both entail that *human beings* exist, the existence of human beings does not discriminate between them. This might suggest that we would do better to deflate the two hypotheses even further:

(*ID−*)   An intelligent designer gave organisms their adaptive features.
(*Chance−*) The adaptive features of organisms are due to purely mindless processes.

Maybe the existence of *human beings* discriminates between these two hypotheses. But even if it does, the existence of organisms that have adaptive features does not.

Perhaps the way to solve this problem is not to fixate on a single formulation of the hypotheses of chance and intelligent design but to insist that the formulations be parallel. It is a pyrrhic victory for a defender of intelligent design to compare *ID+* with *Chance* and declare that the fact that human beings have eyes favors the former; it is equally pyrrhic for an Epicurean to compare *ID* with *Chance+* and conclude that the fact that human beings have eyes favors the latter. Better to compare apples with apples, oranges with oranges.

In addition to honing the competing hypotheses that the design argument seeks to evaluate, we also should consider what the observations are that the argument brings to bear on those hypotheses. Which descriptions of the watch and the eye should we use? In the famous passage I quoted in §2.3, Paley says that the watch, but not the stone, has *utility* and lacks *tolerance*; the watch is useful to its owner because it keeps track of time, and the watch would not be useful in this way if any of its parts were altered in shape, size, or location. However, a few sentences later, Paley describes the watch's mechanical details. Should we use these mechanical details, or merely the fact that the watch is useful and intolerant, as the observation that we ask the hypotheses of *ID* and *Chance* to address? I won't try to assess whether Paley was aware of this question. My main goal, as I have said, is to identify the most defensible version of the design argument; given that we are now understanding the design argument as a likelihood inference, the important point is that it is arbitrary to limit the observations we consider just to the fact that the system in question is useful and lacks tolerance. Rather, we should consider *all* the observations. Even if a system's usefulness and intolerance is evidence favoring intelligent design over chance, we should consider the other features the system has as well.

In any event, it is wise not to put too much weight on the fact that a system is useful and intolerant. Recall that Paley recognized that a watch that fails to keep perfect time still should be explained in terms of intelligent design. The same point holds of a broken watch; it is *useless* as a timepiece, but presumably Paley would still want to argue that its features favor intelligent design over chance. If so, usefulness is not necessary. It might be replied that Paley's first criterion should be put in terms of the object's having a *function* (which even a broken watch might be said to have), not in terms of its being *useful*. However, this invites the question of what it means for an object to have a function and whether a plausible definition of function can be spelled out that the design argument can use without begging the question.[16] Similar questions attach to Paley's point about the object's intolerance – that the device would not be useful (or be able to perform its function) if any of its parts were altered in shape, size, or location. It is interesting that engineers often build devices that are highly *redundant*. Since any component might break, machines are often engineered so that the whole system continues to function even when some of its parts malfunction.[17] Suppose Big Ben were controlled not by a single mechanism but by 100 separate watches, with the position of Ben's hands dictated

---

[16] For example, Plantinga (1993) maintains that objects have functions only if they were made by an intelligent designer. His argument for this thesis is thin: that no one has so far managed to satisfactorily define what "function" means while omitting mention of a designer. In any event, if Plantinga is correct, the concept of function cannot be used to describe the observations that the design argument uses to adjudicate between intelligent design and chance. This description of the "observations" is too theory-laden; it begs the question.

[17] Organisms also exhibit redundancy and fallback mechanisms aplenty; witness the fact that people have *two* kidneys, *two* lungs, *two* gonads, and that cell death usually does not cause organs to malfunction.

by an average over those 100 timepieces. Imagine that the average is taken after outlying values are discarded. Surely this mechanical arrangement cries out for explanation in terms of intelligent design at least as much as the watch that Paley found on the heath. However, this system is much more tolerant than a watch whose hands are controlled by a single mechanism.[18] The design argument does not need to begin with the claim that a functional device is intolerant.

A more fundamental reason why the design argument should not focus exclusively on the fact that the system under study is useful and intolerant is that whether a system is intolerant depends on how it is decomposed into parts. Consider the eye. The familiar division of the vertebrate eye into cornea, retina, and so on, leads to the conclusion that the eye is intolerant. But suppose we divide someone's eye into individual atoms and call each atom a "part" of the eye. The system so described is *not* intolerant; change the position of a single atom and the eye still sees. This example might suggest that the idea of intolerance should be defined by saying that a system is intolerant if *some* division of the system into parts satisfies the requirement, not that *every* division must do so. The problem with this suggestion is that it entails that many systems are intolerant when we'd intuitively judge that the opposite is true. Consider the wine bottle. Its function, I take it, is to hold a certain liquid. There is a fine-grained segmentation of the bottle into parts that entails that the bottle is highly tolerant, since shaving a very thin slice off the surface does not impair the bottle's ability to hold liquid. However, there is another division that leads to the opposite conclusion. Just divide the bottle into a number of identically shaped top-to-bottom slices; remove any of *these* parts and the bottle can no longer serve as a container for liquids. A possible response to this problem is to claim that there is a uniquely correct division of a system into parts; however, this raises the question of how that uniquely correct breakdown should be defined and defended. Call this *the wine-bottle problem*.

One natural way to address this problem is to individuate the parts of a system in terms of the processes by which it was assembled. According to this approach, the customary division of the watch into parts is legitimate

---

[18] It might be argued that each of the 100 watches has its own function and that this function would not be performed if any its parts were changed; Big Ben might be highly tolerant even if its components are not. Well, maybe so, but why can't the component watches have their own redundancy? And even if they do not, the argument from design should apply to Big Ben and not just to its components.

because it represents the separate objects that the watchmaker created and manipulated; an atom in the watch does not count as one of its parts because there was no process that treated it as a unit apart from the other atoms in the watch. Evolutionists use a similar conception when they talk about the definition of a trait. When they say, for example, that having five fingers on the left hand is not a different trait from having five fingers on the right, they mean that digits-on-the-left and digits-on-the-right did not evolve independently. Although there is plausibility to this etiological conception of what the parts of a system are, it is not something that the design argument can embrace, since the argument needs to be able to describe the parts of a system in a way that begs no questions about how the system was brought into being.

These problems disappear if we focus, not just on the fact that a system is useful and intolerant but on a more detailed description of its parts. Understood in this way, Paley's comments about usefulness and intolerance should be understood as heuristic: they are intended to point us in the direction of a more detailed set of observations, and it is this set that is said to favor intelligent design over chance.

2.10   THE PRINCIPLE OF TOTAL EVIDENCE

The assessment presented in the previous section of the observations we should consider in the design argument was guided by the *principle of total evidence* (Carnap 1950), which I discussed in §1.2. This is just the idea that we should use *all* the evidence we have to evaluate the hypotheses at hand. It would be a mistake to focus exclusively on the fact that the watch (or the eye) is useful and intolerant. But it also would be wrong to restrict our attention to the mechanical details of how watches and eyes work and neglect the fact that they are useful to their owners. The principle of total evidence says that we need not choose between these two partial descriptions: it is best to use both.

This principle has a special relevance when we consider a long-standing criticism of the design argument. Paley's argument about the vertebrate eye, when formulated within a likelihood framework, consists of two claims:

Pr(the human eye has features $F_1, \ldots, F_n$ | Chance) is tiny (though not zero).
Pr(the human eye has features $F_1, \ldots, F_n$ | ID) is larger.

This argument has been criticized by shifting to a logically weaker observation. Instead of considering a detailed description of the human eye as the fact that requires explanation, there are other, less specific, observations we might consider instead. Here are some of them:

There are organisms in the universe that have eyes with features $F_1, \ldots, F_n$.
There are organisms in the universe that have eyes.
There are organisms in the universe that have adaptive features.
There are organisms in the universe.

The case against Epicureanism weakens as we change the problem from explaining a fact about human beings and their eyes to a proposition farther down the list. After all, it is not *that* improbable that creatures *somewhere in the universe* have eyes, if organisms were produced by a mindless random process. And it is even more probable that there should exist organisms in the universe that have adaptive features of some sort, if Epicureanism is true. With the observations reconfigured in this way, they no longer seem to strongly favor intelligent design over chance. Van Inwagen (1993: 144) presents this objection to the design argument and explains it by way of an analogy. Suppose you toss a coin twenty times and it lands heads every time. You should not be surprised at this outcome if you are one of a million people who each toss a fair coin twenty times. After all, with so many people tossing fair coins, it is all but inevitable that someone will get twenty heads. The outcome you obtained, therefore, was not improbable according to the chance hypothesis.

Van Inwagen's analysis involves shifting from a logically stronger to a logically weaker description of the observations:

- This coin landed heads on each of the twenty tosses.
- Some coin out of the million that were tossed twenty times landed heads on each toss.

The first observation seems to favor the hypothesis that the coin is strongly biased in favor of heads ($p = 0.9$, say) over the hypothesis that the coin is fair ($p = 0.5$), since

(I) Pr(this coin landed heads on each of 20 tosses | this coin is fair)

   $= (0.5)^{20}$.

   Pr(this coin landed heads on each of 20 tosses | this coin is

   strongly biased in favor of heads) $= (0.9)^{20}$.

The bias hypothesis has a likelihood that is $(1.8)^{20}$ times bigger than the fair hypothesis. However, if we shift to the logically weaker description of the observations, *and also change the hypotheses we are considering*, the situation looks very different:

(II) Pr(some coin lands heads on each of 20 tosses | 1 million

fair coins are each tossed 20 times) $= 1 - [1 - (0.5)^{20}]^{1,000,000}$

Pr(some coin lands heads on each of 20 tosses | 1 million coins

strongly biased in favor of heads are each tossed 20 times)

$= 1 - [1 - (0.9)^{20}]^{1,000,000}$.

Now the two likelihoods are both *very* close to unity.

There is a third problem to consider. What are the likelihoods when we consider the observations obtained from this coin relative to hypotheses about the 1 million coins? Here are the likelihoods:

(III) Pr(this coin lands heads on each of 20 tosses | this coin is one of

a million fair coins that each are tossed 20 times) $= (0.5)^{20.}$

Pr(this coin lands heads on each of 20 tosses | this coin

is one of a million coins strongly biased in favor of heads

that each are tossed 20 times) $= (0.9)^{20}$.

Regardless of whether the 1 million coins are all fair or all biased, their tosses are *mutually independent*; what happens to *this* coin when it is tossed is not affected by what the *other* coins are like.[19]

The data we have – that this coin landed heads on all twenty tosses – is telling in two of the three discrimination problems just described. The observations strongly favor the hypothesis that this coin is strongly biased to land heads over the hypothesis that this coin is fair; this was problem I. They also strongly favor the hypothesis that the 1 million coins are all biased to land heads over the hypothesis that they all are fair; this was problem III. In fact, the strength of evidence in these two problems is the same. Matters change if we logically weaken our description of the observations, as we did in problem II, but that should not distract us.

As discussed in §1.3, it is a standard feature of likelihood comparisons that logically stronger and logically weaker descriptions of the observations may differ in their evidential significance. This was the point of

---

[19] See Hacking's (1987) treatment of "the inverse gambler's fallacy."

the example about the breakfast order of toast and eggs represented in Figures 1.6 and 1.7. The principle of total evidence tells us to focus on logically stronger descriptions of the evidence, if doing so makes a difference in our likelihood assessments. Even if Epicureanism says that it is highly probable that there will be *some* organisms that have *some* adaptive features *somewhere* in the universe, this does not show that the theory says that it is highly probable that *human beings* will have *eyes* with features $F_1, \ldots, F_n$. Paley is right to focus on the logically stronger description of the observations. He is following the principle of total evidence. It is true but irrelevant that intelligent design and chance confer very similar probabilities on observations that are logically weaker.

2.11 SOME STRENGTHS OF THE LIKELIHOOD FORMULATION OF THE DESIGN ARGUMENT

Some of Hume's criticisms of the design argument in his *Dialogues Concerning Natural Religion* (published 1779) dissolve once we formulate the argument as a likelihood inference. For example, Hume at one point has the character he calls "Philo" say that the design argument is an argument from analogy and that the conclusion of the argument is supported only very weakly by its premises. His point can be formulated by thinking of Paley's argument as follows:

Watches are produced by intelligent design.
Organisms are similar to watches to degree p.

$p\left[\rule{7cm}{0pt}\right.$

Organisms were produced by intelligent design.

Notice that the letter p appears twice in this argument. It represents the degree of similarity of organisms and watches and it represents the probability that the premises confer on the conclusion. Think of similarity as the proportion of shared characteristics. Things that are 0 percent similar have no traits in common; things that are 100 percent similar have all their traits in common. The analogy argument says that the more similar watches and organisms are, the more probable it is that organisms were produced by intelligent design.

Hume thinks this argument is undermined by the fact that watches and organisms have relatively few characteristics in common: watches are made of metal and glass; organisms metabolize and reproduce, etc. Even if Hume is right about the analogy argument, his objection does not touch the likelihood formulation of the argument from design. With respect to watches, the only relevant question is whether their observed features are made more probable by the hypothesis of chance or by the hypothesis of intelligent design; with respect to the eye, the same comparative question is the only one that matters. Paley's analogy between watches and organisms is merely heuristic. The likelihood argument about organisms stands on its own (Sober 1993b, 2004b).

Hume also has Philo construe the design argument as an inductive argument and then complain that the inductive evidence is weak. Philo suggests that if we are to have good reason to think that our world was produced by an intelligent designer, we must have visited other worlds and observed that all or most of them were produced by intelligent design. But how many other worlds have we visited? The answer is *not even one*. Apparently, the design argument is an inductive argument that could not be weaker; its sample size is zero. This objection also dissolves once we move from the model of inductive sampling to that of likelihood. Observing numerous worlds and seeing how they were brought into being is not essential if the point is just that the two hypotheses about the world we inhabit confer different probabilities on what we observe. We will revisit the idea of inductive sampling in §2.18, but for now it should be clear that the likelihood argument takes the sting out of the fact that none of us has seen an intelligent designer create an organism from nonliving materials.

There is another objection to the likelihood version of the design argument that many philosophers find in Hume's *Dialogues* and think is devastating. This is the point that the design argument does not establish the attributes of the designer. The argument does not show that the designer who gave organisms their complex adaptations is morally perfect, or all-knowing, or all-powerful, or that there is just one of him, or that he also created the universe. Perhaps this undercuts some versions of the design argument, but it does not touch the likelihood argument we are considering. Paley, perhaps responding to this Humean point, makes it clear in Chapter 5 of *Natural Theology* that his argument about the watch and the eye is intended to establish only the *existence* of a designer and that the question of the designer's *characteristics* must be addressed separately. Does this limitation of the design argument make the argument trivial? Not at all. It is *not* trivial to claim that the adaptive features of organisms are due to intelligent design. This supposed "triviality" would be *big news* to evolutionary biologists. And it also would be enough to refute Epicureanism. To concede that the design argument establishes the existence of an intelligent designer is to concede a great deal.

2.12 THE ACHILLES HEEL OF THE LIKELIHOOD ARGUMENT

Paley's likelihood argument about the watch goes as follows:

The watch has features $G_1 \ldots G_n$.

Pr(the watch has features $G_1 \ldots G_n$ | chance) = tiny.

Pr(the watch has features $G_1 \ldots G_n$ | intelligent design) > tiny.

The law of likelihood

_____

The watch's having features $G_1 \ldots G_n$ favors intelligent design over chance.

His argument about the eye has the same logical form:

The eye has features $F_1 \ldots F_n$.

Pr(the eye has features $F_1 \ldots F_n$ | chance) = tiny.

Pr(the eye has features $F_1 \ldots F_n$ | intelligent design) > tiny.

The law of likelihood

_____

The eye's having features $F_1 \ldots F_n$ favors intelligent design over chance.

The arguments are both deductively valid. The first premise of each reports the observations, the last premise invokes an epistemological principle, and the second and third state that the observations are more probable under the one hypothesis than they are under the other. If Paley reasoned cogently about the watch, how can one deny that he did the same when he discussed the eye? This denial may sound implausible, but this is precisely what I propose to argue. There is a deep difference between the two arguments. The argument about the watch has a missing premise that, once acknowledged, does the argument no harm. The argument about the eye also has a missing premise, but it turns out to be indefensible. The difference between the two arguments becomes visible once we scrutinize the third premise of each.

With respect to the watch, it seems undeniable that

Pr(the watch has features $G_1 \ldots G_n$ | intelligent design) > tiny.

But, in fact, whether this claim is true depends on what *goals and abilities* the designer would have if such a being existed. For example, suppose we assume that

A_f: If an intelligent designer made the watch, he would have wanted (above all) to give it features $G_1 \ldots G_n$ and he would have had the ability to ensure that this is so.

Given this assumption, it is true that the intelligent-design hypothesis makes the observations highly probable; with this assumption in place, intelligent design has a higher likelihood than the hypothesis of chance. A_f is thus an auxiliary assumption that is *favorable* to the design hypothesis. However, there are other assumptions that block the conclusion that Paley wants to draw. For example, consider the following

A_u: If an intelligent designer made the watch, he would have wanted (above all) to *prevent* the watch from having features $G_1 \ldots G_n$ and he would have had the ability to insure that the watch fails to have these features.

If we adopt *this* auxiliary assumption, the design hypothesis has a likelihood of a zero. If A_u correctly describes what the designer would have been like if such a being had existed, then the watch has a higher probability of exhibiting the features we observe if it is the result of chance. Given the *unfavorable* assumption A_u, chance has a higher likelihood than intelligent design.

There is an asymmetry between A_f and A_u. For Paley's argument about the watch to go through, there must be a reason to reject A_u. However, it isn't essential, if his argument is to work, that there be a reason to accept A_f. The favorable assumption A_f suffices, but it is not necessary, for intelligent design to have a higher likelihood than chance. All Paley needs is an assumption that ensures that

Pr(the watch has features $G_1 \ldots G_n$ | ID) > tiny

and this can be true even if there is considerable uncertainty as to which goals and abilities the designer would have if there were such a being. This is because

$Pr($the watch has features $G_1 \ldots G_n \mid ID)$

$$= \sum_i Pr(\text{the watch has features } G_1 \ldots G_n \mid ID \ \& \ A_i) Pr(A_i \mid ID).$$

Here $A_1, A_2, \ldots, A_n$ are different auxiliary propositions about the goal–ability pairs that the designer of the watch might have had if there were such a being.

Although Paley's argument about the watch requires assumptions about the goals and abilities the designer would have had if there were such a being, I don't think this leaves the argument in the lurch. Paley is aware that there are many human designers not far from the heath on which he is walking and that these designers know how to make watches and have every inclination to do so. Provided that there is even a small chance that the designer of the watch is a human being of this sort, his argument goes through.

Parallel questions arise for Paley's argument concerning the eye, but they lead to a different verdict concerning whether his argument succeeds. As before, whether intelligent design has a higher likelihood than chance depends on what we are entitled to assume about the goals and abilities that the designer of the eye would have had if such a being had existed. There are favorable assumptions we might make here and unfavorable assumptions are available as well. I pointed out in §2.7 that Gould makes unfavorable assumptions about the designer's goals and abilities when he discusses the panda's thumb; Gould assumes that if an intelligent designer had made the panda he would have chosen *not* to give the panda the spur of bone we call a "thumb" and instead would have given the panda some more efficient device. With this unfavorable assumption, the intelligent-design hypothesis has a likelihood of zero, so the hypothesis of chance and the hypothesis of evolution by natural selection both have higher likelihoods. My criticism of Paley is that his discussion of the eye makes the same mistake that Gould made. Paley assumes that if an intelligent designer created the human eye, the designer would have wanted to give us eyes with features $F_1 \ldots F_n$ and would have had the ability to do so. Paley is no more entitled to adopt these favorable assumptions than Gould is entitled to embrace his unfavorable assumptions. What is required, whether we are talking about the panda's thumb or the vertebrate eye, is an *independent reason* for believing assumptions about goals and abilities (Kitcher 1983; Pennock 1999; Shanks 2004; Sober 1999b).[20]

It is no good arguing as follows: "Look, the eye was created by an intelligent designer and the eye has features $F_1 \ldots F_n$. Therefore, the designer in question probably wanted the eye to have features $F_1 \ldots F_n$ and had the ability to achieve this goal." This line of reasoning *assumes* that the intelligent-design hypothesis is true; however, this is just what the design argument is trying to establish, and so it can't serve as a premise in the argument. What is needed is information about goals and abilities that we can know is correct without already needing to have an opinion as to whether the intelligent design or the chance hypothesis is true.

When hypotheses are tested against each other, each is asked to say how probable the observations are. It is here that a point that Pierre Duhem (1914) emphasized in connection with theories in physics becomes relevant: Theories rarely make predictions on their own; rather, auxiliary assumptions need to be brought to bear. For example, the general theory of relativity, by itself, does not make predictions about when eclipses will occur or what features they will have. However, if auxiliary information about various celestial bodies is taken into account, the general theory of relativity does make predictions about these matters. Duhem's point holds for most of the hypotheses that the sciences consider,[21] and it also holds when we recognize that prediction rarely involves deduction. Duhem's idea is that the usual pattern in science is that the hypothesis *H* does not entail whether the observation statement *O* will be true; rather it is *H&A* that will have this kind of entailment, for suitably chosen auxiliary assumptions *A*. The likelihood version of this Duhemian point is that the value of $Pr(O \mid H)$ is rarely well defined but that the value of $Pr(O \mid H\&A)$ can be, again for suitably chosen auxiliary assumptions *A*.

This raises the question of which auxiliary assumptions we should use to render a theory testable. What makes an auxiliary assumption "suitable?"

---

[20] As noted in §1.7, scientists often use auxiliary propositions to test theories that include *harmless idealizations*. These are simplifying falsehoods that lead the theory to make the same, or nearly the same, predictions it would make if true auxiliary propositions were used instead (McMullin 1985; Hausman 1992). Assuming that a planet is spherically symmetrical or that a population is infinitely large can often be justified in this fashion. In this case, the demand for independently justified auxiliary propositions is not the demand that those propositions be shown to be *true*; rather, what is required is independent evidence that the idealization is harmless. However, to know whether an auxiliary proposition is a harmless idealization, one would have to know what the truth is, and this is precisely what Paley and Gould do not know with respect to the auxiliary assumptions they adopt.

[21] Quine (1953) generalized Duhem's thesis, claiming that *all* theories, not just *physical* theories, have this character. This thesis is now often referred to as the Duhem–Quine thesis.

The point against Paley and Gould is that, in testing H_1 against H_2, you must have a reason to think that the auxiliary proposition A is true that is independent of whatever you may already believe about H_1 and H_2. For example, suppose you are on a jury. Jones is being tried for murder, but you are considering the possibility that Smith may have done the deed instead. Evidence is brought to bear: A size 12 shoe print was found in the mud outside the house where the murder was committed, as was cigar ash, and shells from a Colt .45 revolver. Do these pieces of evidence favor the hypothesis that Smith is the murderer or the hypothesis that Jones is? It is a big mistake to answer these questions by *inventing* assumptions. If you assume that Smith wears a size 12 shoe, smokes cigars, and owns a Colt .45 and that Jones wears a size 10 shoe, does not smoke, and does not own a gun, you can conclude that the evidence favors Smith over Jones. If you make the opposite assumptions, you can draw the opposite conclusion. Surely it would be wrong simply to commit to assumptions that help convict Smith or to assumptions that help acquit him, as is your whim. What is needed is *independently attested information* about Smith's and Jones's shoe sizes, smoking habits, and gun ownership.

I hope it is obvious that if you want to use the observation O to test hypothesis H_1 against hypothesis H_2, that the auxiliary assumptions you make must not depend for their justification on assuming that H_1 is true or on assuming that H_2 is true. What is perhaps less obvious is that the auxiliary assumptions must be justified without assuming that O is true. Here is why that additional constraint is needed: If O is true, so is the disjunction "either H_1 is false or O is true." If you use this disjunction as your auxiliary assumption A_1, then it turns out that the conjunction $H_1 \& A_1$ entails O. This allows H_1 to make a prediction about O even when H_1 has nothing at all to do with O. The same ploy can be used to obtain auxiliary assumptions A_2 so that the conjunction $H_2 \& A_2$ also entails O. Using propositions A_1 and A_2 as auxiliary assumptions leads to the conclusion that the two hypotheses H_1 and H_2 both have likelihoods of unity. The way to prevent this is to insist that the auxiliary assumptions used to bring the hypotheses H_1 and H_2 into contact with the observation O must be justified without assuming H_1 or assuming H_2 or assuming O.

The objection I have described to the organismic design argument applies to a design hypothesis that postulates an otherwise unspecified designer and also to a hypothesis that says that this designer is God himself. If we suppose that God, if he exists, is omnipotent, omniscient, and omni-benevolent, it still isn't clear that Pr(the human eye has features $F_1 \ldots F_n \mid$ the eye was made by God) > tiny. The supposition that God is omnipotent of course ensures that God has the *ability*; he could give human beings an eye with the features we observe if he chose to do so. The question is whether God would have had the *desire*. As noted earlier, we must be careful not to beg the question. We can't reason that since the eye was made by God, that God must have wanted human beings to have eyes with the features we observe. What is needed is evidence about what God would have wanted the human eye to be like, where the evidence does not require a prior commitment to the assumption that there is a God and also does not depend on looking at the eye to determine its features. The supposition that God is omnipotent, benevolent, and omniscience does not provide this information. There are many architectures for eyes, and they are distributed in an interesting way across different organisms. For example, human beings and other vertebrates have camera eyes, and so does the octopus. The octopus eye does not have a blind spot, but ours does. Scorpions have eyes equipped with "sunglasses," but we do not. It seems pretty clear that the mix of features that the human eye has cannot be predicted from the supposition that God made those eyes.

Does this problem disappear if we retreat to some logically weaker description of the observations that need to be explained, as contemplated in §2.9? For example, instead of trying to explain the details of the vertebrate eye, what if we simply try to explain why it is useful and intolerant? Now the problem of blindspots and sunglasses disappears, but the main problem remains in place. The fact that the eye is useful is explained by the fact that organisms with eyes live in environments of certain kinds – roughly, in environments in which useful information about the environment is packaged in light. The question may then be put as to why these organisms live in such environments rather than in darkness, and this returns us to questions about the designer's goals and abilities. The same puzzle arises when we turn to the task of explaining why the eye is intolerant. Even if we waive the wine-bottle problem (§2.9), it is puzzling why an intelligent designer would have constructed the eye so that removing or changing any of its parts causes the organism to stop seeing. Why did the putative designer choose to make the eye so fragile?

The criticism I have made of the design argument is one that Descartes endorses in *The Principles of Philosophy*:

when dealing with natural things we will, then, never derive any explanations from the purpose which God or nature may have had in view when creating them, and we shall entirely banish from our philosophy the search for final causes. For we should not be so arrogant as to suppose that we can share in God's plans. (I, 28)[22]

It is interesting to put this passage side by side with others in which Descartes plainly thinks it legitimate to explain *human* behavior by describing the beliefs and desires that human beings have. Apparently, invoking *human* purposes to explain a set of observations is one thing, invoking *God's* purposes is another. The failure to heed this distinction is the mistake that undermines Paley's argument.

2.13 PALEY'S STONE

I have focused on Paley's discussion of the watch and the eye, but it is worth revisiting what he says about his first example: the stone found on the heath. He introduces this example only to brush past it dismissively; perhaps, he says, "it had lain there forever; nor would it, perhaps, be very easy to show the absurdity of this answer." The stone, unlike the watch and the eye, does not provide a compelling argument for intelligent design. But why not? The likelihood representation of the design argument suggests an answer. It is a consequence of the law of likelihood (§1.3) that

$$Pr(O \mid ID) > Pr(O \mid \text{Chance}) \text{ if and only if}$$
$$Pr(notO \mid ID) < Pr(notO \mid \text{Chance}).$$

If an object's being useful and intolerant favors intelligent design over chance, then its *failing* to be both useful and intolerant must have the opposite evidential significance. This may be why Paley does not offer the stone as a compelling argument for intelligent design. The watch and the eye fill the bill; the stone does not.

Although restricting our evidence about the eye and the watch to the conjunction "useful and intolerant" and restricting our description of the stone to the negation of that conjunction makes everything fall into place, the situation becomes less tidy once we recognize that we know more about all three objects. It then emerges that my objection to what Paley says about the eye also applies to what he says about the stone. If the

---

[22] See also *Principles* (III, 2) and his replies to Gassendi (Descartes 1985: II, 241–77).

stone's specific features do not favor intelligent design over chance, this must mean, within the likelihood framework, that

$$Pr(\text{the stone has features } S_1 \ldots S_n \mid ID)$$
$$\leq Pr(\text{the stone has features } S_1 \ldots S_n \mid \text{chance}).$$

But why should we think that this inequality holds? If Paley gets to help himself to assumptions about the goals and abilities of the putative designer that are favorable to the design hypothesis in the case of the eye, why should he abstain from doing so in the case of the stone? We can easily describe assumptions about the putative stone-maker that render design more likely than chance. And, of course, there are other assumptions that have the opposite effect. The design argument has no more basis for claiming that design is the better supported hypothesis in the case of the eye than it has for saying that chance is the better supported hypothesis in the case of the stone.[23]

2.14   TESTABILITY

My criticism of the design argument might be summarized by saying that the design hypothesis is untestable, but it is important that this summary be understood in the right way. It involves no endorsement of Popper's proposal that testability should be glossed understood in terms of his concept of falsifiability, a proposal that I criticized in §2.8. It is perfectly clear that "an intelligent designer gave vertebrates their eyes" entails that vertebrates have eyes; therefore, this intelligent-design hypothesis is falsifiable. What is less clear is that this hypothesis can be tested *against* the Epicurean hypothesis that a mindless chance process gave vertebrates their eyes (or, for that matter, against the evolutionary hypothesis that the process of evolution by natural selection did the work). In addition, my analysis takes seriously the Duhemian insight that testable theories typically do not make predictions (whether deductive or probabilistic) all by themselves but need to make use of additional auxiliary propositions. This point should make us cautious about claiming that the design hypothesis is, *in principle*, untestable. A statement may *now* be untestable because we don't at present have the independently attested auxiliary propositions that are needed to bring it into contact with observations, but the situation may change as knowledge grows. I do not expect this to happen in

---

[23] I am grateful to Susanna Rinard for drawing my attention to this point about the stone.

the case of the design argument for the existence of God, but my criticism of the design argument does not depend on making forecasts. To say that it will *never* be possible to test a given hypothesis requires considerable knowledge about what the future of inquiry may bring. The fact that we now can't imagine how a given hypothesis could be tested may be due to the fact that the hypothesis is in principle untestable, but, alternatively, it may simply reflect our limited powers of imagination.

In the course of trying to develop a theory of what makes a statement empirically testable, the logical positivists developed a number of proposals, but one after another turned out to be flawed (Hempel 1951, 1965a). Many philosophers took this to show that there can be no criterion of testability.[24] For them, the lesson learned is that the problem is *impossible*, not just that it is *difficult*. I am less pessimistic. The logical positivists employed a limited set of tools. They tried to define testability by using the concept of an observation statement and the resources of deductive logic. What would happen if probability theory were added to deductive logic in this enterprise? And what if we recognize that testability is an epistemic concept and isn't strictly logical at all? And how is the problem affected by taking on board the fact that testing is typically a contrastive enterprise?

Although many philosophers of science now seem to take the past failures of the positivist project to show that nothing much can be said to clarify what testability means, they frequently hold that testing is important in science and that philosophers of science have the job of clarifying what it means to test a theory. Their implicit view is that testing makes sense but testability does not. This two-part position is peculiar. What would we think if chemists took the view that there is lots to say about what happens when salt dissolves in water, but that there is no hope of providing a theory of water-solubility? Testing is to testability as dissolving is to solubility. Philosophers should never have walked away from the concept of testability (Sober 1999b).

Hempel developed his pessimistic assessment of the prospects for a criterion of testability against the backdrop of the assumption that testability and "cognitive meaningfulness" are one and the same property.[25] In retrospect, it seems clear that meaningfulness and testability are different. I suppose that the sentence "undetectable angels exist" is untestable, but the

---

[24] Justus (2007) provides a useful review of these of trials and errors and argues that one of Carnap's later proposals remains unrefuted.

[25] More precisely, most positivists thought that a meaningful statement must either be empirically testable or analytic (its truth value being a logical consequence of definitions). This entails that nonanalytic statements must be empirically testable if they are meaningful.

sentence is not meaningless gibberish. We know what it says, what logical relations it bears to other statements, and we can discuss whether it is knowable; none of this would be possible if the string of words literally made no sense. However, once testability and meaningfulness are separated, the condition of adequacy that Hempel uses in his discussion of the problem becomes highly questionable when it is applied to the concept of testability. Hempel (1965a:102) says that if S is a meaningless string of words, then the same is true of all statements in which S occurs as a (truth-functional) part; if S is meaningless, so is the negation of S, and so are all disjunctions and conjunctions in which S occurs. Hempel's condition of adequacy may be right as a claim about what it is for a string of words to lack meaning, but it is wrong as a constraint on what it is for a statement to be untestable. To see why, let's begin with a natural Bayesian understanding of this concept.

In §1.2, we examined how Bayesianism defines confirmation, disconfirmation, and evidential irrelevance. For Bayesians, "O is evidentially irrelevant to H" means that $Pr(H\,|\,O) = Pr(H)$. There is a natural generalization of this idea, one that clarifies what it means for H to be impervious to observational test:

> H cannot be observationally tested if and only if, for all observation
>
> statements O, $Pr(H\,|\,O) = Pr(H)$.

I do not claim that this is a flawless Bayesian definition of testability,[26] and, of course, I have already registered my reservations about Bayesianism in Chapter 1. But notice how the Bayesian definition of evidential irrelevance leads naturally to a concept of testability. And notice also how it violates Hempel's criterion of adequacy. Suppose that the statement "undetectable angels exist" is untestable in this Bayesian sense. It does not follow that the same must be true of conjunctions in which that sentence occurs. For example, if you draw twenty balls at random from an urn (with replacement) and observe that all twenty are green, this raises the probability that all the balls in the urn are green. The same observation also raises the probability of a conjunction: "All the balls in the urn are green and undetectable angels exists." The conjunction is more probable, given the observation, than it was before. Maybe the conjunction $H \& X$ is meaningless if X is, but that does not entail that $H \& X$

---

[26] Skyrms (1984) discusses this Bayesian proposal, as does Creath (1991). See also Reichenbach's "probability theory of meaning" (1938: 46–57).

must be untestable just because *X* is. If Hempel's condition of adequacy is wrong, his article can hardly be regarded as the nail in the coffin that it is often taken to be.

Here, then, is my suggestion for how testability should be understood. Let us begin by considering what it means for a statement to have deductive implications about observations once suitable auxiliary propositions are added. The problem is to understand what "suitable" means. To see the kind of problem that the positivists encountered, consider the following proposal:

Proposition *P* has observational implications if and only if there exists an auxiliary assumption *A*, and an observation statement *O*, such that *P&A* entails *O* but *A* by itself does not entail *O*.

The trouble with this criterion is that it is too broad; it has the consequence that every proposition *P* has observational implications. Just let *A* be the statement "if *P* then *O*." Can this criterion be repaired by requiring that the auxiliary assumption *A* be true? The problem now is that if *O* is true, so is the statement "*notP or O*" for any proposition *P* you care to name. Let this disjunction be the auxiliary proposition *A*. Notice that *A* is true, *P&A* entails *O*, and *A*, by itself, does not entail *O*. As the positivists came to see, the problem is tricky. With some fear that I am stumbling into the same old quagmire, I offer the following proposal:

Proposition *P* now has observational implications if and only if there exist true auxiliary assumptions *A*, and an observation statement *O*, such that (i) *P&A* entails *O*, but *A* by itself does not entail *O*, (ii) we now are justified in believing *A*, and (iii) the justification we now have for believing *A* does not depend on believing that *P* is true (or that it is false), and also does not depend on believing that *O* is true (or that it is false).

The word "now" marks the fact that whether a proposition has observational implications depends on the rest of what we are justified in believing, and that can change.

This definition of what it means for a statement to have observational *implications* does not, by itself, provide an account of *testability*. For one thing, it ignores the fact that probability statements (like the claim that a coin is fair) are testable though they do not deductively entail what we will observe (§2.8). For another, it follows from the above proposal that contradictions have observational consequences (assuming, as I do, that classical logic is correct in saying that contradictions entail *all* statements), but this is not enough to show that contradictions are empirically testable. Bayesians often recommend that contradictions be assigned a probability of 0. This assignment entails that no observation can change their probability. The reason is that the probabilities 0 and 1 are *sticky*; if $Pr(H) = 0$ (or 1) then $Pr(H|E) = 0$ (or 1), no matter what E is. Bayesians who adopt this recommendation will therefore agree that contradictions are not empirically testable, though, if they accept classical logic, they will grant that contradictions have observational consequences.

Although the above definition does not capture what testability means, we can use it as a model for how likelihoodism should understand the latter idea. The key is to recognize that testing a proposition often involves probabilistic rather than deductive relations to observations and that the concept of testing needs to be understood contrastively:

Hypothesis H_1 can now be tested against hypothesis H_2 if and only if there exist true auxiliary assumptions A and an observation statement O such that (i) $Pr(O|H_1 \& A) \neq Pr(O|H_2 \& A)$, (ii) we now are justified in believing A, and (iii) the justification we now have for believing A does not depend on believing that H_1 is true or that H_2 is true and also does not depend on believing that O is true (or that it is false).

Notice that this definition makes use of the concept of justified belief and thus requires a concept that I set to one side in §1.1. Notice also that Hempel's criterion of adequacy is no more plausible within a likelihood framework than it is within Bayesianism. Likelihoodists, recall, are happy to grant that specific scientific theories like the general theory of relativity have well-defined likelihoods, but they deny that this is true of catchall hypotheses like the general theory of relativity's negation (§1.3). For a likelihoodist, *GTR* can be tested against a specific alternative, but *notGTR* cannot be.

It remains to say something about what an observation statement is. Like the concept of testability itself, the concept of observation has come in for heavy criticism. A rallying cry of 1960s philosophy of science was the slogan that "all observations are theory-laden." Kuhn (1962), Hanson (1969), and Feyerabend (1974) promoted this idea, as did many others. If the idea of testability requires there to be an observation language that is *absolutely* theory-neutral, then I concede that there are no observations, provided that we are prepared to be fairly relaxed about what a "theory" is. Every statement has the following fairly trivial property: You have to know *something* if you are going to (justifiably) affirm or deny that the statement is true. For example, to assert or deny that "this is an apple," you have to know what an apple is (or, at least, what counts as evidence for and against something's being an apple). Call this your "theory of apples." *QED*. Still, it would be a mistake to take this simple point to show that the concept of an observation should be abandoned. The above definition of what it takes for H_1 to be testable against H_2 contains the idea we need. For a statement O to count as an observation in this testing problem, it must be possible to tell whether O is true without assuming that H_1 is true or assuming that H_2 is. An observation in a testing problem must be *relatively* theory neutral; it need not be *absolutely* theory neutral (Sober 2008a). Scientists often use theories to formulate their observations; for example, the outputs of measuring devices are interpreted by using theories, but that does not prevent these interpreted outputs from counting as observation reports. Suppose you use a radioisotope dating machine to estimate the age of a dinosaur fossil. There is nothing wrong with your saying "I see that the fossil is 80 million years old" as you gaze at the machine's output screen. You then can use this observational report in the context of testing hypotheses about why dinosaurs went extinct. A statement can count as an observational report in one problem even though it is a hypothesis under test in another.[27]

Although I earlier expressed my doubts about our ability to tell that a given statement is untestable in principle, this does not mean that there should be doubts about what that concept means. By replacing the word "now" with an existential quantifier in the proposed criterion of testability, we can define what it means for there to exist a time at which it is possible to test H_1 against H_2. We then can further modalize the concept by defining what it means for it to be possible that there exists a time at which H_1 can be tested against H_2. Of course, there are a variety of concepts of possibility to think about here – logical, nomological, and so on. However, this should not discourage us: testability is, in this respect, like the concept of solubility.

The fact that the above definition of testability is stated within a likelihood framework means that it will need to be modified if the likelihood framework is inadequate. For example, the proposal ignores how model selection criteria (§1.7) permit models to be compared, though it is easy enough to see how "AIC testability" might be defined along the same lines.[28] Despite this limitation, I believe that this

[27] There is more to the concept of an observation statement than the requirement of epistemic independence described here. For example, we know that $5 + 7 = 12$ is true without needing to have an opinion on whether relativity theory or Newtonian theory is true, but that does not mean that the arithmetic proposition is an observational report. I won't try to complete the analysis here; see Sober (2008a) for further discussion.

[28] The requirement that auxiliary assumptions should be epistemically independent of the hypotheses under test also makes sense in the context of a model selection criterion like AIC. Akaike discovered assumptions that suffice for AIC to be an unbiased estimate of a model's

likelihoodist conception of testability is a step forward from the failed proposals of the logical positivists. The problem of developing a criterion of testability is inseparable from the task of understanding the concept of evidence. It should be addressed, not set aside as if it were a pseudo-problem.

2.15 THE RELATIONSHIP OF THE ORGANISMIC DESIGN ARGUMENT TO DARWINISM

To see why the design argument is defective, there is no need to have a view as to whether Darwin's theory of evolution is true. In spite of what the no-designer-worth-his-salt argument suggests (§2.7), we are not in a situation in which the design hypothesis makes one set of predictions and evolutionary theory another. The problem with the hypothesis of intelligent design is not that it makes inaccurate predictions but that it doesn't predict much of anything. Rather, the design hypothesis merely allows our observations – whatever they turn out to be – to be folded inside a simple formula. If the human eye has one set of features, we can construct the hypothesis that an intelligent designer brought this about; but if the eye turns out to have a different set of features, that outcome also can be accommodated within the framework of intelligent design. From the point of view of Duhem's thesis, the problem with the design hypothesis is that we have no independent knowledge of the goals and abilities that the designer of organisms would have if such a being existed.[29]

2.16 THE RELATIONSHIP OF PALEY'S DESIGN ARGUMENT TO CONTEMPORARY INTELLIGENT-DESIGN THEORY

Michael Behe's (1996) book, *Darwin's Black Box*, is one of the most influential works of the contemporary intelligent-design movement. The main claims of the book are these:

(1) The various biochemical adaptations that Behe describes are "irreducibly complex."

---

predictive accuracy. If this is part of what justifies our using AIC to compare LIN and PAR, then we must have reason to think that those assumptions (or some other assumptions that suffice for the theorem to hold) are true. This had better not depend on our already having an opinion as to whether LIN is more predictively accurate than PAR.

[29] Johnson, a leading intelligent-design theorist, agrees that God's purposes are "inscrutable" (1991: 67) and "mysterious" (1991: 71).

(2) Irreducibly complex adaptations cannot evolve by the gradual Darwinian process of evolution by natural selection.

(3) If an irreducibly complex adaptation cannot evolve by the process of Darwinian gradualism, then it is plausible to think that it was designed and created by an intelligent being.

Behe (1996, 2005) has emphasized that his argument for the existence of an intelligent designer is not formulated as an argument for the existence of God. The identity of the designer – God or a team of extraterrestrials – is left unspecified.[30] Behe personally believes that God is the designer in question, but he admits that this conviction goes beyond what he has been able to establish scientifically, at least so far.

How is Behe's framework related to Paley's? Behe says that an irreducibly complex system is "a single system composed of several well-matched interacting parts that contribute to the basic function, wherein the removal of any one of the parts causes the system to effectively cease functioning (1996: 39)." Behe's notion of irreducible complexity is very close to the concepts of usefulness and intolerance that Paley (1802) described. Both require that the system have a function or purpose or that it be useful. However, there is a modest difference between Paley and Behe that concerns how they spell out the second requirement. Behe asks whether *removing* any part would result in the system's malfunctioning, whereas Paley asks whether *changing the shape, size, or placement* of any part would do so. In spite of this small difference, the wine-bottle problem, discussed in §2.10 in connection with the concept of intolerance, applies, with devastating results, to Behe's concept. Whether a system that has a function counts as irreducibly complex depends on how the system is divided into parts. Another difference between Paley and Behe concerns the examples they discuss; Behe's focus is on biochemical systems such as the bacterial flagellum and the mechanisms that cause blood to coagulate; these details were, of course, unknown to Paley. But of greater import is the fact that Behe aims to refute evolutionary theory whereas Paley's target was Epicureanism. Behe's critique of evolutionary theory raises some new issues.

We have already seen why Epicureanism is logically compatible with the existence of adaptive complexity. Monkeys pounding at random on typewriters *can* produce the works of Shakespeare, and a hurricane

---

[30] His position, when supplemented with independently plausible additional assumptions, entails the existence of a *supernatural* intelligent designer, or so I argue in Sober (2007a).

whirling through a junkyard *can* produce a functioning airplane. These outcomes are not impossible; what is true is that they have extremely small probabilities. We saw in §2.5 that Epicureanism and Darwinian evolution are different; the first describes a purely random process, whereas the second involves a two-part process that is partly random and partly not. Even so, the same point holds: evolution *can* produce adaptive features that are irreducibly complex.

Evolutionary biology describes evolution as a probabilistic process. Evolution isn't like the motion of Newtonian billiard balls. In Newtonian physics, the initial conditions of a system plus the laws that govern it uniquely determine what its future will be. This is what it means for the theory to be deterministic. In a probabilistic theory, the initial conditions plus the relevant laws provide a probability distribution; some futures are more probable and others are less, but there are multiple possible futures.[31] To apply this distinction between deterministic and probabilistic theories to evolutionary biology, we need to come to grips with the fact that evolutionary theory contains a number of different models of the evolutionary process. The theory recognizes that there are different possible causes of evolution. There are simple models that describe what will happen when just one of these causes is in operation and more complicated models that describe the outcome of multiple causes interacting (Sober 1984).

Although biology conceives of evolution as a probabilistic, not a deterministic, process, it also is true that biology contains deterministic models of the evolutionary process. This may sound like a contradiction, but it is not. The reason it is not is that these deterministic models are *idealizations*. Like models in Newtonian mechanics that assume that the Earth is a perfect sphere or that an inclined plane is frictionless, idealized models in evolutionary biology make false assumptions. In evolutionary biology, deterministic models are false because they assume that the populations being described contain infinitely many individual organisms. To understand why finite versus infinite population size makes a difference to a model's predictions, consider the following simple example. Suppose the organisms in a population reproduce asexually. The organisms either have trait *A* or trait *B*, and suppose that inheritance has perfect fidelity, with offspring always having the same traits as their parents. Suppose further that *A* individuals have a higher probability of surviving to reproductive age than *B* individuals have. That is, the two types of organism differ in their *fitness*. Let's suppose, to make things

---

[31] For a more detailed analysis of what determinism involves, see Earman (1986).

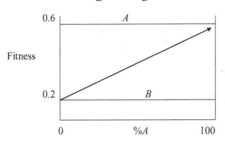

Figure 2.3 If *A* individuals have a fitness of 0.6 and *B* individuals have a fitness of 0.2, no other evolutionary forces impinge, and the population is infinitely large, trait A must evolve to 100-percent representation.

concrete, that *A* individuals have a probability of 0.6 of surviving and that *B* individuals have a probability of surviving of 0.2. Again to keep things simple, let's suppose that generations do not overlap; in each generation, the individuals who survive to reproductive age all produce a single off-spring at the same time and then they die.

How will traits *A* and *B* evolve in this population? If a finite population begins with a few *A* individuals and mostly *B* individuals, it is *possible*, after many generations, that the less fit trait *B* will disappear and the population will become 100 percent *A*. However, there is no necessity in this outcome; this is because, with *finite* population size, there is a non-zero probability that all the *A* individuals in a generation die without reproducing and that this never happens to the *B* individuals. On the other hand, if we assume that population size is *infinite*, then the fitter trait (*A*) *must* go to fixation (meaning 100 percent representation).[32] The evolution of this infinitely large population is depicted in Figure 2.3. In such deterministic models, the fitter trait *must* increase in frequency and the less fit trait *must* decline.

Figure 2.3 represents selection acting on a dichotomous trait. To think about the same set of questions in connection with the evolution of a quantitative trait (like the length of a polar bear's fur), it is useful to consider the trait's *fitness function*. Figure 2.4 depicts three examples; each maps trait values onto fitness values. What must a quantitative trait's fitness function be like if the trait is to evolve from one value to another? For example, if a polar bear population at one time has an average fur length of 10 centimeters and the climate gets colder, will selection lead the population to evolve to an average fur length of 20 centimeters? Darwin

---

[32] More precisely, the probability that *A* goes to fixation in this infinite population is one.

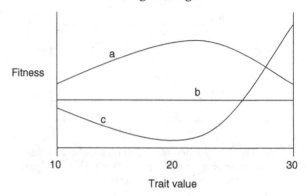

Figure 2.4 A trait that evolves from a value of 10 to a value of 20 by the process of Darwinian gradualism in an infinite population must have a fitness function that monotonically increases from 10 to 20. This is true for (a), but not for (b) or (c).

thought of natural selection as a gradual process; his idea was that the average trait value in the population will increase by small changes. If a fur length of 11 centimeters is fitter than a fur length of 10 centimeters, and a length of 12 centimeters is fitter than 11 centimeters, and so on up to 20 centimeters, the population can be transformed from 10 to 20 centimeters by the process of Darwinian gradualism. Notice the word "can" in the previous sentence. It is here that the distinction between deterministic and probabilistic models becomes relevant. If the population is infinitely large, fitter trait values *must* replace less fit trait values, and so the fitness function depicted in (a) of Figure 2.4 entails that the population *must* evolve from 10 to 20 centimeters. On the other hand, if the fitness function is either (b) or (c) in Figure 2.4, an infinite population *cannot* make this transition. However, these claims of necessity and impossibility are both cancelled if the population is finite. What is true is that a finite population has a *higher probability* of evolving from 10 to 20 centimeters if the fitness function is the one shown in (a), but its probability of making this transition is not zero if (b) or (c) is the fitness function in place. In (b), the population must undergo neutral evolution; in (c), the population must evolve against the tide of natural selection.

We can use the concept of a fitness function to say something about how the irreducible complexity of a trait is related to how (or whether) the trait will evolve. Consider an organ (like the eye) or a biological process (like blood coagulation) that has many parts (p_1, p_2, . . . , p_n). Suppose that if all the parts are present, the organism gains an advantage. In the case of the eye, the organism can see. Suppose further that *a miss is as good*

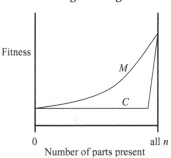

Figure 2.5 If there are *n* parts to an eye, how fit are organisms that have 0, or 1, or 2, ...,
or (*n* − 1), or all *n*? If natural selection in an infinite population is going to cause the eye
to evolve one part at a time, a monotonic fitness function, such as *M*, is required.
However, if the trait is irreducibly complex, the fitness function will resemble *C*.

as a mile; if an organism fails to have all the parts in place, it has the same,
lower, fitness value. This idea is summarized in the nonmonotonic fitness
function *C* shown in Figure 2.5. It contrasts with the monotonic fitness
function *M*, in which each additional part increases the organism's fitness.
If a finite population is to evolve from 0 to 1 to 2 ... to all *n* parts in
stepwise fashion, this progression is more probable under fitness function
M than it is under *C*. For all *n* parts to evolve under the latter fitness
function, the trait must drift by neutral evolution from 0 parts to *n*−1;
only then does natural selection kick in, causing the population to take
the last step, from *n*−*1* to *n*. On the other hand, if the population is
infinite, something stronger can be said: The stepwise evolution *must*
occur if *M* is true but it *cannot* do so if *C* is true.

Chapter 6 of *The Origin of Species* is called "Difficulties of Theory."
Darwin begins the section on "Organs of Extreme Perfection" with this
comment on Paley's well-worn example:

To suppose that the eye, with all its inimitable contrivances for adjusting the
focus to different distances, for admitting different amounts of light, and for the
correction of spherical and chromatic aberration, could have been formed by
natural selection, seems, I freely confess, absurd in the highest degree. Yet reason
tells me, that if numerous gradations from a perfect and complex eye to one very
imperfect and simple, each grade being useful to its possessor, can be shown to
exist; if further, the eye does vary ever so slightly, and the variations be inherited,
which is certainly the case; and if any variation or modification in the organ be
ever useful to an animal under changing conditions of life, then the difficulty of
believing that a perfect and complex eye could be formed by natural selection,
though insuperable by our imagination, can hardly be considered real.   (Darwin
[1859] 1964: 186–7)

Biologists often follow Darwin's lead here, arguing that the eye and other complex adaptations have monotonic fitness functions. Creationists, in contrast, tend to think in terms of fitness functions like C. "What good is 1 percent of an eye?" they ask, thinking that the answer to this rhetorical question is obvious: *It is no good at all*. Behe, like his predecessors, thinks this shows that the trait cannot evolve. I hope that the distinction between finite and infinite population size makes it clear that this does not follow.

There is a second problem with Behe's position on irreducible complexity. Although a system like the camera eye can be segmented into parts in such way that it counts as irreducibly complex, this does not guarantee that the evolution of the system involved a stepwise accumulation of those parts. Consider the horse and its four legs. A horse with zero, one, or two legs cannot walk or run; suppose the same is true for a horse with three. In contrast, a horse with four legs can walk and run, and it thereby gains a fitness advantage. So far so good: the tetrapod arrangement satisfies the definition of irreducible complexity. The mistake comes from thinking that horses (or their ancestors) had to evolve their tetrapod morphology one leg at a time. In fact, the development of legs isn't controlled by four sets of genes, one for each leg; rather, there is a single set that controls the development of appendages (Griffiths et al. 2005: Chapter 18). A division of a system into parts that entails that the system is irreducibly complex may or may not correspond to the historical sequence of trait configurations through which the lineage passed. This point is obvious with respect to the horse's four legs, but needs to be borne in mind when other, less familiar, organic features are considered.

A third problem for Behe's framework concerns the fact that Figures 2.5 and 2.6 have "fitness" on the y-axis and say nothing about the *function* of the system in question. What matters to the process of evolution by natural selection is how a trait affects an organism's probability of surviving and reproducing as compared to the effects that alternative traits have that are also present in the population. It doesn't matter whether the trait always performs the same function throughout its evolution or switches function. This point is not so obvious when we think about the evolution of the eye, since both Darwinians and creationists tend to think about the evolution of the eye by asking how gradual modifications of the eye affect an organism's ability to extract information about the environment that is present in light. We therefore need to examine a new example to see the relevance of the idea of function switching to Behe's views on irreducible complexity.

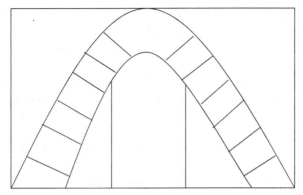

Figure 2.6 An arch surmounted by a keystone satisfies the definition of irreducible complexity. Yet, it can be built by a process of addition and subtraction, one stone at a time (Cairn-Smith 1982).

Consider the wing. A full wing allows the organism to fly. But 1 percent of a wing doesn't permit 1 percent flying. Tiny wings do not provide enough lift for flying or even for gliding. The wing seems to be a more alarming problem for Darwinian gradualism than the eye is. At first glance, it appears that the fitness function for the wing will look like *C*, not *M*, in Figure 2.6. Kingsolver and Koehl (1985) provide an interesting approach to this problem in connection with insect wings. There are insects around now that have tiny wing buds; these buds aren't big enough to permit the insect to fly or glide. Rather, they are useful as thermoregulators; insects turn these buds towards or away from the sun and thereby raise or lower their body temperature. This and other details suggest the following scenario. Insect wings began evolving because wings were useful as thermoregulators. Once these "wings" had evolved to a sufficient size, a new selection process began in which larger wings were selected because they promoted flight. Wings started evolving for one reason but then continued to evolve for another. Function switching is a pervasive theme in evolution.[33] Notice that Behe's concept of irreducible complexity ignores it. He defines irreducible complexity in terms of the function a structure has *now* and whether the structure would be able to perform *that same function* if one of its parts were excised. Behe thinks it is

---

[33] Darwin ([1859] 1964: 190–2) discusses the importance of function-switching in evolution, citing the example of the swimbladder in fish – "an organ originally constructed for one purpose, namely flotation, may be converted into one for a wholly different purpose, namely respiration."

a problem for evolutionary biology if wings now have the function of promoting flight but would not be able to perform that function if even one of their parts were removed. This is a problem only if flying was the name of the game all along.

Cairn-Smith (1982: 93–9) describes a nonbiological example that brings out another important feature of the concept of irreducible complexity. Suppose you come upon a stone arch that frames an open doorway and is surmounted by a keystone. The arch has been constructed without mortar. You notice that if any of the stones in the arch were removed, it would collapse. You naturally assume that the arch has the function of providing a doorway through the surrounding wall and conclude that the arch satisfies the definition of irreducible complexity. Of course, you have no doubt that the arch is the product of human intelligent design. But could the arch have been built one stone at a time? If you consider just the stones in the arch you see before you, the answer seems to be *no*. But consider Figure 2.6, which depicts a wall that can be built by adding stones one by one, after which three stones can be removed one at a time, with the result that only the arch set in the wall remains. Behe may reply that irreducibly complex structures like the arch can, of course, be built by intelligent designers; his point is that evolution can do no such thing. But Cairn-Smith's arch, built by stepwise addition and subtraction with irreducible complexity resulting at the end, provides a nice model for a mindless evolutionary process that does precisely the same thing.

The example may be developed further by bringing it into contact with the idea of function-switching. Suppose the builders add stones one at a time to make a solid wall, but that once the wall is built, a new purpose comes to mind – they want to create a passageway through the wall, so they remove the three stones. It may seem far-fetched that builders would revise their plan like this, but that does not matter to the point at hand. When the wall was on the way up, each rock was *useful*, but once the wall was completed and the three stones were removed, the remaining rocks in the arch became *indispensable*. This is the pattern that Orr (1996, p. 7) sees in the evolution of lungs:

> The transformation of air bladders into lungs that allowed animals to breathe atmospheric oxygen was initially just advantageous: such beasts could explore open niches – like dry land – that were unavailable to their lung-less peers. But as evolution built on this adaptation (modifying limbs for walking, for instance), we grew thoroughly terrestrial and lungs, consequently, are no longer luxuries – they are essential. The punch-line, is, I think obvious:

although this process is thoroughly Darwinian, we are often left with a system that is irreducibly complex.

Behe (2001, pp. 692–693) replies to Orr by denying that lungs and swim bladders are single systems, which means that the definition of "irreducible complexity" does not apply. The reason Behe offers for this verdict is that organs "contain so many active, unknown components . . . that one is dealing with a 'black box' whose capacities are substantially obscure." Notice that this is a point about our *knowledge;* it hardly follows that lungs aren't irreducibly complex. However, if Behe's point is that we are in no position to say whether lungs are irreducibly complex because there are open questions about one or more of their components, the same will be true of the examples that Behe thinks are clear cases of irreducible complexity – the bacterial flagellum and the biochemical processes of blood coagulation. In any event, whether lungs are, in fact, irreducibly complex isn't the main point that Orr is making. His point about lungs is the same one that Cairn-Smith made about the arch – *if* a system is irreducibly complex, this doesn't rule out its evolving by an incremental Darwinian process.

These points can be tied together by considering the concept of *epistasis* used in population genetics. Suppose an organism's fitness is influenced by its genotype at two loci, as shown in Figure 2.7. The numbers represent an individual's probability of surviving from egg to adult. This set of fitness values is epistatic because the fitness ordering of the three genotypes at one locus depends on what genotype is present at the other. Notice that in an infinite population in which the *a* gene remains at fixation it will be impossible for selection to transform the

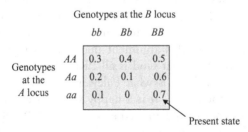

Figure 2.7 Hypothetical example of epistatic fitness relationships. Cell entries are fitness values (probabilities of surviving to reproductive age) that depend on the organism's genotype at two loci. The population can evolve from *aabb* to *aaBB* without crossing a fitness valley.

population from 100 percent *b* (i.e., everyone is a *bb* homozygote) to 100 percent *B* (i.e., everyone is *BB*). To do so would involve traversing a fitness valley; if a single mutation from *b* to *B* occurs in a population of *aabb* individuals, the individual affected will be *aaBb* and selection will eliminate the novel allele. In a population in which everyone is *aa*, the *BB* genotype satisfies the definition of irreducible complexity: Remove a copy of *B* (or replace it with the *b* allele) and the organism is dead. If we observe an *aaBB* population and discover that *aaBb* is lethal, we might be inclined to think that some process other than selection was responsible for *B*'s replacing *b*. But considering both loci shows that there is a path of increasing fitness that takes the population from 100 percent *b* to 100 percent *B* without crossing a valley. If the population starts at *aabb*, it will evolve to *AAbb*, and then to *AABB*, and then to *aaBB*, with fitness increasing every step of the way. This analysis of how genes at the two loci evolve does not depend on what phenotypes are produced by what gene combinations. Nor does the word "function" play any role. Maybe there is function switching as the population evolves. It is curious that the trajectory of gene frequencies in Figure 2.7 resembles the curve of the arch in Figure 2.6. Just as the arch can be built by adding and subtracting stones, so the *aaBB* phenotype can be built from *aabb*, first by adding *AA* and later by taking it away. In both cases, the initial impression that "you can't get there from here" by a series of small steps is an illusion.

2.17   THE RELATIONSHIP OF THE DESIGN ARGUMENT TO THE ARGUMENT FROM EVIL

The *argument from evil* is an argument for atheism; it asserts that the kinds and quantities of evil that exist in our world show that there is no God, where God is assumed to be omnipotent (all-powerful [P]), omniscient (all-knowing [K]), and omni-benevolent (entirely good [G]). One easy reply to this argument that is open to the theist is to abandon the assumption that God is all-PKG. Another is to deny that there is so much evil. This is what Paley maintains in Chapter 26 of *Natural Theology*; for Paley, "it is a happy world after all. The air, the earth, the water, teem with delighted existence" (1809: 498). Paley also records that "few diseases are fatal" and offers as evidence a hospital record that summarizes the outcomes for patients over a six-year period (p. 458):

Admitted 6,420
Cured 5,476
Dead 234

He argues that the preponderance of pleasure over pain shows that God is benevolent.[34] The problem of evil is more difficult for theists who decline to see the world through rose-colored glasses and who wish to retain their belief that God exists and is all-PKG.

Atheists who agree with my criticism of the design argument should reconsider whether they think the argument from evil is convincing. If complex adaptations cannot be said to favor intelligent design over chance, since we have no independent knowledge of what the putative designer's goals and abilities would be, how can we be so sure that an all-PKG God would not have had his reasons for allowing evils to exist in the quantity that we observe? Perhaps these evils are necessary correlates of greater goods, where some of these correlations are completely beyond our ken. If we don't know what God would have wanted to achieve if he built a panda, perhaps we also don't know what amount of evil there would be if God created the entire world. The design argument and the argument from evil reach opposite conclusions: The former is a proof of the existence of God whereas the latter is a proof that God does not exist – but maybe they have a common flaw.

The problem of evil is not that evil exists, but that there is so much of it. *Apparently*, there is more evil in the world than is consistent with the existence of an all-PKG being. But perhaps the appearance can be explained away. The project of attempting to explain how an all-PKG God could permit so much evil to exist is called *theodicy*. For example, it is sometimes pointed out that some of the evils that exist are soul-building (Hick 1978); sometimes, when we experience adversity, the experience makes us stronger and also helps others to become better people by witnessing our fortitude. If the benefit of having moral fortitude exceeds the cost we pay when we gain our fortitude by suffering through adversity, then soul-building may explain why an

---

[34] Was Darwin ([1859] 1964: 62) replying to Paley when he wrote that "we behold the face of nature bright with gladness, we often see superabundance of food; we do not see, or we forget, that the birds which idly singing round us mostly live on insects or seeds, and are thus constantly destroying life; or we forget how largely these songsters, or their eggs, or their nestlings, are destroyed by birds and beasts of prey; we do not always bear in mind, that though food may be now superabundant, it is not so at all seasons of each recurring year"?

all-PKG being would permit *some* evil to exist. But there is so much more. Another familiar argument in the project of theodicy concerns free will. It is often claimed that some evils exist because human beings have free will and sometimes freely choose actions that are wrong. Free will is supposed to be such a wonderful thing that a benevolent God would have given us this great benefit even though it brought with it a considerable cost. Like a number of other philosophers, I don't see why having free will rules out always freely choosing to do the right thing.[35] If a sinner can have free will, why can't a saint? But even if we grant for the sake of argument that free will and soul-building account for *some* of the evils that there are, there still appears to be vastly more evil than we have so far explained. There are evils that don't build souls and that aren't due to anyone's exercising their free will. How can this excess be reconciled with the existence of an all-PKG God?

I think it fair to say that no one so far has explained this. But that does not *prove* that atheism is true (even granting the assumption that God, if he exists, is all-PKG). You can't *deduce* the nonexistence of an all-PKG God from the fact that there is more evil than human beings have been able to explain. But perhaps something more modest can be claimed. Maybe the existence of so much evil is *evidence* against the existence of such a being. When a coin is tossed 1,000 times and lands heads every time, that does not allow one to *deduce* that the coin isn't fair, but surely these observations *favor* the hypothesis that the coin is biased over the hypothesis that it is fair. For the evil we observe to favor atheism over the hypothesis that an all-PKG God exists, the following likelihood inequality would have to be true:

$$Pr(E \mid \text{there is no God}) > Pr(E \mid \text{an all-PKG God exists}).$$

In this inequality, E is a detailed description of the evils that exist. This is a likelihood representation of the thought behind the *evidential* argument from evil (Rowe 1979).[36] By using likelihoods, it is possible to consider competing hypotheses that are exclusive but not exhaustive;

---

[35] Compatibilism is the philosophical position that freedom and causal determinism are not in conflict. On this view, freedom does not require the absence of determinism but only that our actions be caused in a certain way; the compatibilist philosopher then has the job of describing what that way is. Hume was a compatibilist and so are a number of contemporary philosophers.

[36] Howard-Snyder (1996) is a useful anthology of recent work on the evidential problem of evil. Draper (1989, 1996) represents the argument in terms of likelihoods.

for example, I have left aside the possibility that there is a God who is not all-PKG. And there is no need to consider the prior probabilities of the two hypotheses, either.

Wykstra (1984: 87–9) argues against the evidential argument by saying that it is committed to the following assumption:

If an all-PKG God had a reason for permitting horrendous evils to exist, human beings would know what those reasons are.

Wykstra (like Plantinga 1974) denies that this proposition is true. The likelihood formulation of the evidential argument from evil shows why the argument does not require so strong a premise. The evidential argument is consistent with our having considerable uncertainty about what God's motives would be in allowing some horrendous evils to exist. The question is whether E's probability *increases* if there is no God.

In an earlier publication (Sober 2004b), I took the view that the organismic design argument and the argument from evil are precisely on a par in that both require assumptions that we are not entitled to make. Now I am not so sure. Perhaps the design argument requires more knowledge of the designer's goals than the argument from evil does. Maybe the hypothesis that an all-PKG God exists makes predictions concerning how much evil there should be even though it does not predict the architecture of the vertebrate eye or the panda's paw. If so, atheists can consistently accept my criticism of the design argument and still think there is something to the evidential argument from evil. What seems less plausible is the thought that the organismic design argument is successful, while the evidential argument from evil requires assumptions that we have no reason to believe. This is not the place to pursue these issues further. Suffice it to say that theists who reply to the argument from evil by saying that God's goals are inscrutable should have little sympathy with the organismic argument from design.

2.18 THE DESIGN ARGUMENT AS AN INDUCTIVE SAMPLING ARGUMENT

I have argued that the design argument is unsuccessful because we have no way to evaluate

Pr(the eye has features $F_1 \ldots F_n$ | the eye was made by an

intelligent designer).

My point is not that we don't know what the point value is of this probability but that we can't even judge whether it is greater or less than

Pr(the eye has features $F_1 \ldots F_n$ | the eye was the result of a mindless random process).

The value of this second probability is very low, but it is not zero. As we have seen, auxiliary propositions can be invented about the putative designer's goals and abilities that insure that the likelihood of the intelligent-design hypothesis is very high, but it is equally true that auxiliary propositions can be invented that insure that the likelihood of the intelligent-design hypothesis is zero. What is needed is not the invention of auxiliary propositions (whether they help or hurt the design hypothesis) but the identification of auxiliary information that is independently supported. Paley did not provide this information, and the same is true of modern defenders of the design argument.

I now want to consider an objection to my criticism of the argument from design. It contends that there are clear cases in which we have ample reason to infer that an intelligent designer is responsible for what we observe even though we have no independent information as to what goals and abilities this putative designer would have if he existed. The thought is that there are observations O that give $Pr(ID\,|\,O)$ a high value, even though we don't know whether $Pr(O\,|\,ID)$ is high or low. True, when the design argument is given a *likelihood* formulation, information about goals and abilities is needed. But the suggestion is that this requirement does not apply if we formulate the design argument as a *probability* argument based on inductive sampling.[37] After all, there are

---

[37] Behe (1996: 197) says that "the conclusion that something was designed can be made quite independently of knowledge of the designer." It isn't clear to me what kind of argument Behe (1996) thinks his version of the design argument is, though he (1996: 285–286) praises Dembski's (1998) framework for detecting design; see Fitelson et al. (1999) for criticisms of Dembski's approach. More recently, Behe (2006) has suggested that the design argument is inductive. Commenting on his testimony at the *Kitzmiller* v. *Dover Area School District* trial, he says:

as I testified, the intelligent design argument is an induction, not an analogy. Inductions do not depend on the degree of similarity of examples within the induction. Examples only have to share one or a subset of relevant properties. For example, the induction that, *ceteris paribus*, black objects become warm in the sunlight holds for a wide range of dissimilar objects. A black automobile and a black rock become warm in the sunlight, even though they have many dissimilarities. The induction holds because they share a similar relevant property, their blackness. The induction that many fragments rushing away from each other indicates a past explosion holds for both fire-crackers and the universe (in the Big Bang theory), even though firecrackers and the universe have many, many dissimilarities. Cellular machines and machines in our everyday world share a plenty of cases in which we can estimate the value of $Pr(H \mid O)$ but have no clue as to the value of $Pr(O \mid H)$. Consider, for example, the recent discovery of two genes (*BRCA1* and *BRCA2*); most women who have one or the other of these genes develop breast cancer; from this frequency data, we can infer that $Pr(S$ will develop breast cancer $\mid S$ has *BRCA1* or *BRCA2*) is high. But knowing the frequency with which women who have these genes develop breast cancer leaves completely unspecified how often women with breast cancer have these genes and, thus, no basis for estimating $Pr(S$ has *BCA1* or *BCA2* $\mid S$ has breast cancer). As it happens, among women who have breast cancer, fewer than 10 percent have these genes, but that is a separate fact.

The thought that the design argument does not require information about the goals and abilities of the putative designer can be developed in connection with some examples that Hume discusses in his *Dialogues Concerning Natural Religion*. In Book III, Cleanthes replies to Philo, who criticized the design argument for being a weak inductive or analogical inference, a criticism we considered in §2.11. "Suppose," says Cleanthes,

that an articulate voice were heard in the clouds, much louder and more melodious than any which human art could ever reach: suppose, that this voice were extended in the same instant over all nations, and spoke to each nation in its own language and dialect: suppose, that the words delivered not only contain a just sense and meaning, but convey some instruction altogether worthy of a benevolent Being, superior to mankind: could you possibly hesitate a moment concerning the cause of this voice? and must you not instantly ascribe it to some design or purpose? Yet I cannot see but all the same objections (if they merit that appellation) which lie against the system of Theism, may also be produced against this inference.

Cleanthes' point, applied to the criticism of the likelihood formulation of the design argument that we are considering, is this: If we have no independent information about the goals and abilities of the putative designer in the case of the vertebrate eye, then we also have none in the case of the voice from the clouds. But surely that should not stop us from inferring an intelligent designer in the latter case. And so it should not do

relevant property – their functional complexity, born of a purposeful arrangement of parts – and so inductive conclusions to design can be drawn on the basis of that shared property. To call an induction into doubt one has to show that dissimilarities make a relevant difference to the property one wishes to explain.

so in the former. After introducing the example of the voice from the clouds, Cleanthes adds the following:

Might you not say, that all conclusions concerning fact were founded on experience: that when we hear an articulate voice in the dark, and thence infer a man, it is only the resemblance of the effects which leads us to conclude that there is a like resemblance in the cause: but that this extraordinary voice, by its loudness, extent, and flexibility to all languages, bears so little analogy to any human voice, that we have no reason to purpose any analogy in their causes: and consequently, that a rational, wise, coherent speech proceeded, you know not whence, from some accidental whistling of the winds, not from any divine reason or intelligence? You see clearly your own objections in these cavils, and I hope too you see clearly, that they cannot possibly have more force in the one case than in the other.

Cleanthes is here seeking to undermine Philo's earlier suggestion, discussed in §2.11, that the design argument requires there to be a strong analogy. Cleanthes says that a strong analogy is present when we hear a voice in the dark and absent when we hear the voice in the clouds, but the two arguments are equally compelling. He then attempts to bolster his position by introducing another example. Suppose we discovered books that reproduce themselves. These books, like the voice from the clouds, would be entirely *unprecedented*; Cleanthes' point is that, even so, we are right, in both cases, to infer the existence of an intelligent designer. I don't see much hope for analyzing these arguments in terms of whether the analogies they use are strong or weak (§2.11). The voice from the clouds is similar to the terrestrial voices we routinely hear in some ways, and, of course, it differs from them in others. But if the argument is not an argument from analogy, what kind of argument is it? The suggestion I wish to explore is that the design argument aims to show that $Pr(ID \mid O)$ is high without addressing whether $Pr(O \mid ID)$ is high or low. Maybe this *probability* argument can succeed even though the *likelihood* argument fails. Let us see.

Cleanthes' two examples about the voices have a common structure. Each begins with the same set of mundane observations:

$f_1 = freq$(an intelligent designer produced noise $X \mid$ noise X is an

English sentence & we hear noise X & we see what

produced X) is high.

When we hear an English sentence (I use this language as a convenient example) and we see what caused it, we almost always see that an intelligent designer (namely, a fellow human being) produced the sentence.

The argument about the voice in the dark takes this frequency data to show what is true when the lights are off:

$p_{1a} = Pr($an intelligent designer produced noise X | noise X is an English sentence & we hear noise X & we do not see what produced X because the lights are off$)$ is high.

The argument about the voice in the clouds takes the same frequency data to show what is probably true when the voice comes from the clouds:

$p_{1b} = Pr($an intelligent designer produced noise X | noise X is an English sentence & we hear noise X & we do not see what produced X because the voice came from the clouds$)$ is high.[38]

Both arguments extrapolate from the mundane cases described in f_1 though they do so to different degrees; the voice in the dark is a modest extrapolation, while the voice in the clouds is more dramatic. These arguments do not require us to say anything concerning how probable it is that an intelligent designer who lives in the dark or in the clouds will produce an English sentence. With respect to the former, the probability is apt to be low unless we're talking about a monolingual speaker of English. And who knows how inclined a celestial intelligence would be to boom English sentences down to Earth?

The organismic design argument can be reformulated on this model. We begin with frequency data:

$f_2 = $freq(an intelligent designer produced X | X is complex and useful & we observe what produced X) is high.

Here I'm using the term "complex and useful" to refer both to the complex features of organisms that help them survive and reproduce and to the complex features of humanly created artifacts (like watches) that make them

---

[38] Cleanthes adds a surprising detail to this example. It isn't just that the voice from the clouds produces an English sentence; rather, the single sequence of sounds that the voice produces is simultaneously heard by each speaker around the world in his or her native language. Cleanthes does not explain how a single sequence of sounds can simultaneously be a sentence in all these many languages (Dye 1988), but that is a detail that does not matter to the epistemology. What we know by observation is that each speaker hears a voice from the clouds speaking in his or her own language, and each draws the conclusion that the sound was produced by an intelligence. We do not observe whether there is one intelligence or many, and whether there is one sequence of sounds or many occurring simultaneously (with each listener hearing only one of them) does not matter.

useful to human beings. From this fact about what we observe, the design argument draws an inference concerning a somewhat different type of case:

$$p_2 = Pr(\text{an intelligent designer produced } X \mid X \text{ is complex}$$

and useful & we do not observe what produced X) is high.

If Cleanthes is right about the voice in the dark and the voice from the clouds, it seems that the design argument is on firm ground as well when it is formulated as an inductive sampling argument. Here again, the argument does not require that we know how inclined or how able an intelligent designer would be to produce the complex and useful devices we see around us.

There are two differences between the *f*s and the *p*s in these arguments. First, *f* is a sample frequency while *p* is a probability. Second, the *f*s and the *p*s contain different conditioning propositions. To knit the premise that *f* is high to the conclusion that *p* is high, we need to think of these inferences as each having two steps. To this end, we need to introduce a new probability, *q*, that "goes between" *f* and *p* in each case. Combining Cleanthes' two arguments about voices into one, we obtain:

(VOICES) $f_1 = freq$(an intelligent designer produced noise X| noise X is an English sentence & we hear noise X & we see what produced X) is high.

Therefore, $q_1 = Pr$(an intelligent designer produced noise X| noise X is an English sentence & we hear noise X & we see what produced X) is high.

Let $p_1 = Pr$(an intelligent designer produced noise X| noise X is an English sentence & we hear noise X & we do not see what produced X because the lights are off or because the voice comes from the clouds).

$q_1 \approx p_1.$

Therefore, p_1 is high.

The inductive sampling version of the organismic design argument also has two steps:

(IND-ID) $f_2 = freq$(an intelligent designer produced X| X is complex and useful & we see X & we see what produced X) is high.

Therefore, $q_2 = Pr$(an intelligent designer produced X| X is complex and useful & we see X & we see what produced X) is high.

Let $p_2 = Pr$(an intelligent designer produced X| X is complex and useful & we see X & we do not see what produced X).

$q_2 \approx p_2.$

Therefore, p_2 is high.

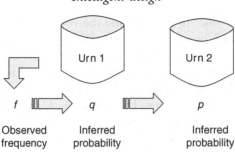

Figure 2.8   If we accept the bridge principle $q \approx p$, we can estimate the value of p by observing the frequency f.

First we infer that q is high from the fact that f is high. This permits us to conclude that p is high as well, provided that we add a *bridge principle* that links the values of the two probabilities. The bridge principle says that $q \approx p$. When we hear the voice in the dark or the voice in the clouds, we can conclude that this was probably due to intelligent design even though we do not see what produced those voices. This is justified because, in other cases in which we *do* see where voices come from, we see that they almost always come from intelligent beings. By parity of reasoning, when we see an object that is complex and useful and do not observe what produced the object, we can conclude that it was probably produced by an intelligent designer. This is justified because, in other cases in which we *do* see where complex and useful objects come from, we see that they almost always come from intelligent designers. Cleanthes is proposing a persuasive analogy, as persuasive in its way as Paley's analogy between the watch and the eye. But does it work?

The structure of these arguments – (VOICES) and (IND-ID) – is captured by the two-urn problem depicted in Figure 2.8. We sample balls from the first urn and observe what the frequency f of green balls is in the sample. From this we infer what value we should assign to q, the probability of a ball's being green if it is drawn from the first urn. Given this estimate of the value of q, we need to say what the probability p is of drawing a green ball if we sample from the second urn. It is obvious that the first urn provides no guidance as to the second unless a bridge principle linking the two probabilities is correct. If we accept the bridge principle that $q \approx p$, we can infer what the probability is of drawing a green ball from the second urn, based on data concerning what we observed when we drew from the first.

I have emphasized the isomorphism of these examples, just as Cleanthes would wish. And I grant that there is nothing wrong with inferring that the voice in the dark and the voice in the clouds were both probably produced by an intelligent being. Yet, I churlishly persist in thinking that the inductive sampling version of the organismic design argument is unsuccessful. Before I explain why, I need to tinker with the first premise in IND-ID, since I think this first premise was known to be false in the past and is known to be false today. Many of us have had the experience of seeing organisms brought into being by the process of biological reproduction. We see what causes a puppy to come into existence when we see its parents copulate. And we see something further – that the puppy has eyes because this feature was transmitted to it by its parents. We do not see an intelligent designer fashioning the puppy and its adaptive features. Of course, defenders of the design argument do not maintain that each organism is the result of a separate act of intelligent planning; their idea is that a designer created one or more organisms initially and subsequent organisms arose because earlier organisms reproduced. This means that the first premise of (IND-ID) needs to be changed to something like the following:

$f_2^* = freq$(an intelligent designer designed and produced $X \mid X$ is complex and useful & we see X & we see what produced X & if X is a member of a lineage of objects that produce copies of themselves, then X is the first member of that lineage) is high.

I take it that the only objects that satisfy the conditioning proposition in this revised premise are humanly designed artifacts. A somewhat similar modification is needed in the first premise of (VOICES), at least for modern readers familiar with radios, televisions, or CD players. With these modifications in place, the first premise in each argument is true.

Given the value of an observed frequency (f), (VOICES) and (IN-IND) then infer the value of a probability (q). I do not fault this inference, though it is well to recall from Chapter 1 that Bayesians think that this inference requires prior probabilities (§1.2), whereas frequentists understand induction in terms of maximum likelihood estimation or some other inference procedure that is justified because of its long-term operating characteristics (§1.5). Neither school thinks of induction as basic and irreducible. Even so, both statistical philosophies can see their way clear in the two arguments to infer that q is high from the fact that f is high. So far, so good.

The next step in the two arguments is where there is a parting of the ways. The bridge principle in (VOICES) is reasonable, but that in (IND-ID) is not. Even if we have never before heard a voice in the dark, our experience gives us reason to think that the English sentence we now are hearing in the dark was produced by an intelligent being. We have heard English sentences in lighting conditions that range from bright sunlight to the dimmest twilight. Across this range of cases, we see that the sounds always or almost always emanate from an intelligent being, namely a fellow human being. Reducing the amount of illumination apparently does not affect the frequency with which the sound of an English sentence comes from a being who has a mind. This is an important reason for thinking that if the lights were entirely off that the probability of an intelligent designer would be approximately the same. The same point holds for the voice in the clouds. Even if we have never before heard a voice from the clouds, both we and our eighteenth-century predecessors have had ample opportunity to observe that the frequency with which English sentences come from intelligent beings is not affected by how high off the ground the sound is. In the 1780s, the Montgolfier brothers and their imitators transported people from earth to sky in hot-air balloons. As the passengers rose, spectators on the ground could see that the sounds coming from above that took the form of English or French sentences were invariably produced by intelligent human beings. And before hot-air balloons, people were accustomed to hearing speeches delivered and songs sung from balconies, towers, and windows above the ground floor. There was ample evidence that elevation above the surface of the Earth does not matter – regardless of elevation, sounds that constitute sentences always, or almost always, issue from the mouths of human beings. The bridge principle in (VOICES), $p_1 \approx q_1$, is reasonable.

What evidence do we (or our eighteenth-century predecessors) have that the difference between living organisms and nonliving artifacts does not matter in the case of (IND-ID)? We considered a continuum of lighting conditions and elevations in (VOICES), so perhaps we should consider a continuum of some other sort in the case of (IND-ID). But what continuous quantity should we contemplate? If there were a continuum between "not being alive" and "being alive," and we had sampled along this continuum, it would be no great leap to conclude that what we found in our sample also applies to unsampled objects that are a little more down the line. Our sampling would assure us that this matter of degree difference does not matter to the question of whether a complex and useful device is the result of intelligent design. But the vital processes we and our eighteenth-century predecessors see in living things do not seem to be like this. My point here is not that the bridge principle in (IND-ID) (that $p_2 \approx q_2$) is false but that there was no sampling evidence in the eighteenth century, nor is there any now, that it is true.

There is a second and perhaps more glaring flaw in the bridge principle in (IND-ID). This inductive sampling argument simply *ignores* theories that say that the complex and useful traits of organisms arose by mindless processes that occurred long ago or at a rate too slow for us to observe them do their work from start to finish. The inductive sampling version of the design argument rejects Epicureanism and Darwinism without even considering them; its first premise focuses exclusively on cases in which we *observe* a process producing a complex and useful trait. You do this when you watch carpenters build a barn in the space of an afternoon, but none of us lived when Epicurean whirling in the void was supposed to have occurred, nor do any of us live long enough to witness a Darwinian process that takes many human lifetimes to produce a result. The way to consider hypotheses of this second kind is to think about what they *predict*, but this leads us back to the likelihood formulation of the design argument discussed earlier. The inductive sampling version of the argument introduces a massive sampling bias into the inference.

My point about the observational evidence for the bridge principle in (VOICES) uses a strategy of inductive reasoning that Galileo made famous. When Galileo pointed his telescope at Jupiter and claimed to see that it has four moons, what basis did he have for thinking that his telescope provided reliable evidence about that distant state of affairs? On the face of it, he had no inductive justification for this confidence, since no one had checked claims made about celestial objects by traveling to the heavens and directly observing whether those claims were in fact correct. This skeptical remark resembles Philo's comment that we have never journeyed to distant worlds and seen intelligent designers making them. The proper reply is that even though Galileo could not cite the testimony of space travelers, he did have sampling evidence of another kind. Galileo trained his telescope on distant buildings and on ships coming over the horizon that appeared to be heading for land; he then sent his assistants to check whether the buildings and the ships entering port really had the features he thought he saw. In this way, Galileo accumulated evidence that distance does not matter. Of course, objects that are very far away may fail to produce any telescopic image at all; but when they do so, the probability that the image is veridical does not depend on how far away the seen object is. This was the basis for Galileo's claim that a telescope trained on Jupiter provides reliable information that the planet has moons (Kitcher 1993, 2001).

It is interesting that the two urns depicted in Figure 2.8 provide a useful model for another of Hume's arguments, his famous argument for inductive skepticism. Let the first urn represent the observations you might make of the present and past. No matter how many times you draw from that first urn, it is impossible to use the sampling data you obtain to say anything at all about the composition of the second urn (which represents the observations you might make of the future) unless a bridge principle is added. This bridge principle is not *a priori* true, nor is it justified solely by the data you drew from the first urn. This was Hume's insight. Hume went further and claimed that all inductive inferences rest on the same bridge principle, the *principle of the uniformity of nature*. This was a mistake. The fact that each induction from past and present to future requires a bridge principle does not entail that there is single bridge principle on which all inductive inferences rely (Sober 1988).[39]

2.19 MODEL SELECTION AND INTELLIGENT DESIGN

The likelihood framework provides one good way to understand the design argument. A second option is the one explored in the previous section: that the design argument is an argument based on inductive sampling. In this section, I want to consider a third approach, one that draws on ideas concerning model selection (§1.7). To begin, let's consider a variant of the example about coin tossing discussed in §2.10. There we examined the principle of total evidence in connection with the observation that a coin landed heads each of the twenty times it was tossed. At first glance, this outcome seems to strongly favor the hypothesis that the coin is biased towards heads over the hypothesis that the coin is fair. But then we are invited to consider the fact that the coin is one of a million coins that each was tossed twenty times. The point is then made that if all these coins were fair, it would be very probable that at least one of them would land heads on all twenty tosses. The likelihood framework, coupled with the principle of total evidence, was then offered as a reason for not allowing this point about the million coins to weaken our conviction that the observation of the first coin's twenty tosses really does favor the hypothesis that the coin is biased. Although I hope my argument was persuasive, I do not dismiss the suggestion that we should think about the one coin by placing it in the wider context of thinking about the 999,999 other coins that were tossed. The discussion in §2.10 did not pursue this idea very far. True, we considered *hypotheses* about other coins, but we did

---

[39] We see here another instance of the birthday fallacy, which I discussed in §2.2 in connection with Aquinas' version of the design argument.

not bring in any *data* about the results of tossing them. How should a wider data set, which describes the result of tossing each of these million coins twenty times, be used to evaluate hypotheses about all these coins?

Let us construct some candidate models. The first of the following two models assumes that each toss of each coin has the same probability of landing heads, but the second is more complicated:

(One)   For each toss of each coin, $Pr(\text{heads} \mid \text{the coin is tossed}) = p$.
(Million)   For each coin i, $Pr(\text{heads} \mid \text{coin } i \text{ is tossed}) = q_i$
            $(1 \leq i \leq 1{,}000{,}000)$.

The (Million) model views each coin as potentially different from all the others; it says that every toss of the first coin has a probability q_1 of landing heads, while every toss of the second coin has a possibly different probability q_2 of landing heads, and so on, for each of the million coins. The (One) model rules out this heterogeneity; it contains just one adjustable parameter. Other models might be considered as well. For example, there is a model that organizes the data by tosses, not by coins:

(Twenty)   For the jth toss of any coin, $Pr(\text{heads} \mid \text{the } j\text{th toss of the}$
            $\text{coin}) = r_j$ $(1 \leq j \leq 20)$.

The (Twenty) model says that all first tosses have a probability r_1 of landing heads, while all second tosses have a possibly different probability r_2 of landing heads, and so on, for the twenty tosses. We also could consider models that say that the outcome of earlier tosses of a coin alters the probabilities that attach to subsequent tosses of the same coin. (One), (Million), and (Twenty) are just the tip of the iceberg.

Bayesianism provides a strategy for assessing these models: We assign them prior probabilities, which we then update by taking the observations into account. The Bayesian goal is to discover which model has the highest probability of being true. As explained in §1.7, model selection criteria like AIC have a different goal: predictive accuracy, not truth or probable truth. We fit models to data by finding maximum likelihood estimates of their adjustable parameters and then ask how accurately those fitted models can be expected to predict new data drawn from the same underlying reality. With respect to our coin-tossing problem, we need to estimate the values of p, $q_1 \ldots q_{1,000,000}$, and $r_1 \ldots r_{20}$, from the data at hand and then ask how well these different fitted models would do in predicting a new experiment in which the million coins are once again each tossed twenty times.

With these different models in mind, let's go back to the single coin that lands heads twenty times. This is the first of the million coins in our data set. Suppose that half the 20 million tosses were heads. This means that the maximum likelihood estimate of the single parameter p in the (One) model will be $\frac{1}{2}$. The (Million) model, on the other hand, treats each coin as a separate inference problem; since the first coin landed heads all twenty times, the maximum likelihood estimate of the parameter q_1 is that it has a value of unity. The probability of the data from the first coin, relative to the two fitted models, is, therefore:

$$Pr[\text{the first coin lands heads on all 20 tosses} \,|\, L(\text{One})] = \left(\frac{1}{2}\right)^{20}.$$

$$Pr[\text{the first coin lands heads on all 20 tosses} \,|\, L(\text{Million})] = (1)^{20}.$$

With respect to the first coin, (Million) beats the pants off (One) as far as the likelihoods of the two fitted models are concerned. What happens when these models address data from the other coins? For each of these other coins, L(Million) will make the observations more probable than L(One) will, except when the coin in question happens to produce exactly ten heads and ten tails; in this circumstance, the two fitted models will tie. The (Million) model has far more adjustable parameters than the (One) model, so it is no surprise that (Million) fits the data far better than (One) does. Relative to all the data, L(One) will therefore have a much lower likelihood than L(Million).

What, then, is wrong with the (Million) model? When I think about these two models with my Bayesian hat on, I am at a loss to give an answer. Since (Million) *allows* the coins to have different probabilities of landing heads without *requiring* that they do so, it's clear that (One) entails (Million). As noted in Chapter 1, this means that $Pr(\text{One}) \leq Pr(\text{Million})$ and that $Pr(\text{One} \,|\, E) \leq Pr(\text{Million} \,|\, E)$, no matter what proposition E happens to be. The defect in the (Million) model becomes visible when we think about making accurate predictions, not when we think about finding models that are probably true. As noted in §1.7, there are a number of model selection criteria now on the market; AIC is just one of them. These different criteria view fit-to-data as one virtue of models, but not the only one. The other virtue is simplicity. The (Million) model gets high marks for fitting the data but low marks for simplicity; the (One) model has the opposite mix of strengths and weaknesses. Which model has the overall better score depends on the data; this can't be determined *a priori*. Indeed, there is another model we might consider in this problem, one that is even more complex than (Million). It assigns a separate parameter to each toss of each coin:

(20 Million)   For each of the k tosses, $Pr(\text{heads} \mid \text{the } k\text{th toss}) = s_k$
                   $(1 \leq k \leq 20{,}000{,}000)$.

This model achieves *perfect* fit; if a toss lands heads, the maximum likelihood estimate of the probabilistic parameter that pertains just to that toss's coming up heads will be assigned a value of unity, and if a toss lands tails, the estimate of that toss's probability of landing heads will be zero. When fitted to the data, (20 Million) has a likelihood than which none greater can be conceived. But that does not mean that it receives high marks when a model selection criterion is applied. Its likelihood score is high, but the penalty it pays for complexity is huge (Hitchcock and Sober 2004).

How does all this help us understand the organismic design argument within the framework of model-selection theory? To begin, we need to think about *models* of intelligent design, where a model is understood as a proposition that contains at least one adjustable parameter. We want the data we have to allow us to obtain maximum likelihood estimates of the parameters in those models. It also must be true that the models apply not just to the data we happen to have, but to new data sets that might be drawn from the same underlying reality. It is easy enough to describe intelligent design models that can be fitted to data. Suppose we have 20 million observations concerning the different traits that different species possess. Here is a model that can be made to fit those observations perfectly:

(ID-40 Million) $Pr(\text{an intelligent designer wanted species } s_i \text{ to have}$
                        $\text{trait } T_j) = g_{ij}.$
                        $Pr(\text{species } s_i \text{ has trait } T_j \mid \text{an intelligent designer}$
                        $\text{wanted species } s_i \text{ to have trait } T_j) = a_{ij}.$

As noted earlier, for an intelligent-design hypothesis to have implications about the traits that different species will probably have, the hypothesis needs to exploit auxiliary propositions about the putative designer's goals and abilities. When the intelligent-design hypothesis is formulated as a *model*, those goals and abilities are represented by adjustable parameters (g_{ij} and a_{ij}, respectively) whose values need to be inferred from the observations. According to (ID-40 Million), each trait of each species is the result of a separate intelligent decision. This model has twice as many parameters as there are observations. For this reason, the model is not identifiable; since there is no unique maximum likelihood estimate of values for its parameters, it makes no sense to talk about its predictive accuracy, which means that AIC does not apply (§1.7). However, the number of parameters in (ID-40 Million) can be cut in half by building into the model the assumption that the hypothesized designer is *omnipotent*:

(ID-20 Million) Pr(an omnipotent intelligent designer wanted species s_i to have trait T_j) = g_{ij}
Pr(species s_i has trait T_j | an omnipotent intelligent designer wanted species s_i to have trait T_j) = 1.0.

This model also fits the data perfectly and it is identifiable. There is a unique likeliest member of this model, and it has a likelihood of unity.

Moving from the likelihood framework to the framework of model selection changes the kind of problem that the design hypothesis encounters. From a likelihood point of view, the problem with the design hypothesis is that we have no independently justified auxiliary propositions that permit the hypothesis to make predictions about the observations (§2.12). This objection disappears when we use a model selection framework. It is easy to estimate the values of the adjustable parameters in (ID-20 Million). However, there are two problems that attach to this ID model. The first is that it is so complex. It may seem that one virtue of the hypothesis of intelligent design is that it is very *simple*. After all, it can postulate a single intelligent designer from whom all features of different species are said to flow. Thinking about intelligent design as a problem in model selection shows that this judgment about its simplicity is misguided. The (ID-20 Million) model views each trait in each species as a separate act of creation; it is exceedingly complex, when complexity is measured by the number of adjustable parameters a model contains. The fact that the model postulates a single *designer* is besides the point.

There is a question and a problem we need to address in connection with this intelligent-design model. The question is: What predictions does this model make about new data when it is fitted to old? Perhaps we should think of ourselves as sampling new organisms from the same species and our goal is to predict the characteristics of those newly sampled organisms based on what we observe in the ones present in the old data set. It is easy to see that (ID-20 Million) may do a poor job in this prediction problem if the organisms described in the old data set exist at one time and one place while those described in the new set exist at another. I don't claim that this flaw must be present in *all* intelligent-design models but merely mention it to highlight the fact that we need to think about the prediction problem we are trying to solve. Another problem is that mention of an intelligent designer in (ID-20 Million) is idle. This model is structurally just like the (20 Million) model about the coins. Again, I do not claim that *all* ID models must be like this. Rather, the point is to realize that there is hard work that needs to be done if we want to have intelligent-design models that are worth considering. Creationists and intelligent-design theorists have not taken even the first step in this direction.

Just as there are multiple possible models for our coin-tossing experiment, there are many models that might be considered as versions of the hypothesis of intelligent design. The ones that are worth thinking about will have fewer parameters than there are observations to explain. In coin tossing, parsimonious models collect different tosses together and view them as reflections of the values of a single set of parameters; similarly, in intelligent-design modeling, we need to collect different observations together and view them as consequences of a single plan that the designer (s) had in mind. How should this be achieved? I don't know; this is a task for intelligent-design theorists to address. But a few *caveats* can be stated nonetheless.

In the coin-tossing example, we began by thinking of a single coin that lands heads on all twenty tosses and then considered that coin as one of a million coins each tossed twenty times. If we think of that single coin in isolation, it may seem very implausible to regard it as fair. But if we think of the entire data set within the framework of model selection, it becomes far more plausible that the (One) model may be the best of the models we're considering. The (One) model *unifies* the 20 million observations whereas the (20 Million) model does not. When fitted to the data, the (One) model says that some coins produced results on their twenty tosses that had very low probabilities; this is something that the (20 Million) model never has to say. But what the (One) Model loses in fit-to-data, it gains because of its simplicity. The two models – one of them unified, the other disunified – are shown in Figure 2.9.

Creationists usually concede that evolutionary theory provides a satisfactory explanation of *micro*-evolutionary processes, but they dig in their heels when it comes to *macro*-evolution. Here, creationists are using the distinction that biologists draw between evolutionary novelties that arise within a species and the appearance of traits that mark the origin of new species. Behe's (1996) version of intelligent-design theory also involves a division of cases; he concedes that evolutionary theory explains the evolution of traits that are not irreducibly complex but holds that intelligent

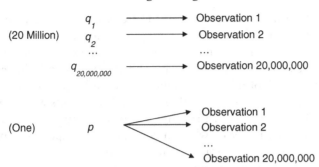

Figure 2.9 The (One) model unifies the 20 million observations; the (20 Million) model treats each toss of each coin as a separate problem and is therefore more disunified.

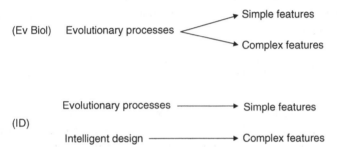

Figure 2.10 Evolutionary biology proposes a unified model of the features that organisms have. Intelligent-design theory proposes a disunified model.

design is needed to explain irreducible complexity (§2.16). Behe's partition may or may not coincide with the older division of micro from macro; they nonetheless have in common the fact that both propose a disunified treatment of the traits that organisms possess. The difference between their approach and that taken by evolutionary biologists is depicted in Figure 2.10. I hope the parallel with the coin-tossing problem is clear. Perhaps evolutionary theory says that some of the features we observe in organisms had very low probabilities of arising; intelligent-design theorists never have to make this concession. The framework of model selection shows why this difference does not mean that the intelligent-design approach is better.

Darwin's most fundamental objection to the doctrine of special creation is that it is empty; it allows each observation to be described *post hoc*, and that is all. What is wrong with such "theories" is that they do not make predictions. The framework of model-selection theory throws light on these methodological ideas. There is no categorical imperative that says that unified theories are always superior to disunified theories (§1.8), although it is true that models that have more parameters than there are items in the data set are beyond the pale. Recall the example, discussed in §1.7, of inferring whether the mean heights in two fields of corn are the same. The model that says the two means are the same is unified, while the model that assigns a different parameter to each field is not. The data get to decide which model should be expected to be more predictively accurate; no *a priori* principle can settle this – not that creationists have developed intelligent-design models that enter into empirical competition with Darwinian theory. The point is that the extreme disunification of (ID-20 Million needs to be seen for the defect that it is. The reason it never has to say that any observation was improbable is that it treats each observation as an isolated phenomenon in its own right.

2.20 THE POLITICS AND LEGAL STATUS OF THE INTELLIGENT-DESIGN HYPOTHESIS

The present legal test in the USA for whether creationist ideas can be taught in public schools asks whether the purpose or effect of doing so would be to promote religion. This means that it is not enough for intelligent-design proponents to drop the G-word from their theories and claim, merely, that an intelligent designer, whose identity remains unspecified, gave organisms the complex adaptations we observe. The word "God" does not appear in this minimalistic assertion, but that does not show that the purpose of including this circumspect proposition in a high-school biology course is not religious.[40]

Given the character of this legal test, it is no wonder that creationism and intelligent-design theory have lost in recent court battles. The reason is that an inquiry into the motives of the people behind these theories, and of the school board members who find these theories attractive, usually leaves no doubt about their ultimate goal. For example, the Discovery Institute in Seattle is the flagship think tank for intelligent-design theorizing, and the "Wedge Strategy" (available at www.antievolution.org/features/wedge. html) is its political manifesto. The document is an internal memo that was

---

[40] For an argument that even a minimalistic version of intelligent-design theory has implications about the existence of supernatural designers, see Sober (2007a).

leaked on the Internet in 2001; the Institute says its goal is to "replace materialistic explanations with the theistic understanding that nature and human beings are created by God." According to the Wedge Strategy, "design theory promises to reverse the stifling dominance of the materialist worldview, and to replace it with a science consonant with Christian and theistic convictions."[41]

The constitutional principle of the separation of church and state raises many important questions about law and public policy, but these have not been the subject of the present chapter. These policy questions have an important *psychological* component: What do policy-makers intend, and how is a policy likely to affect the minds of students and of citizens generally? The subject of the present chapter, in contrast, has been *logical*, not psychological. In examining the hypothesis of intelligent design, my focus has been on the strength of the arguments that have been advanced to support it, not on the motives of those who endorse those arguments. This division of psychological from logical questions reflects a long-standing distinction that philosophers draw; Reichenbach (1938) called it the distinction between the *context of discovery* and the *context of justification*. Reichenbach's point was that scientists get their inspiration from a wide variety of sources. Newton and the other architects of the scientific revolution of the seventeenth century were devout Christians whose science was religiously inspired. The nineteenth-century chemist Kekulé is said to have discovered the structure of benzene by having a hallucination in which he saw snakes whirling around and then grabbing each other's tails. Anything that works works. These different pathways to new ideas – the context of discovery – are matters for psychology and sociology to try to understand. Reichenbach thought that there is a separate question that is nonpsychological in character. Once scientists formulate their theories, are those theories supported or disconfirmed by evidence? It is here that our sources of inspiration do not matter. Formulating a theory is one thing; testing it is another.[42] This is the context in which I have attempted to evaluate the logical strengths and weaknesses of the design argument.

---

[41] Phillip Johnson, one of the principal architects of the intelligent-design movement, discusses the wedge strategy in his 1997 book.

[42] Reichenbach's distinction does not deny that scientists are influenced in what they believe, as well as in what theories they consider, by nonevidential considerations. Furthermore, the distinction, as I understand it, allows for the possibility that facts about how a hypothesis was generated are sometimes relevant evidence as to whether the hypothesis is true; see the discussion of the so-called "genetic fallacy" in Sober (1993b).

2.21 DARWINISM, THEISM, AND RELIGION

"[A]ny confusion between the ideas *suggested* by science and science itself must be carefully avoided." (Monod 1971: xiii)

Was Darwin a theist? In the *Autobiography* (1958: 92–3), written at the end of his life as a personal document for his family, Darwin refers to

the extreme difficulty or rather impossibility of conceiving this immense and wonderful universe, including man [...] as the result of blind chance or necessity [...] I feel compelled to look to a First Cause having an intelligent mind in some degree analogous to that of man; and I deserve to be called a theist.

But on the next page, Darwin refers to himself as an "agnostic," by which he says he means someone "who has no assured and ever present belief in the existence of a personal God or of a future existence with retribution and reward." In his letters, Darwin several times describes himself as "in a muddle"; although he finds the cosmological argument for the existence of God irresistible, he often doubts that he is thinking about the argument in the right way (Brown 1986: 25, 29). This worry found its way into the *Autobiography* as well, where Darwin wonders whether a mind that evolved from the "lowest animal" is equipped to draw "such grand conclusions" (1958: 93). Darwin concludes, "the mystery of the beginning of all things is insoluble by us; and I for one must be content to remain an Agnostic" (1958: 94).

In addition to finding the cosmological argument dizzying, Darwin was deeply troubled by the problem of evil (§2.17). Darwin's response to suffering was visceral. It forced him to stop studying medicine (his father's profession) at the University of Edinburgh. It also led him to refuse to impale live worms on hooks when he went fishing; instead, he euthanized them in brine (Darwin 1958: 27); he also eventually gave up hunting (1958: 55). When Darwin was launched as an evolutionist, examples of horrible suffering in nature were constantly placed before him. One example that particularly revolted him is a parasitic wasp that inserts her eggs in a caterpillar she first paralyzes; the offspring hatch and eat the caterpillar while it is still alive. In a letter to the American biologist Asa Gray, Darwin wrote that he could not persuade himself that "a beneficient and omnipotent God would have designedly created" this arrangement (Burkhardt 1993: 224). In Hume's *Dialogues*, Philo says that a theist who looks at nature objectively cannot avoid concluding that God is indifferent to suffering. This thought intruded on Darwin's consciousness again and again. Indeed, Darwin's theory gave the problem of evil a new twist. In the theory of evolution by natural selection, death is not a minor detail but is central to the process that generates the organic diversity we observe. Why would an all-PKG God use the deaths of billions of organisms as his method for bringing new species into existence? There is grandeur in the *product* of this process, but the process itself is gruesome.[43]

Whether or not Darwin remained a theist, he gradually ceased to be a Christian. After abandoning his medical studies, he went to Cambridge to study for the clergy. Darwin read Paley with enthusiasm and lived in the same rooms in Christs College that Paley had occupied. In the *Auto-biography*, Darwin describes himself as having had, as a young man, "no wish to dispute any dogma" (1958: 57). But by middle age he had left his Christianity behind. A crucial moment in this dispiriting odyssey was the death of his beloved nine-year-old daughter Annie. Darwin's bitterness about her pitiful suffering drove a wedge between him and religion (Desmond and Moore 1991). There was no sudden crisis of faith; he says in the *Autobiography* that "disbelief crept over me at a very slow rate" (Darwin 1958: 87). In the same passage, Darwin describes Christianity as a "damnable doctrine" because it says that his brother, father, and grandfather must suffer everlasting punishment for their lack of belief. This sticking point, which helped drive Darwin from the religion of his youth, did not stop his wife, Emma Wedgwood, from remaining a Christian; Emma simply dismissed the doctrine of eternal damnation as unchristian (Darwin 1958: 238; Brown 1986: 32). At the start of their marriage, Emma had expressed to Charles her concern that he did not share her faith that they would be together in Heaven forever. Darwin was advised by his father to keep his religious doubts to himself so as not to trouble his new wife. Darwin followed this advice and, more generally, was careful not to attack religion in his public life.

When I argued at the start of this chapter that Darwinian theory and theism are *logically consistent*, I did not say that theistic evolutionism is *plausible*. This is a question with two parts. There is, first, the nature of the evidence for *evolution*, which will be the subject of the rest of this book. The second part of theistic evolutionism is *theism*; the question of its plausibility leads to a wide range of issues. Evolutionary theory does not prove that there is no God; like other scientific theories, it is neutral

---

[43] Chapter 3 of *The Origin of Species* is called "The Struggle for Existence." The incessant theme is that nature is "battle within battle" ([1859] 1964: 79). Yet, the chapter ends with a surprising sentence: "when we reflect on this struggle, we may console ourselves with the full belief, that the war of nature is not incessant, that no fear is felt, that death is generally prompt, and that the vigorous, the healthy, and the happy survive and multiply."

on this question. Even if you are not impressed with the argument from design, you still need to consider other arguments for the existence of God – for example, the cosmological argument that Darwin sometimes found compelling. There also is the prior question of whether you should rely exclusively on evidence to decide what your theistic convictions ought to be. Must *all* your beliefs be dictated by the evidence you have and by nothing else? This is the question that William James and W. K. Clifford debated – "The Will to Believe" versus "The Ethics of Belief" (§1.1).[44] Just as scientific theories are silent on the question of whether God exists, they also have nothing to say about this question of ethics.

2.22   A PREDICTION

It was clear to Paley and to other defenders of the organismic design argument that the intelligent designer who built organisms must have been far more intelligent and efficacious than any human being could ever be. This is why the organismic design argument was for them an argument for the existence of *God*. I predict that it will eventually become clear that the organismic design argument should never have been understood in this way. This is because I expect that human beings will eventually build organisms from nonliving materials. This achievement will not close down the question of whether the organisms that human beings observe were created by intelligent design or by mindless natural processes; in fact, it will give that question a practical significance, since the organisms we will see around us will be of both kinds.[45] However, it will be abundantly clear that the fact of organismic adaptation has nothing to do with whether *God* exists. When the Spanish conquistadors arrived in the New World, several indigenous peoples thought these intruders were gods, so powerful was the technology that the intruders possessed. The locals were mistaken; they did not realize that these beings with guns and horses were merely *human* beings. The organismic design argument for the existence of God embodies the same mistake. If my prediction is correct, our descendants will someday look back on Paley and see him and Montezuma in the same light.

---

[44] Plantinga (2000) argues that it is rational to believe propositions that are "properly basic" even if you can produce no good argument that they are true. Plantinga holds that "God exists" is properly basic.

[45] Human beings have been modifying the characteristics of animals and plants by *artificial selection* for thousands of years. This means that some traits of some of the organisms around us *now* are due to intelligent design while others are not (Dennett 1987a: 284–5). Even so, the organisms that human planners have deliberately modified were not *created* by designers working just with nonliving raw materials.

CHAPTER 3

Natural selection

In the previous chapter, I explored what happens when Duhem's point about the need for auxiliary propositions in testing is brought into contact with a likelihood formulation of the design argument. Pre-Darwinian versions of the argument did not consider Darwin's theory of evolution but instead considered Epicureanism as the alternative to intelligent design, arguing that

Pr(the eye has features $F_1 \ldots F_n$ | an intelligent designer made the eye)

$> Pr$(the eye has features $F_1 \ldots F_n$ | the eye was produced by

mindless chance).

I granted that the probability on the right is very small (though it is not zero) but complained that we have no basis for assigning a value to the probability on the left. The reason is that the hypothesis that an intelligent designer fashioned the vertebrate eye predicts nothing about what features the eye should exhibit until further propositions are added about the putative designer's goals and abilities. It does no good *inventing* these needed further propositions; rather, they need to be *independently attested*.

The very same standards apply to evolutionary theory – *what is sauce for the goose is sauce for the gander*. Can the theory measure up? If the only thing that evolutionary biologists do is go around saying "that's due to natural selection" when they examine the complex and useful traits that organisms have, they are engaged in the same sterile game that creationists play when they declare "that's due to intelligent design." Assumptions about natural selection of course can be *invented* that allow the hypothesis of natural selection to fit what we observe. But that is not good enough; the question is whether there is *independent evidence* for those auxiliary propositions.

189

An important theme from Chapters 1 and 2 was that testing is *contrastive*: To test a theory, you need to test it *against* alternatives.[1] This seems to entail that testing evolutionary theory means testing it against creationism. If so, why is creationism so rarely discussed in scientific publications? Are biologists willfully burying their heads in the sand? In the last chapter I argued that the hypothesis that an intelligent designer made the complex and useful traits that we observe organisms to have is untestable. However, if intelligent design is a sorry excuse for a scientific theory, where is evolutionary theory to find a more worthy opponent? If testing is contrastive, there must be one. This line of questioning has a false presupposition. Evolutionary theory describes a number of possible causes of evolution. There is natural selection but there also is mutation, migration, random genetic drift, recombination, linkage, inbreeding, as well as others. The theory allows for the possibility that different traits in different lineages might evolve for different reasons; there is no presupposition that one size fits all. What evolutionary biologists spend their time doing is testing one evolutionary hypothesis against another. For example, an important project in population genetics involves using data on the DNA sequences present in different species to test selection against drift, an undertaking I'll discuss in §3.9. Contrastive testing occurs *within* evolutionary biology, not *between* evolutionary biology and something outside. To talk about testing evolutionary theory is a bit like talking about testing chemistry. Evolutionary biology, like chemistry, is a *field* or *discipline* that contains many theories; evolutionary biologists test evolutionary hypotheses *against each other*.

Does this mean that there are presuppositions internal to evolutionary biology that never get tested? For example, when population geneticists use sequence data to test selection against drift, they usually assume that the species considered all derive from a common ancestor. Not only that, they usually assume a specific phylogenetic tree, one that describes which species are closely related to each other and which are related only more distantly. It is true that population geneticists usually make these phylogenetic assumptions, but it is false that those "assumptions" are *merely* assumptions. As we will see in the next chapter, the hypothesis of common ancestry can be tested, and the same is true of more detailed claims about phylogenetic relationship. It is important to distinguish the

---

[1] This isn't true for theories that (together with independently justified auxiliary propositions) have deductive implications about observations; however, for theories that merely confer nonextreme probabilities on observations, the *dictum* is correct.

question of whether *scientists* make assumptions from the question of whether *the science* they work within contains undefended assumptions. Science is a cumulative enterprise; scientists constantly build on results obtained by others. If each scientist had to start from scratch, science would never get anywhere. So *of course* scientists make assumptions. But a proposition that serves as an assumption in one testing problem can be one of the propositions under test in another. Hypotheses about genealogy can be tested, and a lot of work in evolutionary biology goes into testing them.

Historians who study Darwin's work often say that he thought that natural selection is analogous to an agent (see, for example, Ospovat 1981 and Young 1985). Of course, it isn't literally true that natural selection is "trying" to do anything or that it "chooses" who shall live and who shall die. Selection is a mindless process. But Darwin found it useful, and so have ⸳evolutionary biologists down to the present, to think about what natural selection will achieve by thinking about what agents would achieve if they had certain aims and if their choices were limited to a given set of feasible options.[2] What fur length should we expect natural selection to produce in polar bears? It does no harm to think about this by asking what an intelligent designer would do for polar bears if he had the goal of helping them to survive and reproduce in their environment and was limited in his choices to a certain set of options. Although creationists often think of the designer as omnipotent, the analogy of natural selection with a designing agent loses its heuristic value if we assume that natural selection is unlimited in its power. As Maynard Smith (1978) remarked, if natural selection acted without constraint, organisms would live forever and would produce an infinite number of offspring. The reason zebras don't have machine guns with which to repel lion attacks is not that guns would not be useful; rather, this option was not available to them ancestrally (Krebs and Davies 1981). Selection selects only among those options that are actually represented in a population. It is important to remember that what is *conceivable* to an intelligent agent can differ from what is *biologically possible for a species given its history*.

Just as the hypothesis of intelligent design makes predictions only when supplemented with information about the putative designer's goals and abilities, so the hypothesis of natural selection makes predictions only when it is supplemented with information about what selection's "target" is and how selection is "constrained." In what follows I'll try to develop a

---

[2] For discussion of how this "heuristic of personification" can lead one astray, see Sober (1998).

concrete understanding of how "targets" and "constraints" should be understood. I will do this by investigating two simple models of phenotypic evolution. The result will not be a full understanding of the nature of selection or of the nature of constraints, both of which have been discussed at length and have many dimensions.[3] Rather, my goal is to start to fill in the picture of what we need to know if we are to construct meaningful tests of hypotheses about natural selection.

Just as pre-Darwinian creationists pitted the hypothesis of intelligent design against the hypothesis of chance, I want to test hypotheses that invoke natural selection against hypotheses of drift. Creationists think it is ludicrous to suggest that a complex adaptive device like the vertebrate eye arose by chance, and evolutionists emphatically agree. Both may find it tempting to explain why by saying that it would be very improbable for such complex and useful assemblages to come into existence if a chance process were doing the work. This is not an answer that can withstand scrutiny. We need to take seriously the fact that there is no such thing as probabilistic *modus tollens* (§1.4). The fact that a hypothesis says that a set of observations is very improbable is not a good reason to reject the hypothesis. Rather, to understand how a set of observations can favor selection over drift (or have the opposite evidential significance), we need to ascertain what each of these hypotheses predicts. The first step in that direction is to place this problem within a likelihood framework, but model-selection considerations will not be far behind.

3.1 SELECTION PLUS DRIFT (SPD) VERSUS PURE DRIFT (PD)

My goal here is not to discuss the testing of selection against drift in all its complexity but to isolate a fairly simple problem and to analyze the issues it raises. To this end, I don't want to consider a complex adaptive feature such as the vertebrate eye; instead, my example will be an apparently simpler quantitative character – the fact, let us assume, that polar bears now have fur that is, on average, 10 centimeters long. Which hypothesis – selection or drift – confers the higher probability on the trait value we observe polar bears to have?[4]

---

[3] See Antonovics and Van Tienderen (1991) and Schlichting and Pigliucci (1998) for discussion of how evolutionary "constraint" should be understood.

[4] I take it that the probability of the bears' having an average fur length of *exactly* 10 centimeters is zero, on each hypothesis. Rather, we need to talk about a small region surrounding the value of 10 centimeters as the observation that each theory probabilifies. Subsequent discussion should be understood in this way.

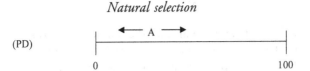

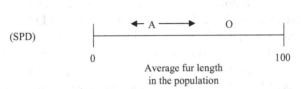

Figure 3.1 The pure-drift (PD) hypothesis can be thought of as a random walk on a line. The selection-plus-drift (SPD) hypothesis can be represented as a biased walk, influenced by a probabilistic attractor, the optimal phenotype. Both processes begin with the lineage in its ancestral state A.

I will assume that evolution in the lineage leading to present-day polar bears takes place in a finite population. This means that there is an element of drift in the evolutionary process, regardless of what else is happening. The question is whether selection also played a role. Thus, our two hypotheses are *pure drift* (PD) and *selection plus drift* (SPD). Were the alternative traits identical in fitness or were there fitness differences among them (and hence natural selection)? I will understand the idea of drift in a way that is somewhat nonstandard. The usual formulation is in terms of random *genetic* drift; however, the example I want to examine concerns fur length, which is a *phenotype*. To decide how random *genetic* drift would influence the evolution of this phenotype, we'd have to know the developmental rules that describe how genes influence phenotypes. I am going to bypass these genetic details by using a purely phenotypic notion of drift. Under the PD hypothesis, a population's probability of increasing its average fur length by a small amount is the same as its probability of reducing fur length by that amount. Average fur length evolves by random walk. This is depicted in Figure 3.1; the PD hypothesis is represented by two arrows of equal size, indicating that the expected amount of change is the same in both directions (note that they sum to zero). Let's suppose that the shortest possible fur length is 0 centimeters and that the maximum possible is 100. If a population happens to land at either of these end points, it isn't bound to stay there; these are not absorbing barriers. I'll assume that mutations always introduce a cloud of variation around the population's average fur length. This means that the population can evolve away from each of these extremes. The SPD hypothesis should be understood in similar fashion. The SPD hypothesis identifies some phenotypic value (*O*) as the optimal phenotype and says that an organism's fitness decreases monotonically as it deviates from that optimum. Thus, if 12 centimeters is the optimal fur length, then 11 centimeters is fitter than 10, 13 centimeters is fitter than 14, etc. Given this singly peaked fitness function, the SPD hypothesis says that a population's probability of moving a little closer to *O* exceeds its probability of moving a little farther away. This is why the arrows that depict the SPD hypothesis in Figure 3.1 are of unequal size; a population in state *A* has a higher probability of moving towards the optimum than away from it. The SPD hypothesis says that *O* is a *probabilistic attractor* in the lineage's evolution.[5]

A natural mathematical model for pure drift is Brownian motion (Harvey and Pagel 1991), according to which the evolution of the population's average phenotype obeys the same rules that govern a molecule moving at random to the right or to the left on a line with a reflecting barrier at each end. A natural formulation of the SPD hypothesis is provided by the Ornstein–Uhlenbeck model (Lande 1976; Hansen 1997; Butler and King 2004). Here the appropriate analogy is with a rubber band stretched between two pins, one above the other. If you hold the band at its center and pull it left or right, the farther you pull the band, the stronger the restoring force is. If the optimal fur length is 12 centimeters, then a population with a value of 7 centimeters experiences a stronger force pulling it towards 12 centimeters than a population at 10 centimeters experiences. The force declines as the population gets closer to its target. The Ornstein–Uhlenbeck model has a selective and a stochastic part:

$$dX(t) = a[\theta - X(t)]dt + \sigma dB(t).$$

The equation describes how much change you should expect to occur in a population's trait value between time *t* and time *t* + *dt*. The first addend on the right describes the effect that selection would have if the

---

[5] Some may prefer to define selection and drift so that they are mutually exclusive; the first involves variation in fitness while the latter means that there is no such variation. This choice of terminology would make the idea of SPD a contradiction. I am using a different terminological convention, but there is no need to fuss over this here, since there is a neutral way to describe the two hypotheses I want to consider: SPD postulates a process of selection in a finite population, and PD says that there is no variation in fitness (and hence no process of selection) in that finite population.

population were infinite and so there is no drift. *X(t)* is the population's trait value at time *t* and *θ* is the optimum. The parameter *a* describes the change that selection can be expected to effect per unit deviation from the optimum. So, for a fixed value of *a*, selection can be expected to produce a bigger change in trait value the more the optimum and the present trait value differ. The second addend describes random fluctuations, whose magnitude is represented by *σ*; *dB(t)* is a vector of independent and identically distributed normal random variables. To apply this equation to a population that now is in a given state, you use the first addend to calculate how far towards the optimum selection would move the population if there were no drift; then you draw a bell-shaped curve around that new value, indicating the uncertainty that is introduced by the fact that the population is finite. The Ornstein–Uhlenbeck equation describes the SPD process, but it includes the case of pure drift as a special case; if there is no selection the first addend is zero and evolution is governed just by the second.

To understand the meaning of the parameter *a* in the Ornstein–Uhlenbeck model, which represents the *expected* response to selection per unit deviation from the optimum, it is useful to consider an idea from quantitative genetics called the *breeder's equation* (Falconer and Mackay 1996). As the name suggests, this part of quantitative genetics was developed as a theoretical foundation for artificial selection, but it applies to natural selection as well. Suppose the polar bears in a given generation differ in fitness because they have different fur lengths. Individuals in this generation reproduce (with fitter individuals being more reproductively successful than less fit individuals), and their offspring then grow to adulthood. How much should we expect these two generations to differ in their average fur length? The breeder's equation says that

Response to selection = heritability × intensity of selection.

If the heritability is zero, then selection will not produce any change.[6] And for a fixed nonzero heritability, there will be a greater response to selection the more intense the selection is.[7] But what does "intensity" (or

---

[6] The breeder's equation reflects the fact that natural selection is described in evolutionary theory as a cause and also as an effect – "intensity of selection" describing the former, "response to selection" the latter. This poses a challenge to philosophers who deny that the theory of natural selection describes a cause of evolution; see, for example, Walsh et al. (2002) and the response of Shapiro and Sober (2007).

[7] There are two kinds of heritability described in quantitative genetics: broad and narrow. It is the narrow sense (meaning the additive genetic variance) that is relevant to the breeder's equation. The

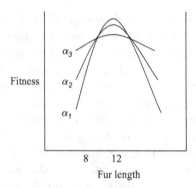

Figure 3.2 Three fitness functions that have the same optimum ($\theta = 12$).

"strength") of selection mean? This refers to how much variation in fitness there is in the population and to the extent to which fitness differences correlate with phenotypic differences for the character in question.[8] Consider, for example, the three fitness functions represented in Figure 3.2. The functions agree on which fur length is the best one for a polar bear to have (i.e., they agree that $\theta = 12$). They disagree about how much a bear's fitness suffers if the organism deviates from that optimum by a fixed amount. Imagine three populations p_1, p_2, and p_3 characterized by the fitness functions a_1, a_2, and a_3, respectively. Suppose that the average fur length in the three populations is the same, say 8 centimeters, that each has the same amount of phenotypic variation around this mean, and that the trait has the same heritability in all three populations. The breeder's equation says that p_1 is expected to move a larger distance towards the optimal value of 12 centimeters than p_2 is, and that p_2 should experience a larger displacement towards 12 centimeters than p_3 does.[9]

The dynamics of SPD are illustrated in Figure 3.3, which comes from Lande (1976). At the beginning of the process, at t_0, the average phenotype in the population has a sharp value. The state of the population at various later times is represented by different probability distributions. Notice

---

additive genetic variance might be regarded as measuring the "evolvability" of a trait subject to natural selection; see Hereford et al. (2004) for further discussion of this point and also of how terms in the breeder's equation should be scaled.

[8] Intensity of selection refers to the *covariance* of fitness and phenotype.

[9] There is a disconnect between the Ornstein–Uhlenbeck equation, which postulates a linear relationship between departure from the optimum and response to selection and the curved fitness functions shown in Figure 3.2. Harmony can be restored by using fitness functions that look like pointed gables or by replacing the linear equation with one that is quadratic. I'll do neither in what follows, for the sake of simplicity. If the curvature is slight, the linear model is a good approximation.

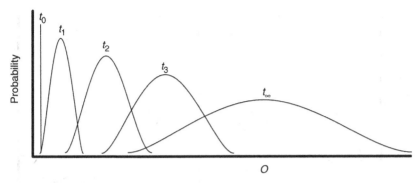

Figure 3.3 According to the SPD hypothesis, a population that has a given trait value at t_0 can be expected to move in the direction of O, the optimal trait value. As the process unfolds, expected values get closer to the optimum but the uncertainty surrounding those expected values increases.

that as the SPD process unfolds, the mean value of the distribution moves in the direction of the optimum. The distribution also grows wider, reflecting the fact that the population's average phenotype becomes more uncertain as more time elapses. After infinite time (at t_∞), the population will be centered on the putative optimum. The speed at which the population moves towards this final distribution depends on the trait's heritability and on the strength of selection. How wide the different distributions at different times are depends on the effective population size N; the larger N is, the narrower the bell curve. In summary, the SPD hypothesis says that trait evolution involves *the shifting and squashing of a bell curve*.

Figure 3.4 depicts the process of PD, which involves just *the squashing of a bell curve*. Although uncertainty about the trait's future state increases with time, the mean value of the distribution remains unchanged. In the limit of infinite time, the probability distribution of trait values is flat, indicating that all average fur lengths for the population are equiprobable. The rate at which the PD process squashes the bell curve depends on N, the effective population size; the smaller N is, the faster the squashing.[10]

---

[10] The case of infinite time in the PD model makes it easy to see why an explicitly genetic model can generate predictions that substantially differ from the purely phenotypic models considered here. Under the process of pure random *genetic* drift (with no mutation), each locus is homozygotic at equilibrium. In a one-locus two-allele model in which the population begins with each allele at 50 percent, there is a 0.5 probability that the population will eventually evolve to 100 percent *A* and a 0.5 probability that it will evolve to 100 percent *a*. In a two-locus two-allele model, again

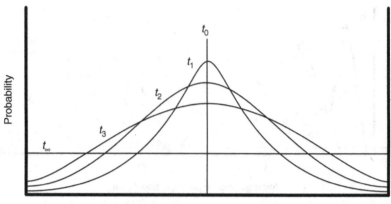

Figure 3.4 According to the PD hypothesis, a population that has a given trait value at time t_0 has that initial state as its expected value at all subsequent times, though the uncertainty surrounding that expected value increases.

The SPD hypothesis as I have formulated it constitutes a relatively simple conceptualization of natural selection in a finite population. The hypothesis assumes that the fitness function is singly peaked and that fitnesses are *frequency independent* – whether it is better for a bear to have fur that is 9 centimeters long or 8 centimeters does not depend on how common or rare these traits are in the population. I also have conceptualized the SPD hypothesis as specifying an optimum that remains unchanged during the lineage's evolution; the optimum is not a moving target. Indeed, the hypothesis assumes that there is a fur length that is optimal for all bears, regardless of how they differ in other respects.[11] My reason for constructing the SPD hypothesis with these features is not that I think they are realistic. My goal is to construct a simple example that makes it clear what information you need to have if you want to say whether SPD or PD has the higher likelihood. Informational requirements do not decline when models are made more complex; rather, they increase.

with each allele at equal frequency at the start, each of the four configurations *AABB*, *AAbb*, *aaBB*, and *aabb* has a 0.25 probability. Imagine that genotype determines phenotype (or that each genotype has associated with it a different average phenotypic value) and it becomes obvious that a genetic model can predict a nonuniform phenotypic distribution at equilibrium. The case of SPD is the same in this regard; there are genetic models that will alter the picture of how the phenotype evolves. See Turelli (1988) for further discussion.

[11] I also am assuming that a lineage that shifts its average fur length does so by a change in gene frequencies; this ignores the possibility that fur length is *phenotypically plastic*.

To visualize what the SPD and PD hypotheses each predict, it may be helpful to think about what each says will happen in 1,000 replicate populations that all begin evolving with the same initial average fur length and all evolve for the same length of time. If the 1,000 populations each experience SPD, we expect them to exhibit different average fur lengths; these different average phenotypes should form a distribution that approximates the theoretical distribution depicted in Figure 3.3 that corresponds to the amount of time that has elapsed. The same is true if the 1,000 replicate populations all experience PD. The PD and the SPD hypotheses both describe a single population by saying that there are different average fur lengths that it might evolve, and that these different possibilities have the different probabilities represented by the relevant curve.

3.2 COMPARING THE LIKELIHOODS OF THE SPD AND PD HYPOTHESES

We now are in a position to analyze when SPD will be more likely than PD. Figure 3.5a depicts the relevant distributions when there has been finite time since the lineage started evolving from its ancestral state (*A*). The SPD curve has moved in the direction of what it claims is the optimal trait value (*O*); the PD curve remains centered on *A*. During this finite interval of time, the PD curve has become more flattened than the SPD curve has; selection impedes spreading out. Figure 3.5b depicts the two distributions when there has been infinite time. The SPD curve is now

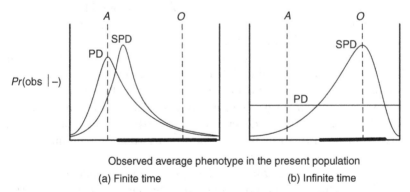

Observed average phenotype in the present population

(a) Finite time   (b) Infinite time

Figure 3.5 The likelihoods of the SPD and the PD hypotheses. SPD has the higher likelihood when the observed value is "close" to the optimum *O* postulated by the SPD model. *A* is the ancestral state of the lineage. The phenotypic values that count as "close" are marked with a solid line.

Ordering of A, P, and O		Which hypothesis is more likely?
(a) The present state coincides with the putative optimum.	$A \rightarrow P = O$	SPD
(b) The population evolves away from the putative optimum.	$P \leftarrow A \qquad O$	?
(c) The population overshoots the putative optimum.	$A \qquad O \overset{\frown}{\longrightarrow} P$	?
(d) The population undershoots the putative optimum.	$A \rightarrow P \qquad O$	?

Figure 3.6 The population must evolve from its ancestral state A to its present state P. How these two states are related to the optimum (O) postulated by the SPD hypothesis influences whether SPD is more likely than PD.

centered at the optimum it postulates while the PD curve is flat. Whether finite or infinite time has elapsed, the fundamental fact about the likelihoods is the same: *The SPD hypothesis is more likely than the PD hypothesis precisely when the population's present value is "close" to the optimum specified by the SPD curve.* I put the word "close" in quotation marks because its meaning depends on further details; compare the range of darkened *x*-values in Figure 3.5a with those in 3.5b. How close the population has to be to the optimum postulated by the SPD hypothesis for that hypothesis to have the higher likelihood depends on how much time has elapsed between the lineage's initial state and the present, on the intensity of selection, on the trait's heritability, and on N, the effective population size. For example, if infinite time has elapsed (Figure 3.5b), the SPD curve will be more tightly centered on the optimum, the larger N is. If 10 centimeters is the observed value of our polar bears, but 11 centimeters is the optimum, SPD may be more likely than PD if the population is small, but the reverse will be true if the population is sufficiently large.

The criterion of "closeness to the putative optimum" suggests that there are just two possibilities that need to be considered in deciding whether SPD is more likely than PD. Either the population's present state is "close enough" or it isn't. This is correct (as long as we remember that how close is close enough depends on further details), but, nonetheless, it is useful to distinguish the four possibilities that are summarized in Figure 3.6. In each, an arrow points from the population's ancestral state (A) to its present state (P); O is the optimum postulated by the SPD hypothesis. The first case (a) is the most obvious; if the optimum (O) turns out to be identical with the population's present trait value (in our example, fur that is 10 centimeters long), we're done: SPD has the higher likelihood. However, if the present trait value differs from the optimum value, we need more information. There are three more cases to consider, which differ in how A, P, and O are related to each other. In possibility (b), the population evolved *away* from the putative optimum. In (c), the population has overshot the putative optimum, whereas in (d) there is undershooting. In all three of these cases, we need to know not just the values of A, P, and O, but other biological facts as well, if we are to say which of SPD and PD has the higher likelihood. This is perhaps not so obvious in case (b). If a population has evolved away from the optimum, isn't that enough to conclude that we have evidence against SPD and for PD? To see that this is not always true, suppose that $P = 10$ centimeters, $A = 10.1$ centimeters, $O = 10.2$ centimeters, and that the population has been evolving for a very long time. The lineage has evolved away from O, but it's still close. If there is only weak selection pushing the population towards 10.2 centimeters, it isn't that surprising that it exhibits a trait value of 10 centimeters. On the other hand, if the PD hypothesis is true and the population evolves for a long time, the observed trait value of $P = 10$ centimeters is far less probable. Outcomes (c) and (d) are likewise inconclusive; after all, a population may undershoot or overshoot the putative optimum by a lot or a little. If there has been a lot of time and strong heritability, a population's evolving from $A = 2$ centimeters to $P = 10$ centimeters may be evidence against SPD, if that hypothesis says that the optimal trait value is $O = 50$ centimeters and that there has been strong selection for that trait value. However, if there has been much less time in the lineage, weaker heritability and weaker selection, this modest shift in the direction of the optimum may be evidence in favor of the SPD hypothesis.

3.3 FILLING IN THE BLANKS

Given the observed present trait value (P) of polar bears, answering the question of whether SPD is more likely than PD depends on what the value is of O (the trait value that would be optimal if there were natural selection), on what the value is of A (the ancestral state of the lineage), and on other details. How should we fill in these blanks? One possibility is to simply *invent* assumptions that allow our pet hypothesis to win the likelihood competition. For example, if you are an adaptationist and want SPD to triumph over PD, perhaps you should assume that the observed trait value of 10 centimeters also happens to be the optimal fur length. On the other hand, if you are a neutralist and want PD to beat SPD, perhaps you should assume that the lineage's present trait value is miles away from the one that would be optimal if the SPD hypothesis were true. As Bertrand Russell (1919: 71) once said in another context, the method of postulation has all the advantages of theft over honest toil. The mere invention of assumptions is an empty exercise – the same one we examined in the previous chapter in connection with the problem of testing intelligent design against chance. We must do better. Within a likelihood framework, the approach we need to pursue is to find auxiliary propositions that are *independently supported*. Once these are in place, we can see whether the observed fur length of polar bears favors SPD over PD.

The optimal trait value O postulated by the SPD hypothesis

As discussed in §2.12, the requirement of "independent justification" says that the auxiliary propositions used in a testing problem must be justified and that their justification should not depend on assuming the truth of any of the hypotheses that are under test. How does this idea apply to the fitness function used by the SPD hypothesis? The PD hypothesis asserts that all fur lengths have the same fitness. The SPD hypothesis asserts that the fitness function has a single peak. For the SPD hypothesis to make a prediction, what is needed is information about where that peak is. But how can a proposition that says where the optimal value O is located be justified independently of assuming that SPD is true? After all, if PD is the right model, then there *is* no such optimum. The answer is to recognize that what needs independent justification is a conditional that has the following form:

If the SPD hypothesis is true, then the optimal trait value is $O =$ ___ .

The requirement is that we fill in the blank (with a point value, or a value range) in a way that does not depend on assuming the truth of either SPD or PD. You don't have to believe that the SPD hypothesis is true to see that a conditional proposition of this form is justified. In the 1988 movie *Midnight Run* (dir. Martin Best, 1988), the actors Charles Grodin and Robert De Niro have a memorable dialogue:

GRODIN: If I were your accountant, I'd have to strongly advise you
          against –
DE NIRO: But you're not my accountant.
GRODIN: I realize I'm not your accountant. I said that *if* I were your
          accountant, I'd have to –
DE NIRO: But you're not my accountant.

For future reference I will call this the De Niro fallacy. Do not confuse a conditional with its antecedent (or with its consequent).[12] What is needed is evidence for the *conditional* that does not depend on deciding which of SPD and PD is true.

There are two broad strategies that evolutionary biologists use to fill in the blank in the above conditional. The first is more *observational* while the second is more *theoretical*.

If, as we are assuming, there is variation in fur length in the present population, we can observe whether bears with one fur length survive and reproduce more successfully than bears with another. We also can run an experiment – shaving some polar bears, fitting parkas onto others, and leaving still others unmodified. Observing the results provides evidence about the fitness function that characterizes *contemporary* polar bears in their *present* environment.[13] The two italicized words point towards the next step we need to take. We are interested in identifying the fitness function that would apply to a lineage (if that lineage experienced selection on fur length) that began sometime in the past and extends up to the polar-bear populations we now observe. How do observations of the present population allow us to draw a conclusion about the selective regime that was in place ancestrally?

There are two kinds of question to answer here. First, if ambient temperature is relevant to determining which fur lengths are selectively advantageous, we need information about the temperatures that the lineage experienced in the past. Second, the reason one fur length is better than another for a bear in a given physical environment is that the bear has certain other characteristics. For example, the optimal fur length for a bear in a given environment depends on how big the bear is. This raises the question of whether ancestral bears were about the same size as present-day bears. In short, we need information about the past physical environment and also about the biology of ancestral bears if we are to apply the fitness function we infer from data on present-day bears to the lineage as it evolved in the past.

Climatologists can help answer the first question, which concerns the history of weather. As for the second, one source of information about body size in ancestral populations is provided by fossils. This is obvious

---

[12] So that no undue aspersions will be thought to have been cast, let me state categorically that it was the character portrayed by De Niro, not De Niro himself, who makes this mistake. De Niro plays Jack Walsh and Grodin plays Jonathan "the Duke" Mardukas.

[13] In this vein, Baum and Larson (1991: 12) mention painting beetles to test a hypothesis about Batesian mimicry and trimming the toe fringes of lizards to see if this impairs their locomotion.

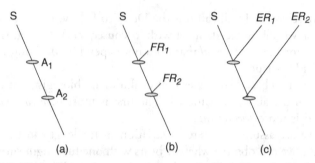

Figure 3.7 The body size of ancestors of current polar bears (S) can be (a) observed, or inferred from (b) fossilized relatives (FR_1 and FR_2), or from (c) extant relatives (ER_1 and ER_2).

when the fossils are ancestors of present-day polar bears (Figure 3.7a), but it also can be true if the fossils are relatives, not ancestors, as shown in Figure 3.7b. And if *fossilized* relatives can provide evidence about the state of ancestors, so can *living* relatives (organisms in other species that are closely related to polar bears), as shown in Figure 3.8c. The only relevant difference separating the three cases depicted in Figure 3.7 is the amount of time between the ancestors of polar bears and the objects we actually observe. The more time there is, the more uncertainty there is about the inference. The fact that (b) provides evidence as well as (a) is important, owing to the fact that biologists usually have no way to tell whether the fossils they observe are ancestors or just close relatives of an extant species whose characteristics they are trying to explain (Patterson 1981). Fossils do not have their genealogies written on their sleeves. The fact that (c) provides evidence as well as (a) and (b) is important too, since fossils are sometimes unavailable and even when they are, the traits in which we are interested often do not fossilize. Body size can be inferred from fossil traces, but maybe fur length cannot.

    In addition to doing experiments on present-day polar bears to see how changing their fur length affects their survival and reproduction, there is a second approach to the problem of identifying the optimal fur length, one that is less observational and more theoretical. Suppose there is an energetic cost associated with growing fur. We know that the heat loss an organism experiences depends on the ratio of its surface area to its volume. We also know that there is seasonal variation in temperature. Although it is bad to be too cold in winter, it also is bad to be too warm in summer. We also know something about the abundance of food. These and other considerations might allow us to construct a model that identifies what the optimal fur length is for organisms that have various other characteristics. Successful modeling of this type does not require the question-begging assumption that the bear's present trait value is optimal or close to optimal. This methodology has been applied to other traits in other taxa (Alexander 1996); there is no reason why it should not be applicable in the present context.

I have not found an optimality model for insulation against ambient temperature in the biological literature, so I'll illustrate the kind of approach I have in mind here by shifting to another example. Many organisms face the question of "when to give up" (Alexander 1996: 71–5). The larvae of ladybirds (a kind of beetle) feed on aphids, extracting the soft tissues and leaving the exoskeleton behind. The more food they extract, the less good it does them to persist in extracting more. There are diminishing returns. At what point should a ladybird stop eating the aphid at hand and set off to find a new one? That is, what is the optimal amount of time to spend on a food item? Dung fly males face a similar problem, although here the resource in question is not food that might be eaten but the eggs in a female dung fly that the male might fertilize. A male dung fly fertilizes more of a female's eggs the longer he copulates with her, though there are diminishing returns. On the other hand, the longer a male remains with one female, the less time there is for copulating with others. What copulation time is optimal for a male? Cook and Cockrell (1978) addressed the problem about ladybird feeding, and Parker and Stuart (1976) the problem about dung fly copulation, and they did so in the same way. They begin by identifying, as optimality modelers must (Maynard Smith 1978), a phenotypic correlate of fitness. For ladybirds, the optimal time to spend eating one aphid before moving on to the next is the amount of time that maximizes their rate of food intake; for male dung flies, the optimal copulation time is the one that maximizes their rate of egg fertilization. These are plausible postulates, though of course they might be wrong. The next step is to derive the trait value that maximizes fitness, so construed.

Both studies use a simple and pleasing graphical method. Cook and Cockrell starved ladybirds for twenty-four hours and then allowed them to start eating an aphid (whose weight they had measured beforehand), interrupting the ladybirds every so often to measure what was left. Ladybirds extract less food in each successive interval of time. This is not because their appetites wane; a starved ladybird does worse on a partially eaten aphid than it does on one that has not been touched. The amount of food (f) that a ladybird obtains by continuing to consume a single aphid is represented in the curve shown in Figure 3.8; notice that f increases, though its rate of increase declines, as t, the amount of time

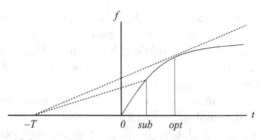

Figure 3.8 The solid curve represents Cook and Cockrell's (1978) estimate of how the amount of food (f) a ladybird obtains from eating an aphid depends on the amount of time (t) spent feeding on it. This curve allows one to calculate the value of t that maximizes the rate of food consumption, namely the ratio $f/(T+t)$, where T is the amount of time it takes a ladybird to find a new aphid. The optimal and a suboptimal value for t are shown.

spent eating the aphid, increases. If it takes a ladybird T units of time to find a new aphid, what value of t maximizes the rate of food consumed, namely the quantity $f/(t + T)$? The answer is given by finding the line that is tangent to the curve. The value of t associated with that point of tangency is the optimal feeding time; a suboptimal value of t is also shown in the figure. My point in describing this example is not to argue that the analysis is beyond question but to draw the reader's attention to the fact that the reasoning involves no assumption that the amount of time that ladybirds actually spend feeding on single aphids is optimal or close to optimal. The optimum derived in the model may turn out to be close to the current trait value, but it also may be far away. Notice also that the model's optimal trait value is not obtained by running an experiment in which ladybirds spend different amounts of time eating an aphid before they are forced to move on, the question being which ladybird gets the most to eat at the end of the day.

The style of argument that Cook and Cockrell deploy is very different from the one in which we shave some polar bears and give parkas to others. Still, there is a similarity. Regardless of whether the optimal trait value is obtained by an experiment or by a more theoretical derivation, what one obtains, in the first instance, is an estimate of the optimum that is based on facts about the *current* environment and the *current* biology of the organisms under study. This is what an experiment on current polar bears reveals, and it also is what one gets from an optimality model that uses information about the amount of time that ladybirds currently spend to find aphids to eat. And yet, the SPD hypothesis is a claim about *history*. The optimal trait value that is relevant to this historical hypothesis is not a timeless quantity; rather, the hypothesis claims that the lineage experienced selection that pushed it towards an optimum. As such, the relevant optimum is the trait value that would have been optimal *then* if there had been selection on the character in question; this may or may not be the trait value that is optimal *now*. The ideal solution to this problem would be to have information about the past with which to work. Information about the history of weather and about the history of polar-bear body size might allow one to calculate the fur length that would have been optimal ancestrally. Failing that, a reasonable fallback position is to calculate the fur length that is optimal now and then to consider a version of the SPD hypothesis that claims that this was the optimum that governed the trait's evolution. Both these strategies are big improvements over the mere assumption that the trait value one currently observes is optimal.

The ancestral state of the lineage

Figure 3.6 indicates that if the population's present state differs from the optimum postulated by the SPD hypothesis, then we need further information if we are to test SPD against PD. One element that matters is the population's ancestral state A. But how are we to determine what the lineage's initial state was?

One method that biologists often use to infer the states of ancestors in a phylogenetic tree is known as *parsimony*. I discussed in §1.7 what parsimony means in the context of model-selection theory: There, a model's complexity is measured by the number of adjustable parameters it contains. In the present context, parsimony has a specifically phylogenetic meaning. Here, the idea is that we should assign character states to ancestors so as to minimize the total amount of evolution that must have occurred in the phylogenetic tree. Consider, for example, the problem depicted in Figure 3.9, in which the average fur lengths of polar bears and two of their relatives are recorded. These extant species are the tips of a phylogenetic tree; reading down the page means going back in time, with interior nodes representing common ancestors. Given the fur lengths of the tip species, the most parsimonious assignment of states to the two common ancestors is $A_1 = 10$ and $A_2 = 7.5$.[14] So as to avoid confusing these two conceptions of parsimony, I'll refer to them in what follows as *model-selection parsimony* and *phylogenetic* (or *cladistic*) *parsimony*.

---

[14] Although no assignment is more parsimonious than $A_1 = 10$ and $A_2 = 7.5$, others are equally parsimonious, for example $A_1 = 10$ and $A_2 = 9$. This tie will be broken if we define parsimony in terms of the net amount of *squared* change; more on this wrinkle below.

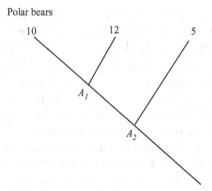

Figure 3.9 Given the trait values of present-day polar bears and their relatives, the principle of parsimony provides estimates of the character states of the ancestors A_1 and A_2.

Figure 3.9 shows that our question about whether the observed fur length of polar bears favors SPD over PD is ambiguous; there are many SPD and PD hypotheses, not just one of each. This is because present-day polar bears have multiple ancestors, and different ancestors may well have had different fur lengths. The problem of explaining why polar bears now have fur that is 10 centimeters long therefore decomposes into a number of subproblems: Why the fur length of A_2 had the value it did; why the fur length exhibited by A_2 evolved to the length found in A_1; and why A_1's fur length evolved to the present value of 10 centimeters (to mention just the two ancestors of polar bears that are depicted in the figure; there were more, of course). SPD may be a better answer than PD for some of these transitions while the reverse might be true for others. Similarly, suppose we pose our question about SPD versus PD by focusing on just one of the ancestors, asking what happened in the lineage connecting that ancestor to the polar bears of today. The answer may be different, depending on whether we use the ancestor A_1 or the ancestor A_2 as our starting point. The question of whether SPD or PD is the more likely explanation of an observed trait value thus needs to be relativized to a choice of ancestor.

Phylogenetic parsimony provides an estimate of the character state of an ancestor in the lineage leading to polar bears, but is this estimate to be trusted? Since we are trying to test SPD against PD, this question about parsimony has a very specific meaning: Is our justification for trusting what the principle of parsimony says about the character state of an ancestor *independent* of the hypotheses under test? Unfortunately, the answer is *no*. We observe that polar bears now have an average fur length of 10 centimeters, so the most parsimonious assignment of character state

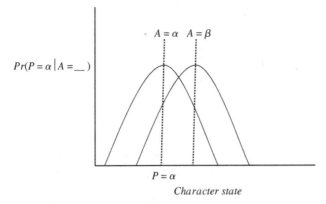

$Pr(P = \alpha \,|\, A = \_\_\,)$

$A = \alpha \quad A = \beta$

$P = \alpha$

Character state

Figure 3.10 If $P = a$ is the present trait value and the lineage has experienced pure drift, the maximum likelihood estimate of the trait value of the ancestor is $A = a$. The alternative estimate $A = \beta$ confers on the observation a lower probability.

to the ancestor is, of course, 10 centimeters. This most parsimonious assignment of trait value to the ancestor (where parsimony means minimizing the *squared* amount of change) is also the assignment of maximum likelihood if the PD hypothesis is true (Maddison 1991); see Figure 3.10.[15] But parsimony and likelihood can fail to coincide if a directional selection process is at work (Sober 2002b). For example, if 12 centimeters is the optimal fur length towards which selection has been pushing the lineage, the most likely assignment of trait value to the ancestor is some value *less than* 10 centimeters (see Figure 3.11); how much less depends on the amount of time separating ancestor and descendant, the intensity of selection, and the heritability. To see why this is so, consider the following analogy: You observe a log floating at the bank of a river and think that it floated there from the other side. But where on the other side did the log begin? You could ask where the log *probably* began, or, more modestly, you could ask which location on the other side is *best supported* by your observation of where the log is now. The latter question poses a problem of maximum likelihood estimation. What location on the other side would maximize the probability of the log's present location? If the river always flows in the same direction, the answer is that the most likely starting position is *upstream* – and the stronger the current, the farther upstream.

[15] Of course, even if $A = 10$ is the maximum *likelihood* estimate of the ancestral character state, given the observation that the descendant is in the state $D = 10$, this does not mean that $A = 10$ is very *probable* (§1.2).

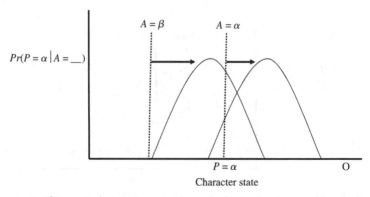

Figure 3.11 If $P = a$ is the present trait value and selection has been pushing the lineage towards the optimal value O, the maximum likelihood estimate of the trait value of the ancestor is not $A = a$; note that $A = \beta$ confers on the descendant's present character state a higher probability.

If parsimony does not provide estimates of ancestral character states that are *independent* of the SPD and PD hypotheses we wish to test, what should we do? To make headway on this problem we need to recall the lesson from the previous section concerning the De Niro fallacy. Although it would be nice to know what the true value was of the ancestral state (A) of the lineage leading to polar bears, this is not necessary as far as the likelihood comparison of SPD and PD is concerned. Rather, what we need to do is fill in the blanks in the following two statements:

If the PD hypothesis is true, the ancestral condition was $A = \underline{\quad}$.

If the SPD hypothesis is true, the ancestral condition was $A = \underline{\quad}$.

The requirement of independent evidence is that we fill in each blank without assuming that PD is true or assuming that SPD is true. Since both hypotheses are probabilistic in character, we can't enter a point value in either blank. Rather, what we need is a probability distribution of the different values A might have (or the expected value thereof), conditional on each hypothesis.

Here's a strategy for getting SPD and PD to each fill in its own blank. The two hypotheses do not say that the processes they postulate suddenly began after the lineage's "initial" state A. Rather, we should think of each as describing what occurred in the lineage before and after A occurs. Both hypotheses say that the lineage moves towards an equilibrium distribution, regardless of what the lineage's ancestral condition was. These equilibrium distributions are depicted in Figure 3.5b. The equilibrium distribution for SPD is centered on the optimum *O*; the equilibrium distribution for PD is flat. The proposal I want to consider is that these equilibrium probabilities should be viewed as providing priors for the state of the ancestor *A*. If this proposal is adopted, we can stop talking about the ancestor *A* entirely. The reason is that once an ancestor *A* is characterized by the equilibrium distribution, so are its descendants. This allows us to pose our question about the likelihoods of SPD and PD in the form depicted in Figure 3.5b. The question is simply whether the present state *P* is "close" to the optimum postulated by the SPD hypothesis. It is no longer relevant what character state the ancestor occupied. The idea that the distribution for the ancestor *A* is the equilibrium distribution is equivalent to the idea that the lineage has been evolving for an infinite amount of time. This is clearly false, but it may be a harmless idealization. Perhaps the lineage has been evolving long enough that the equilibrium distribution is a good approximation.

Even though we do not need to puzzle over what the state of the ancestor *A* was if we are to test SPD against PD, the same is not true for other biological parameters. If the SPD hypothesis says that selection is very strong and that population size is very large, the observed trait value must be very close to the optimum trait value for SPD is to be more likely than PD. With weaker selection and smaller population size, the demands are less stringent. To put the point a bit paradoxically: Weaker selection hypotheses are stronger in their competition against drift than stronger selection hypotheses are. The SPD hypothesis needs to identify the location of the optimal trait value *O*; it must say how strong the selection was that allegedly pushed the lineage towards that optimum; and it must provide a picture of the effective population size. Only then do we know "how close is close enough" – how far the present trait value can be from the optimum specified by the SPD hypothesis if the observed trait value is to favor SPD over PD.

This analysis has implications concerning how the fit between a well-motivated optimality model and an observed trait value should be interpreted. Consider, for example, the work on dung-fly copulation time described earlier. Parker and Stuart (1976) constructed a model that entails an optimal copulation time of 41 minutes. The observed value, they report, is 35 minutes. Is this 6-minute difference small enough that we can conclude that the observations favor the hypothesis that selection has pushed the dung-fly lineage in the direction of that optimal trait value? Or is the 6-minute difference sufficiently large that we should conclude that the evidence tells against the selection hypothesis? If testing must be contrastive (§1.3), there is no answering these questions until we formulate an alternative hypothesis. If the alternative is drift, we still have work to do if we want to say whether 35 minutes is close enough to 41 minutes for this observation to favor SPD over PD (Sober and Orzack 2003).

3.4   WHAT IF THE FITNESS FUNCTION OF THE SPD
      HYPOTHESIS CONTAINS A VALLEY?

Figure 3.6 suggests that testing the SPD against the PD hypothesis gets complicated when the observed fur length of present day polar bears (P) differs from the optimal fur length (O) postulated by the SPD hypothesis. But at least there seems to be clear sailing when $P = O$; if the present fur length is *identical* with an independently identified optimal fur length, surely that suffices for SPD to be more likely than PD! This straight-forward conclusion does hold for the kind of fitness function we have taken the SPD hypothesis to postulate. However, if we abandon the assumption that fitness declines monotonically as a bear's fur length departs from a single optimal value, complications arise even when $P = O$. The vertebrate eye furnishes a nice example that illustrates this point, but not because it is a "complex" trait while fur length is "simple."

Fur length has a fairly obvious "transformation series." If a population is to evolve from an average fur length of 4 centimeters to an average of 10 centimeters by a series of small changes, it must pass through an average that is around 7 centimeters. But consider the evolution of the camera eye in the vertebrate line. If we trace this lineage back far enough, we will find an ancestor that has no eye at all. Again assuming that changes must be small, we can ask what the intermediate stages were through which the lineage must have passed as it evolved from no eye to a camera eye. A more general approach would be to conceive of this problem probabilistically; there may be more than one possible trans-formation series, with different probabilities attaching to different pos-sible changes in character state. The reconstruction of this transformation series is a nontrivial evolutionary problem.

There are nine or ten basic eye designs found in animals, with many variations on those themes. In broad strokes, this variation can be described as follows: Vertebrates, squid, and spiders have camera eyes; most insects have compound eyes (as do many shallow water crustacea); the *Nautilus* has a pinhole eye; the clam *Pectem* and the crustacean *Gigantocypris* have mirror eyes; and flatworms, limpets, and bivalve molluscs have cup eyes. When biologists place these features at the tips of an independently inferred phylogenetic tree (and use parsimony to infer the character states of ancestors), they conclude that these and other basic designs evolved somewhere between forty and sixty-five times in different lineages (Salvini-Plawen and Mayr 1977; Nilsson 1989).[16] For each monophyletic group of tip species that share a given eye design (e.g., vertebrates with their camera eyes), we can ask whether that design favors SPD over PD.

To figure out what the SPD hypothesis predicts, we need to know what trait value that hypothesis says is optimal for the species or taxonomic group in question. In the case of polar-bear fur length, we considered a simple experiment that would provide information about this. Is there a similar experiment for the case of eye design? Even if present technology makes eye transplantation unfeasible, information is now available concerning the optical properties of different eye designs. For example, Nilsson (1989: 302) agrees with Land's (1984) contention that "if the *Nautilus* had a camera-type eye of the same size, it would be 400 times more sensitive and have 100 times better resolution than its current pinhole eye." He has similar praise for camera eyes as compared to compound eyes:

if the human eye was scaled down 20 times to the size of a locust eye, image resolution would still be an order of magnitude better than that of the locust eye. Diffraction thus makes the compound eye with its many small lenses inherently inferior to a single-lens eye. (Nilsson 1989: 306)

If the SPD hypothesis we are considering says that the camera eye is *globally* optimal, then isn't it obvious that a lineage has a higher probability of evolving that trait if the SPD hypothesis is true than would be the case if the PD hypothesis were true? The answer is *no*. Suppose that a lineage that evolved from a compound eye to a camera eye would have to pass through the state of having a cup eye. And suppose that the fitness function of these three architectures is the one given in Figure 3.12. Evolving from compound to camera involves traversing a fitness valley if selection governs the lineage's evolution. So which hypothesis, SPD or PD, make it more probable that a lineage now has a camera eye? That depends on the state of the ancestor. If the ancestor has a cup eye, SPD

---

[16] Geneticists working in "evo-devo" (evolutionary developmental biology) have recently discovered that, although camera eyes evolved independently in different lines, some of the genes that help build those eyes are ancient and homologous; see Gehring (2002).

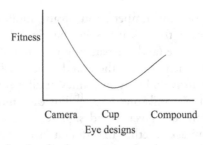

Figure 3.12 A fitness function for the camera, cup, and compound eye that has a valley.

makes the outcome more probable than PD does. But if the ancestor has a compound eye, the reverse may be true.[17]

The fitness function depicted in Figure 3.12 is not merely hypothetical. If the camera eye is fitter than both the pinhole and the compound eye, why don't all organisms with eyes have the camera eye? Spiders and squid are as lucky as we are, but bees and the *Nautilus* are not. Why not? Nilsson (1989: 306) suggests that

> at an early stage of evolution, the simple eye would be just a single pigment cup with many receptors inside [...], whereas the compound eye would start as multiple pigment cups with only a few receptors in each [...] At this low degree of sophistication, neither of the two designs stands out as better than the other. It is only later, when optimized optics have been added, that the differences will become significant. But then there is no return, and the differences remain conserved.

The idea is that organisms with compound eyes are trapped on an adaptive peak that is locally, but not globally, optimal (Salvini-Plaven and Mayr 1977; Nilsson and Pelger 1994). The point of importance for our assessment of SPD and PD is that when there is a fitness valley and a descendant has the globally optimal phenotype, it is not inevitable that the SPD hypothesis has the higher likelihood. More information is needed to say whether this is so.

This problem does not disappear just by avoiding the De Niro fallacy. When the SPD hypothesis postulates a singly peaked fitness function, the hypothesis itself entails an equilibrium distribution that we can use as a prior for the state of the ancestor *A*. But the fitness function depicted in

---

[17] The probability that a population will cross an adaptive valley depends on how high and steep the hills are, the effective population size, the amount of time there is between the initial state and the present, the heritability, and the width of the valley (Lande 1985).

Figure 3.12 does not indicate what state the ancestor probably occupied. When we observe that some present species *P* has a camera eye, it is true that

$$Pr(P = \text{camera} \,|\, \text{SPD} \,\&\, A = \text{cup})$$
$$> Pr(P = \text{camera} \,|\, \text{SPD} \,\&\, A = \text{compound}).$$

This inequality follows from the fitness function in Figure 3.12. However, the fitness function does not entail that

$$Pr(A = \text{cup} \,|\, \text{SPD}) > Pr(A = \text{compound} \,|\, \text{SPD}).$$

In fact, just the opposite conclusion seems to be true, since cup eyes are an unstable equilibrium according to Figure 3.12; a lineage that begins with a cup eye can be expected to move to either a camera or a compound eye, given enough time.

3.5 SELECTION VERSUS DRIFT FOR A DICHOTOMOUS CHARACTER

Our exploration of the SPD and PD hypotheses as possible explanations for why polar bears now have an average fur length of 10 centimeters has uncovered a number of difficulties; there are things you need to know before you can say which hypothesis has the higher likelihood. Perhaps these epistemological difficulties would vanish if we reconfigured the problem. Instead of asking why polar bears now have fur that is 10 centimeters long, perhaps we should ask why they have "long" fur rather than "short." Isn't it clear that polar bears are better off with long fur than they would be with short? If so, long is the *optimal* value for this dichotomous character. Doesn't this allow us to conclude without further ado that SPD has a higher likelihood than PD? Apparently, you don't need to know any further biological details to make this argument.

To make this suggestion more precise, I want to present the standard Markov model for thinking about the evolution of a dichotomous character (Parzen 1962: 293–5). I'll also use this model in the next chapter. It resembles the Ornstein–Uhlenbeck model described earlier for quantitative characters. At any moment in time, a lineage is in one of two states (call them 0 and 1). Within a very small period of time (an "instant"), the population has a small probability (u) of changing from state 0 to 1 and a possibly different probability (v) of changing from 1 to 0. A lineage's probability of ending in state j, if it begins in state i ($i, j = 0, 1$) and there have been t units of time in between, $Pr_t(i \rightarrow j)$, is a function of these instantaneous probabilities and the amount of time. Here are the equations for these lineage transition probabilities:

$$Pr_t(0 \to 1) = \frac{u}{u+v} - \frac{u}{u+v}(1-u-v)^t$$

$$Pr_t(0 \to 0) = \frac{v}{u+v} + \frac{u}{u+v}(1.-u-v)^t$$

$$Pr_t(1 \to 0) = \frac{v}{u+v} - \frac{v}{u+v}(1-u-v)^t$$

$$Pr_t(1 \to 1) = \frac{u}{u+v} + \frac{v}{u+v}(1-u-v)^t.$$

Note that the first two probabilities sum to one, as do the third and fourth. In each of these equations, the first addend fails to mention the amount of time t between the lineage's start and finish. The second addend does, and it shrinks towards zero as t increases. This means that the first addend describes the *equilibrium* probability, the probability that obtains when there is an infinite amount of time in the lineage. When time is short, the value of these transition probabilities is mainly determined by the lineage's initial state; if the lineage begins in a given state, it will almost certainly end in that same state. For example, if $t = 0$, $Pr(1 \to 1) = Pr(0 \to 0) = 1$ and $Pr(1 \to 0) = Pr(0 \to 1) = 0$. As the duration of the lineage is increased, the process plays a progressively larger role in determining the probability of the final state and the initial condition of the lineage is steadily forgotten. The Markov model entails that a "backwards inequality" holds true: regardless of the values of u, v, and t, $Pr_t(j \to j) > Pr_t(i \to j)$. Compare the first and fourth equations above (and also the second and third). The Markov model says that if a descendant is in a given state, the most *likely* hypothesis about its ancestor is that the ancestor was in the same state. Don't confuse this with the "forwards inequality" $Pr_t(j \to j) > Pr_t(j \to i)$; the equations leave it open whether stasis is more probable than change in a lineage.[18]

---

[18] The backwards inequality provides a likelihood justification for preferring the most parsimonious assignment of character state to an ancestor when the trait in question is dichotomous and one has observed the character state of a single descendant; if P is a present species and A its ancestor, it is both more parsimonious and more likely that A was in the same state that P occupies. This holds regardless of whether there was selection or drift in the lineage. It is interesting that this unconditional relationship between likelihood and parsimony holds for dichotomous characters, but not when traits are quantitative (§3.3). We will further consider the use of parsimony to infer the character states of ancestors in §3.11.

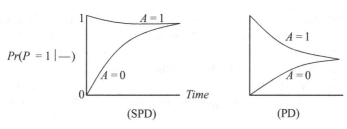

Figure 3.13 The SPD and PD hypotheses differ in the probabilities they specify for a lineage's ending in the state $P = 1$. In both cases, the probability depends on the lineage's ancestral state A and on the amount of time between A and P. The SPD hypothesis is here understood to say that there is strong selection for character state 1.

How can this simple model be used to represent selection and drift when the trait is dichotomous? Selection for state 1 (long fur length) means that $u > v$; if selection favors long fur over short, the lineage has a larger probability of evolving from short to long than from long to short. Drift, on the other hand, can be represented by the claim that $u = v$; the process is unbiased, with both changes having the same probability. The two hypotheses are represented in Figure 3.13. We now can evaluate the likelihoods of the SPD and PD hypotheses when the trait observed in our polar bears is dichotomous; the bears have "long" fur. First, we solve the problem of estimating the lineage's initial state by using the equilibrium value that each hypothesized process provides. If the SPD process is running for a long time before the ancestor A appears, then SPD says that A was, with very high probability, in state 1; PD, on the other hand, says the probability that $A = 1$ is 0.5. This has the consequence that the SPD hypothesis says that the present state $P = 1$ had a very high probability, while PD says that the probability of this observation is 0.5. In other words, when we observe that polar bears now have "long" fur, the hypothesis that this was due to selection favoring long fur over short will have a higher likelihood than the hypothesis that says that fur length evolved by pure drift. This is not surprising; there is no fitness valley when a lineage is characterized in terms of a dichotomous trait.

It is interesting how often informal reasoning about natural selection focuses on dichotomous characters. For example, sociobiologists often discuss why human beings "avoid incest," not why they avoid incest to the degree they do. The selection hypothesis says that outbreeding has a higher fitness value than inbreeding. This hypothesis renders the observed "avoidance of incest" more probable than does the hypothesis that trait evolution was governed by a pure drift process. Of course, the problem gets more difficult if we estimate how much inbreeding there is in human populations and then ask whether that quantitative value is more probable under the SPD or the PD hypothesis. But why shouldn't we simply admit that this quantitative problem *is* more difficult but still maintain that the simple likelihood argument just described solves the qualitative problem?

Although testing SPD against PD is straightforward once we have a dichotomous character, a problem arises in connection with understanding how this dichotomous character is related to the underlying continuous reality. For the quantitative character of fur length, there is an intermediate optimum. This means that for many of the cutoffs we might draw to separate "long" fur from "short," it will *not* be true that any fur length longer than the cutoff is fitter than any that is shorter. In what sense, then, is it true that "long" fur is fitter than "short?" Inbreeding involves a different situation if we suppose that the less of it the better. If a cutoff is drawn to separate "no-inbreeding" from "inbreeding," it *will* be true that any value on one side of the cutoff is fitter than any value on the other. But where should this line be drawn? Maybe the answer is that a population counts as "not inbreeding" precisely when mates are on average no more closely related to each other than they would be if pairs formed at random. But then the fact of the matter is that human beings *do* engage in a nonzero degree of inbreeding. In terms of the dichotomous character, the population has the *sub*optimal trait value. If we choose instead to draw a line that separates "little or no inbreeding" from "more," human populations can, if we wish, be placed in the former and fitter category. But how much satisfaction is there in then concluding that SPD is likelier than PD in this instance? After all, a different cutoff would lead to the opposite conclusion.

Even when a dichotomous character makes sense, there is a mistake we need to avoid. The mistake is to assume that the straightforward argument concerning the dichotomous character automatically carries over to the continuous trait. It seems natural to say "since the fact that polar bears have long fur rather than short favors SPD over PD, the same must be true of the fact that their fur is, on average, 10 centimeters long." This does not follow. Remember the point from §1.3 concerning the principle of total evidence: Logically strengthening the description of the data can affect which hypothesis has the higher likelihood. This point will surface yet again in the next chapter.

3.6 A BREATH OF FRESH AIR: CHANGE THE *EXPLANANDUM*[19]

I have gone into detail about the comparison of selection and drift for a single character because so much discussion (especially informal discussion) of natural selection focuses on a single trait in a single species or taxon. We behold the vertebrate eye and find the conclusion irresistible that this complex and useful trait must have been produced by natural selection. The idea that it could have been produced by drift seems ludicrous. These reactions, if they are to be anything more than intuitive feelings, must be defended by an argument. I have tried to show that there is more going on here than may first meet the eye.

I am guessing that some biologists who have read thus far are impatiently tapping their feet. They are thinking that the problem I have addressed has been misconceived from the outset. The thought is that we should not attempt to explain the fact that a given taxon has this or that characteristic. Rather than seeking to explain why polar bears now have an average fur length of about 10 centimeters, we should try to account for the fact that bears that live in colder climates tend to have longer fur than bears that live in warmer climates. What demands explanation is a cross-species *correlation*, not the fact that a single species has a single *trait value*. I emphatically agree. The preceding analysis, with its catalog of difficulties, was intended to describe a problem. Shifting the *explanandum* to a cross-species correlation is part of the solution. This is another example of the keys and the lamppost (§1.7).

To think about the interpretation of this sort of cross-species correlation between ambient temperature and fur length, let's consider the hypothetical data set given in Figure 3.14. This data set seems to provide strong evidence for the hypothesis that fur length is an adaptation for coping with ambient temperature; natural selection seems to have moved the different species towards their optimal trait values. This optimality line, let us suppose, is derived from experiment or theory in the manner sketched in §3.3. The fact that the data points show a downward trend and are tightly clustered around the optimality line seems pretty compelling. If fur length and ambient temperature were causally independent – e.g., if fur length had evolved by drift – we'd expect there to be no association between observed fur length and ambient temperature, in which case we'd

---

[19] The main argument of this section is drawn from Sober and Orzack (2003).

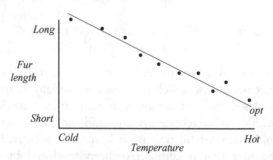

Figure 3.14 The observed fur lengths for different bear species show a downward trend and are closely clustered around an independently motivated optimality line. This seems to be strong evidence favoring selection over drift.

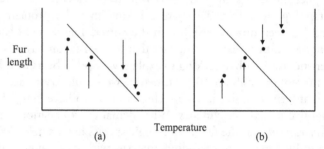

Figure 3.15 Two scenarios in which selection causes bear lineages to evolve in the direction of an optimality line. Depending on the ancestral states of those lineages, the result can be a set of extant species (represented by dots) that exhibit either (a) a downward trend or (b) an upward trend.

expect the best-fitting regression line drawn through the data to have a slope indistinguishable from zero.

As intuitively compelling as this reasoning appears to be, it nonetheless needs to be scrutinized. Let's begin with the question of why the selection hypothesis predicts a downward trend in the data. It is true that selection pushing lineages in the direction of an optimality line that has a downward slope can result in a data set that also has a downward slope. But it is equally true that this type of selection can generate a data set that has an upward slope. Both possibilities are depicted in Figure 3.15. Why, then, should selection lead us to expect the one pattern in the data, but not the other?

The hypothesis we are considering describes an optimality line and claims that this line is a probabilistic attractor; a lineage that exists in an environment with a given temperature is inclined to evolve towards the trait value that is optimal for that environment, as the Ornstein–Uhlenbeck

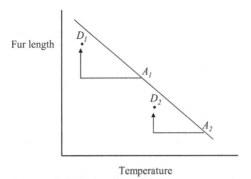

Figure 3.16 Suppose the ancestors A_1 and A_2 both have optimal trait values and their environments then get colder. If the lineages stemming from those ancestors evolve in the direction of the optimality line, a line through the descendants D_1 and D_2 that those ancestors produce can be expected to have a negative slope.

model claims. The process is probabilistic; the lineage is not *bound* to evolve in that direction. Given this hypothesis, why expect the first pattern in Figure 3.15 any more than the second? One justification for this expectation can be extracted from the assumption that the lineages have been evolving for an infinite amount of time. If this were the case, the upward trend in the data in Figure 3.15b would not be observed. This pattern in the data might exist for a while if the ancestral states of the different lineages were arrayed just so, but it would be washed away if the lineages were allowed to evolve longer. Given enough time, pattern (a), rather than pattern (b), is what the selection hypothesis predicts. Another possible justification for holding that the selection hypothesis predicts a downward rather than an upward pattern in the data is provided by the assumption that the ancestors of the lineages all had optimal trait values. This idea is depicted in Figure 3.16. If ancestors have optimal fur lengths and the climate changes, with the result that their descendants initially have suboptimal trait values and then evolve in the direction of the optimality line, the descendants of those ancestors should exhibit a downward trend.

There is a third possible rationale for thinking that the selection hypothesis predicts a downward, not an upward, trend in the data. If the lineages all trace back to a common ancestor, and if they have heritabilities that are approximately the same, then the pattern of data that will probably arise is the one depicted in Figure 3.17. Here is a fact that is obvious but highly significant: *when two descendant lineages stem from a common ancestor, they must, of necessity, begin in the same state.* If both are

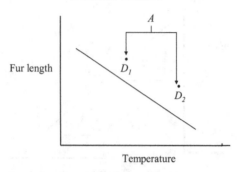

Figure 3.17 If two descendant lineages stem from a common ancestor A and then evolve in the direction of an optimality line that has a negative slope, the expectation is that a line through D_1 and D_2 will also have a negative slope, if the trait's heritability is approximately the same in the two lineages.

attracted to an optimality line that has a negative slope, how far can each be expected to evolve in the direction of that line? That depends on both the strength of selection and the heritability. We know that selection is stronger, the more distant a lineage is from its optimal value; this is a consequence of the Ornstein–Uhlenbeck model (§3.1). And we also know that two lineages stemming from a common ancestor begin with the same heritabilities; their *expected* heritabilities are thus the same, though they may subsequently diverge by chance. The prediction is therefore that a line drawn through the end points (D_1 and D_2) of those evolving lineages will have a negative slope, mirroring the slope of the optimality line towards which the lineages have been attracted. Common ancestry is an auxiliary proposition that allows the selection hypothesis to make a pre-diction about observed trait values.

A similar analysis is available for understanding why the hypothesis of pure drift predicts that there will be no association between fur length and ambient temperature. If the different lineages do not share common ancestors and evolve for an infinite amount of time, a regression line drawn through the descendants of those lineages can be expected to have zero slope. But what if time is finite? If there is common ancestry, the drift hypothesis predicts that there will be no association between fur length and temperature in the descendants, as shown in Figure 3.18.[20] What if there are more than two species? If fur length evolves on a phylogenetic

---

[20] The analysis offered here concerning how common ancestry leads selection and drift to make different predictions provides a rationale for the sign test that Burt (1988) proposed for testing selection hypotheses.

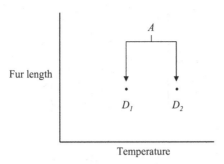

Figure 3.18 If two descendant lineages stem from a common ancestor A and then evolve by drift, the expectation is that a line through D_1 and D_2 will have zero slope.

tree by random drift, we expect that more closely related tip species will be more similar in their fur lengths: The predictor of fur length should be propinquity of descent, not ambient temperature.

In the first part of this chapter, we considered the likelihood competition between SPD and PD, where each hypothesis is assessed by considering the present state of a single species (in our running example, that polar bears now have an average fur length of 10 centimeters). We considered how the different relationships that might obtain among the lineage's ancestral state (A), its present state (P), and the optimal state (O) postulated by the SPD hypothesis affect the likelihood comparison. One issue that emerged from that discussion was the problem of overshooting. Suppose a lineage starts evolving with a fur length of 5 centimeters and the optimum described by the SPD hypothesis is 12 centimeters. If the lineage's present state is a fur length of 25 centimeters, is this evidence that favors drift over selection? The answer depends on the values of several biological parameters. A similar question arises in the new setting of the problem we are now considering in which it is the correlation among species, not the absolute trait value of a single species, that is used to test selection against drift. Overshooting now poses the problem depicted in Figure 3.19.

A final tweak in our formulation of the problem of testing selection against drift can solve this puzzle. In many biological studies, the exact location of the optimality line that the SPD hypothesis should endorse is unknown. Even on the assumption that bears evolve their fur length as an adaptive response to ambient temperature, it is very difficult to say exactly what the optimal fur length is for a given temperature regime. But consider the hypothesis that the optimality line has a negative slope – that the optimal fur length for bears in colder climates is greater than the

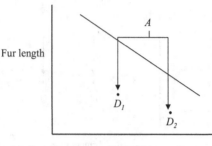

Temperature

Figure 3.19 If two descendant lineages stem from a common ancestor *A* and then overshoot the optimality line postulated by the adaptive hypothesis, does this count as evidence favoring drift over selection?

optimal length for bears in warmer climates. If the selection hypothesis is stated in these terms, with no commitment as to which fur lengths are optimal for which temperatures, the problem of overshooting disappears. If the selection hypothesis says that the optimality line has a negative slope (but does not specify its exact location), then the assumption of common ancestry has the consequence that the selection hypothesis and the drift hypothesis make different predictions. Selection predicts that a line through the trait values of tip species will have a negative slope while the drift hypothesis predicts that the slope will be zero. There is no need to estimate the character states of ancestors. Nor do we need to infer the precise optimal trait value for polar bears or for any other species. And the curvature of the fitness function around the optimum, the heritability, and the effective population size do not matter, either.[21]

This shift to a new *explanandum* and to a formulation of the adaptive hypothesis in which no optimal trait values are postulated has more to be said for it than that the hypothesis becomes easier to test. The hypotheses of selection and drift each purport to describe the *causal processes* that produced our present observations. The point of importance is that *causation is difference-making.* To see whether smoking is a difference-maker in the production of lung cancer, we need to see whether smokers get cancer more frequently than nonsmokers who are otherwise the same.[22] The

---

[21] The shift in *explananda* provides another advantage. When SPD and PD are assessed relative to the single trait value that a single species exhibits, there is an arbitrariness that enters into setting the upper limit the trait value can have. We assumed that polar-bear fur length must fall between 0 and 100 centimeters, but why not say 0 and 11 centimeters, or 0 and 1,000 centimeters?

[22] I add the phrase "otherwise the same," since we need to control for other potential causal factors; failing to do this risks conflating causation and correlation.

relevant data are comparative; the absolute value of the frequency of lung cancer among smokers is not relevant. Fur length is to lung cancer as ambient temperature is to smoking. The selection hypothesis says that ambient temperature causally contributes to longer fur. The claim is that lineages living in colder climates have a higher probability of evolving longer fur (or have a higher expected fur length) than lineages living in warmer climates that are otherwise the same. If this is what the selection hypothesis says, it is perfectly clear why the *absolute* fur length of polar bears is not relevant. What matters is whether they have *longer* fur than bears in warmer climates. Selection hypotheses attempt to identify positive and negative causal factors that influence the trait values of populations. They don't claim that populations are optimal or close to optimal. Nor do they say how probable a given evolutionary outcome is.

I began this chapter by considering why polar bears have fur that is 10 centimeters long and then shifted to the question of why fur length and ambient temperature are negatively associated in a data set covering several bear species. In between this starting and end point, there is a description that we did not pause to consider, one that gives the fur lengths and environmental temperatures for the several species. This detailed enumeration is *logically stronger* than the modest statement that fur length and temperature are positively associated. Have we therefore violated the *principle of total evidence* (§1.3, §2.10)? Not at all. The data you are obliged to consider depend on the hypotheses you wish to test. This simple point was visible in the coin-tossing example discussed in §1.3; if you wish to test $p = 0.25$ against $p = 0.75$ where both hypotheses assume that tosses are independent of each other, the order of heads and tails in the data does not matter; all you need to consider is the frequency. The situation is different if you want to test the hypothesis that tosses are independent against the hypothesis that they are not. You can discard information about the order of heads and tails in the former problem, but not in the latter. The shift from one description of the data to another in testing selection against drift is justified by the character of the hypotheses we wish to test. The selection hypothesis says that the lineages stemming from a common ancestor evolved under the influence of an optimality line with a negative slope; the drift hypothesis says that they evolved from their common ancestor by random walk. The association in the data is sufficient to test the two hypotheses. If adding information about the point values of tip species would change one's judgment as to which of these hypotheses is more likely, then this additional information should be considered. But very often this is not the case.

Thus far, the main thesis of this chapter has been that hypotheses about natural selection should be formulated so that they predict correlations among species, not the absolute trait value of any species.[23] The argument has been somewhat abstract. A less abstract argument for the same conclusion is suggested by the following catalog of examples from *The Origin of Species* (Darwin 1859: 197):

> If green woodpeckers alone had existed, and we did not know that there were many black and pied kinds, I dare say that we should have thought that the green colour was a beautiful adaptation to hide this tree-frequenting bird from its enemies [...] A trailing bamboo in the Malay Archipelago climbs the loftiest trees by the aid of exquisitely constructed hooks clustered around the ends of the branches, and this contrivance, no doubt, is of the highest service to the plant; but [...] we see nearly similar hooks on many trees which are not climbers [...] The naked skin on the head of a vulture is generally looked at as a direct adaptation for wallowing in putridity; and so it may be [...] but we should be very cautious in drawing any such inferences, when we see that the skin on the head of the clean-feeding male turkey is likewise naked. The sutures in the skulls of young mammals have been advanced as a beautiful adaptation for aiding parturition, and no doubt they facilitate, or may be indispensable for this act; but [...] sutures occur in the skulls of young birds and reptiles, which have only to escape from a broken egg.

It takes no great sophistication to recognize that the frequency of cancer among smokers, even if it is high, does not show that smoking causes cancer. It is even more obvious that lung cancer in a single smoker does not suffice to justify the causal claim. Somehow these obvious points seem less so when we think about adaptation, but they are just as true and just as vital.

3.7 MODEL SELECTION AND UNIFICATION[24]

In modern industrial societies, women on average live longer than men. One might suspect that this is a recent phenomenon, a result of improved medical care that reduces the risk of dying in childbirth. The data available suggest otherwise. In eighteenth-century Sweden, women lived longer than men, and this inequality has continued right down to the present, though there has been a steady improvement in the longevities of both sexes. The same is true of the Ache, a hunter-gatherer group now living in Paraguay. Indeed, in twentieth-century societies around the

---

[23] The use of "species" here is unnecessarily restrictive. The point is to look at comparative data that covers a set of objects, not at the trait value of a single object.

[24] This section draws on material in Lang et al. (2002).

world, women almost always have a higher life expectancy than men. Is this fact about human beings to be explained in terms of some constellation of causes that is unique to our species? Or is the pattern of longevity in human beings due to factors that apply to a more inclusive set of organisms? The choice here is between a unified model, in which the difference between human males and females is explained as part of a more general pattern of cross-species sexual dimorphism, and a disunified model, in which each species is furnished with its own special explanation.[25]

Allman et al. (1998) propose a unified model to explain the facts just described concerning human beings. Their hypothesis concerns the evolution of anthropoid primates: when one sex provides more parental care than the other, selection favors reduced mortality in the sex that makes the larger contribution. For example, if females provide more parental care than males, selection on the two sexes should have the result that females take mortality risks less readily than males, because those risks bring with them larger costs if they don't work out; mothers are more in danger of having none of their offspring reach reproductive age than fathers are if they die when their offspring are young. This difference in the selective regimes faced by the two sexes might result in differences between the behaviors of men and women (see §2.2 on Arbuthnot's 1710 remark that men "seek their food with danger"), but it also might take the form of morphological and physiological differences as well. Allman et al. take their hypothesis about selection in the past to predict that, among present-day species, there should be a correlation: the more fathers help rear their offspring, the longer they should live compared with the life spans of mothers. The authors present the data shown in Figure 3.20. It turns out that human beings fall in the middle of this data set, with males and females having fairly similar life spans and males providing a middling level of parental care. The logic behind this test is contrastive. The negative correlation of female/male survival ratio and paternal care is what the selection hypothesis predicts. On the other hand, if parental care and life span were causally independent, we'd expect there to be zero correlation. Notice that the argument does not require the selection hypothesis to describe what the *optimal* survival ratio is for a given amount of male and female parental care. Nor is there any need to

---

[25] There is, of course, a third possibility: That all species except human beings are subject to one causal process, while human beings have evolved by other rules entirely. There are other possibilities as well.

Primate	Female/Male survival ratio	Male care of offspring
Chimpanzees	1.418	Rare or negligible
Spider monkey	1.272	Rare or negligible
Orangutan	1.203	None
Gibbon	1.199	Pair-living, but little direct role
Gorilla	1.125	Protects, plays with offspring
Human (Sweden 1780–1991)	1.052–1.082	Supports economically, some care
Goeldi's monkey	0.974	Both parents carry offspring
Siamang	0.915	Carries offspring in second year
Owl monkey	0.869	Carries infant from birth
Titi monkey	0.828	Carries infant from birth

Figure 3.20 Survival ratios and male care of offspring in anthropoid primates (from Allman et al. 1998).

estimate the character states of the common ancestors that the species in the data set share.

The two hypotheses that Allman et al. consider are both *unified*. Neither makes an exception of human beings or of any other species. But what is wrong with disunified models? Why not view each species as obtaining its trait values by a separate and independent deterministic process? The answer cannot be that disunified models fail to fit the data; quite the contrary: a disunified model can be formulated that fits the data *perfectly*. If such models are defective, the epistemological framework that describes this defect must go beyond likelihoods (since, under standard assumptions, fit-to-data is a measure of likelihood). It is here that a model selection framework is fruitful (§1.7 and §2.19). Instead of evaluating hypotheses in terms of how probable they say the data are, we evaluate them by estimating how accurately they'll predict new data when fitted to old. Suppose that the species in the Allman et al. data set were drawn at random from the anthropoid primates. We fit different models to that data set and then ask how well the fitted models can be expected to predict the survival ratio in a *new* species, given the amount of parental care the two sexes in that species provide. As already noted, model-selection criteria do not automatically favor unified over disunified models; whether they do so depends on the data. If *Homo sapiens* were an *outlier* relative to the other species in the data set, it might make sense to prefer a disunified model, one that treats nonhuman species one way and human beings another. The data, in fact, do not support this, but the possibility cannot be ruled out *a priori*.

In the Allman et al. study, each species is a data point; the project is to characterize and explain variation *among* species, not *within* them. The study therefore does not prejudge whether there is variation in male/female survival ratios among human populations, and, if there is, whether that variation is associated with variation in the amount of paternal care. If there is population variation within our species, this opens the question of whether that variation should be explained in the same way that cross-species variation should be explained. This means that there are two questions about whether the traits of human beings should be explained by the same theories that account for the traits of nonhumans. Models can be doubly unified, singly unified, or not unified at all, as shown in Figure 3.21.[26] Here again, the question cannot be judged *a priori*; the data must be allowed to speak. The framework of model selection provides a useful vehicle for framing this problem.

The pattern of argument exemplified by Allman et al.'s study has wide applicability. Here's another example that can be analyzed in the same way. As human beings age, they tend to sleep less well. Is this due to psychological changes? Koh et al. (2006) find that fruit flies exhibit the same pattern. This result is logically consistent with the disunified hypothesis that postulates that human beings and fruit flies exhibit the same phenotype for different reasons; it also is consistent with the unifying hypothesis that the underlying mechanism is the same. To decide which model is better, there is no need to make it a first principle of scientific reasoning that "to the same natural effects we must, as far as possible, assign the same causes."[27] Model-selection parsimony is not an end in itself; rather, it is a means to an end, that of finding models that make more accurate predictions.

---

[26] Within-species variation, like cross-species variation, could be due to genetic variation, environmental variation, or both; all three possibilities are consistent with the selection hypothesis (Sober 1993b). There need be no commitment to "genetic determinism."

[27] This is Newton's second of four "rules of reasoning in philosophy." See Sober (1988: 51–5) for discussion.

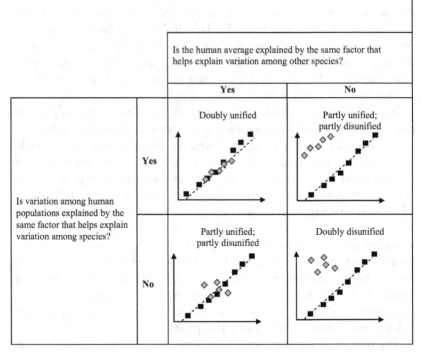

Figure 3.21 Possible explanations of patterns of variation, all for hypothetical data. Each gray diamond represents the average value for a human population. Each black square represents the average value for a nonhuman species. The line is the best-fitting regression line for the non-human species (from Lang et al. 2002).

3.8 REICHENBACH'S PRINCIPLE OF THE COMMON CAUSE

Allman et al.'s study follows an inferential pattern that is typical of many studies in evolutionary biology. A selection hypothesis says that Y evolved as an adaptive response to X. This hypothesis is taken to predict that X and Y should be correlated in data drawn from a set of extant species. In contrast, the hypothesis of causal independence is taken to predict that there will be no correlation. A conclusion is then drawn as to whether the data favor the selection hypothesis. True, the observed correlation does not discriminate between "X causes Y" and "Y causes X"; maybe differential parental care is an adaptive response to differential mortality, not vice versa. And the correlation does not distinguish the hypothesis that X

causes *Y* from the hypothesis that *X* and *Y* are joint effects of a common cause. Still, if the argument merely pits the hypothesis that *X* causes *Y* against the hypothesis that *X* and *Y* are causally independent, it makes sense, and the law of likelihood explains why.[28]

I now want to examine a different approach to this testing problem. It appeals to an idea that Hans Reichenbach (1956) called the *principle of the common cause*. This principle says that if the variables *X* and *Y* are correlated, then either *X* causes *Y*, *Y* causes *X*, or *X* and *Y* are joint effects of a common cause.[29] These three possibilities define what it means for *X* and *Y* to be *causally connected*. Reichenbach's principle has been central to the Bayes net literature in computer science; it is closely connected with the *causal Markov condition* (see Spirtes et al. 2001 and Woodward 2003). Although Reichenbach's principle and the likelihood approach I have taken may seem to be getting at the same thing, I think there is a deep difference. In fact, if the likelihood approach is right, then Reichenbach's principle must be too strong. The likelihood approach does not say that *X* and *Y* *must* be causally connected if they are correlated; it doesn't even say that they *probably* are. The most that the law of likelihood permits one to conclude is that the hypothesis of causal connection is better supported by correlational data than is the hypothesis of causal independence.

To delve deeper into the principle of the common cause, let's begin with an example that Reichenbach used to illustrate it. Consider an acting troupe that travels around the country presenting plays. We follow the company for several years, recording on each day whether the leading man and the leading lady have upset stomachs. This data allow us to see how frequently *each* of them gets sick and how frequently *both* of them get sick. Suppose the following inequality is true:

(1) f(Actor 1 gets sick & Actor 2 gets sick)

        $> f$(Actor 1 gets sick)f(Actor 2 gets sick).

---

[28] The complaint of Leroi et al. (1994) that the comparative method does not get at the causal basis of selection (because it fails to pry apart selection-of from selection-for, on which see Sober 1984) needs to be understood in this light.

[29] Reichenbach additionally believed that when *X* and *Y* are correlated and neither causes the other, not only does there exist a common cause of *X* and *Y*; in addition, if *all* the common causes affecting *X* and *Y* are taken into account, they will *screen off X* from *Y*, meaning that the completely specified common causes will render *X* and *Y* conditionally probabilistically independent of each other. Results in quantum mechanics pertaining to the Bell inequality have led many to question this screening-off requirement (see, for example, Van Fraassen 1982).

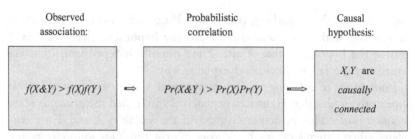

Figure 3.22 Although the principle of the common cause is sometimes described as saying that an "observed correlation" entails a causal connection, it is better to divide the inference into two steps.

Here $f(e)$ means the frequency of days on which the event (type) e occurs. For example, maybe each actor gets sick once every twenty days, but the frequency of days on which both get sick is greater than $\frac{1}{400}$. If this inequality is big enough and we have enough data, our observations will license the inference that the following probabilistic inequality is true:

(2) For each day i, Pr(Actor 1 gets sick on day i &

Actor 2 gets sick on day i)

$>Pr$(Actor 1 gets sick on day i)Pr(Actor 2 gets sick on day i).

It is important to be clear on the difference between the observed association described in (1) and the inferred correlation stated in (2); this distinction was discussed in §2.18 in connection with the inductive sampling formulation of the argument from design. The association is in our data. However, we do not *observe* probabilities; rather, we *infer* them.[30] Once our frequency data permit (2) to be inferred, the principle of the common cause kicks in, concluding that there is a causal connection between one actor's getting sick on a given day and the other's getting sick then too. Perhaps the correlation exists because the two actors eat in the same restaurants; if one of them eats tainted food on a given day, the other probably does too. The two-step inference just described is depicted in Figure 3.22.

How could X and Y be associated in the data without being correlated? Perhaps the sample size is too small. If you toss a pair of coins ten times, it is possible that heads on one will be associated with heads on the other, in

---

[30] In this inference from sample frequencies to probabilities, Bayesians will claim that prior probabilities are needed while frequentists will deny that this is necessary. Set that disagreement aside.

the sense that there is an inequality among the relevant frequencies. But this may just be a fluke; the tosses may in fact be probabilistically independent of each other. One way to see whether this is so is to do a larger experiment. If the association in the ten tosses *is* just a fluke, you expect the association to disappear as sample size is increased.

The principle of the common cause sounds like a sensible idea when it is considered in connection with examples like Reichenbach's acting troupe. But is it *always* true? Quantum mechanics has alerted us to the possibility that X and Y might be correlated without being causally connected; maybe there are stable correlations that are just brute facts. However, it is not necessary to consider the world of micro-physics to find problems for Reichenbach's principle. Yule (1926) described a class of cases in which X and Y are causally independent though probabilistically correlated. A hypothetical example of the kind of situation he had in mind is provided by the positive association of sea levels in Venice and bread prices in Britain over the past 200 years (Sober 2001). Since both have increased monotonically, higher than average values of the one are associated with higher than average values of the other. This association is not due to sampling error; if yearly data were supplemented with monthly data from the same 200 years, the pattern would persist. Nor is this problem for Reichenbach's principle restricted to *time* series data. Variables can be *spatially* rather than *temporally* associated, due to two causally independent processes each leading a variable to monotonically increase across some stretch of terrain. Suppose that bread prices on a certain date in the year 2008 increase along a line that runs from southeast to northwest Europe. And suppose that songbirds on that day are larger in the northwest than they are in the southeast. If so, higher bread prices are spatially associated with larger songbirds. And the association is not a fluke, in that the pattern of association persists with larger sample size. But still, songbird size and bread prices may well be causally independent.

Reichenbach's principle is too strong. The probabilistic correlation between X and Y *may* be due to the fact that X and Y are causally connected. However, to evaluate this possibility, we must consider alternatives. If the alternatives we examine have lower likelihoods, relative to data on observed frequencies, this provides evidence in favor of the hypothesis of causal connection. On the other hand, if we consider an alternative hypothesis that has the same likelihood as the hypothesis of causal connection, then the data do not favor one hypothesis over the other, or so the law of likelihood asserts. There is no iron law of metaphysics that says that a correlation between two variables *must* be due to their being causally connected. Whether this is true in a given case should be evaluated by considering the data and a set of alternative hypotheses, not by appealing to a principle.

Those who accept Reichenbach's principle invariably think that it is useful as well as true. They do not affirm that correlation entails causal connection only to deny that we can ever know that a correlation exists.[31] Reichenbach's treatment of the example of the two actors is entirely typical. The data tell you that there is a correlation, and the correlation tells you that there is a causal connection. This readiness to use Reichenbach's principle to draw causal inferences from observed associations suggests the following argument against the principle. Take a data set that you think amply supports the claim that variables X and Y are probabilistically correlated. If you believe Reichenbach's principle, you are prepared to further conclude that X and Y must be causally connected. But do you really believe that the data in front of you could not possibly have been produced without X and Y being causally connected? For example, take Allman et al.'s data set (§3.7). Surely it is *not impossible* that each primate species (both those in the data set and those not included) came to its values for X and Y by its own special suite of causal processes. I do not say that this is true or even plausible, only that it is possible. This is enough to show that Reichenbach's principle is too strong.

Although the example I have considered to make my argument against Reichenbach's principle involves a data set in which two variables monotonically increase with time, the same point holds for a data set in which the variables each rise and fall irregularly but in seeming synchrony. If a common cause model is plausible in this case, this is not because a Reichenbachian principle says that it must be true. Rather, its credentials need to be established within a contrastive inferential framework, whether the governing principle is the law of likelihood or a model selection criterion like AIC. For a monotonic data set, a fairly simple common-cause model and a somewhat more complex separate-cause model each fit the data well, in which case the former will have a slightly better AIC score than the latter. When the data set is a lot more complex, a common-cause model that achieves good fit will have far fewer adjustable parameters than a separate-cause model that does the same, in which case the difference in their AIC scores will be more substantial. It does not much strain our

[31] It is tempting to argue that Venetian sea levels and British bread prices really aren't correlated because if enough data were drawn from times *outside* the 200-years period, the association would disappear. Well, maybe it would, but so what? Why must real correlations be temporally (and spatially) unbounded?

credulity to imagine that the steady rise in British bread prices and Venetian sea levels is due to separate causes acting on each; the strain may be more daunting for two time series that have lots of synchronous and irregular wiggles. But this difference is a matter of degree and the relevant inferential principles are the same. Strong metaphysics needs to be replaced by more modest epistemology.[32]

3.9 TESTING SELECTION AGAINST DRIFT WITH MOLECULAR DATA

The shift from the task of explaining a single trait value in a single species (§3.1–§3.5) to that of explaining a correlation that exists across species (§3.6) renders the problem of testing selection against drift more tractable. In the former case, you need to know the location of the optimal trait value towards which selection, if it occurs, will push the lineage. In the latter, all that is needed is information about the slope of the optimality line. Instead of needing to know what the optimal fur length is for the polar bear lineage, it suffices to know that, if selection acts on fur length, bears in cold climates have a longer optimal fur length than bears in warm.

A great deal of work in population genetics attempts to get by with even less. Geneticists often test selection against drift by comparing DNA sequences drawn from different species; a number of statistical tests have been constructed for doing this (see Page and Holmes 1998, Kreitman 2000, and Nielsen 2005 for reviews). Scientists carry out these tests with little or no information about the roles that different parts of these sequences play in the construction of an organism's phenotype. If these tests are sound, they require no assumptions concerning what the optimal sequence configuration would be; in fact, they don't even require assumptions concerning how the optimum in one species is related to the optimum in another. If selection can be tested without this type of information, what does the hypothesis of natural selection predict about what we observe?

The parallel question about what drift predicts is easier to answer. The predictions are probabilistic, not deductive. They take the following form: If the process is one of pure drift, then the probability of this or that observable result is such-and-such. If the observations turn out to deviate

---

[32] Hoover (2003) proposes a patch for Reichenbach's principle that introduces considerations concerning stationarity and cointegration, but his proposal is still too strong; it isn't true that a data set that satisfies his requirements *must* be due to the two variables' being causally connected. In addition, Hoover's reformulation makes no recommendations concerning some data sets that in fact do favor a common cause over a separate cause model.

from what the drift hypothesis leads you to expect, should you reject it? If you should, and if selection and drift are the only two alternatives, selection has been "tested" by standing idly on the sidelines and witnessing the refutation of its one and only rival. If probabilistic *modus tollens* (§1.4) made sense, this would be fine. But it does not. For selection and drift to be tested against each other, *both* must make predictions. This is harder to achieve for selection than it is for drift, since drift is a null hypothesis (predicting that there should be zero difference between various quantities; see below) whereas selection is a composite hypothesis (predicting a difference but leaving open what its magnitude should be).

A central prediction of the neutral theory of molecular evolution is that there should be a *molecular clock* (Kimura 1983). In a diploid population containing N individuals, there are $2N$ nucleotides at a given site. If each of those nucleotides has a probability μ of mutating in a given amount of time (e.g., a year), and a mutated nucleotide has a probability u of evolving from mutation frequency to fixation (i.e., 100 percent representation in the population), then the expected rate of substitution (i.e., the origination and fixation of new mutations) at that site in that population will be

$$k = 2N\mu u.$$

This covers *all* the mutations that might occur, regardless of whether they are advantageous, neutral, or deleterious. Because μ and u are probabilities, k isn't the de-facto rate of substitution; rather, it is a probabilistic quantity – an expected value (§1.4). If the $2N$ nucleotides found at a site at a given time are equal in fitness, the initial probability that each has of eventually reaching fixation is

$$u = \frac{1}{2N}.$$

I say that this is the "initial" probability since the probability of fixation itself evolves.[33] If we substitute $1/2N$ for u in the first equation, we obtain one of the most fundamental propositions of the neutral theory:

$$(\text{Neutrality}) \quad k = \mu.$$

---

[33] This equality pertains to the $2N$ *token* nucleotides present at the start of the process; some of those tokens may be of the same *type*. It follows from the above equality that the initial probability that a type of nucleotide found at time t will eventually reach fixation is its frequency at time t. This point applies to phenotypic drift models as well as genetic ones; see the squashing of the bell curve depicted in Figure 3.4.

The expected rate of substitution at a site is given by the mutation rate if the site is evolving by drift. Notice that the population size N has cancelled out.

What will happen if a mutation is advantageous? Its probability of fixation (u) depends on the selection coefficient (s), on the effective population size N_e, and on the population's census size N:

$$u = \frac{2sN_e}{N}.\text{[34]}$$

Substituting this value for u into the first equation displayed above, we obtain the expected rate of evolution at a site that experiences positive selection:

$$(Selection)\ k = 4N_e s\mu.$$

If the probability of mutation per unit time at each site remains constant through time (though different sites may have different mutation probabilities), the neutral theory predicts that the expected overall rate of evolution in the lineage does not change. This is the clock hypothesis. It doesn't mean that the *actual* rate never changes; there can be fluctuations around the mean (expected) value. The selection hypothesis is more complicated. If each site's value for $N_e s\mu$ holds constant through time, (Selection) also entails the clock hypothesis. But there is every reason to expect this quantity to fluctuate. After all, N_e is a quantity that reflects the breeding structure as well as the census size of the population (Crow and Kimura 1970) whereas s, the selection coefficient, reflects the ecological relationship that obtains between a nucleotide in an organism and the environment. With both these quantities subject to fluctuation, it would be a miracle if their product remained unchanged. This is why (Selection) is taken to predict that there is no molecular clock.[35]

If we could trace a single lineage through time, taking molecular snapshots on several occasions, it would be easy to test (Neutrality) against (Selection). Although this procedure can be carried out on populations of rapidly reproducing organisms, it isn't feasible with respect to lineages at longer time scales. It is here that the fact of common ancestry comes to the rescue, just as it did in §3.6. We do not need a time

[34] This useful approximation is strictly correct only for small s and large N.
[35] The simple selection model described here does not predict a clock, but more complicated selection models sometimes do. Discussion of these would take us too far afield.

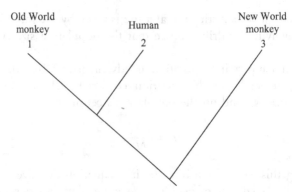

Figure 3.23 Given the phylogeny, the neutral theory entails that the expected difference between 1 and 3 equals the expected difference between 2 and 3 (figure from Page and Holmes 1998: 255).

machine that allows us to travel into the past so that we can observe earlier states of a lineage we now see in the present; rather, we can look at three or more tips in a phylogenetic tree and perform a *relative rates test*. Figure 3.23 provides an independently justified phylogeny of human beings, Old World monkeys, and New World monkeys. We can gather sequence data from these three taxa and observe how many differences there are between each pair. The neutral hypothesis predicts that $(d_{13} - d_{23}) = 0$ (or, more precisely, it entails that the expected value of this difference is zero; here d_{ij} is the number of differences between i and j. We don't need to know how many changes occurred in *each* lineage; it suffices to know how many changes separate one extant group from another. Looking at the present tells you what *must* have occurred in the past, given the fact of common ancestry.

Li et al. (1987) carried out this test and discovered that d_{13} is significantly greater than d_{23}; they didn't look at whole genomes but at a sample of synonymous sites, introns, flanking regions, and a pseudo-gene, totaling about 9,700 base pairs. With respect to these parts of the genome, human beings diverged more slowly than Old World monkeys from their most recent common ancestor (see also Li 1993). This result was taken to favor (Selection) over (Neutrality). In view of the negative comments I made about Neyman–Pearson hypothesis testing in Chapter 1, I want to examine the logic behind this analysis more carefully. Using relative rates to test drift against selection resembles using two sample means to test whether two fields of corn have the same mean height. The null hypothesis says that they are the same; the alternative to the null says that they differ, but it does not say by how much. The Neyman–Pearson theory conceives of the testing problem in terms of acceptance and rejection and requires that one stipulate an arbitrary level for a, the probability of a Type-1 error. I suggested in Chapter 1 that it makes more sense to place this problem in a model-selection framework. In the relative-rates test, the drift hypothesis has no adjustable parameters whereas the selection hypothesis has one. The question is not which hypothesis to reject but which can be expected to be more predictively accurate. The AIC score of the selection hypothesis is found by determining the maximum likelihood value of a single parameter θ, the expected value of $(d_{13} - d_{23})$, taking the log-likelihood when θ is assigned its maximum likelihood value and subtracting the penalty for complexity. The question is whether the selection hypothesis' better fit to data suffices to compensate for its greater complexity.

Bayesians and likelihoodists come at the problem differently. Their framework obliges them to compute the *average likelihood* of the selection hypothesis, which, as noted, is composite. If selection acted on these different parts of the genome, how much should we expect d_{13} and d_{23} to differ? We would need to answer this question without looking at the data. And since different types of selection predict different values for $(d_{13} - d_{23})$, our answer would have to average over these different possibilities. It isn't impossible that empirical information should one day provide a real answer to this question. However, at present, there is no objective basis for producing an answer. I suggest that the model-selection approach is more defensible than both Bayesianism and Neyman–Pearson hypothesis testing as a tool for structuring the relative rate test.

The role played by the fact of common ancestry in facilitating tests of process hypotheses can be seen in another context. If we could look at a large number of replicate populations that all begin in the same state, we could see if the variation among the end states of those lineages is closer to the predictions of neutrality or selection. But why think that each of these lineages begins in the same state? The answer is simple: *If they share a common ancestor, they must have.* The neutral theory predicts that the tips of a tree should vary according to a Poisson distribution. A number of mammalian proteins (e.g., Hemoglobin a and β, Cytochrome c, Myoglobin) were found to be "over dispersed" (Kimura 1983; Gillespie 1986), and this was taken to be evidence of selection. Once again, neutrality is a null hypothesis, and selection is composite.

Like the relative rate test, the McDonald–Kreitman test also relies on an independently justified phylogeny, and there is no optimality line in sight. McDonald and Kreitman (1991) examined sequences from three species of *Drosophila* that all play a role in constructing the protein alcohol dehydrogenase (*Adh*). Fruit flies often eat fruit that is fermented, and they need to break down the alcohol (human beings have the same problem and solve it by way of a different version of the same protein). So it seems obvious that the protein is adaptive. What is less obvious is whether variations in the gene sequences that code for the protein are adaptive or neutral. Perhaps different species find it useful to have different versions of the protein, and selection has caused these species to diverge from each other. And different local populations that belong to the same species may also have encountered different environments that select for different sequences. Alternatively, the variation may be neutral.

The McDonald–Kreitman test compares the synonymous and non-synonymous differences that are found both in different populations of the same species and in different species. A substitution in a codon is said to be synonymous when it does not affect the amino acid that results. For example, the codons UUU and UUC both produce the amino acid phenylalanine, while CUU, CUC, CUA and CUG all produce leucine. It might seem obvious that synonymous substitutions must be caused by neutral evolution, since they do not affect which amino acids and proteins are constructed downstream. This would suggest that the hypothesis that nonsynonymous substitutions evolve neutrally can be tested by seeing if the rates of synonymous and nonsynonymous substitutions are the same. However, it is possible that synonymous substitutions might not be neutral, owing, for example, to differences in secondary structure, for example, having to do with the stability of the molecules (Page and Holmes 1998: 243). What seems safer is the inference that the *ratio* of the rates of synonymous to nonsynonymous substitutions should be a constant if there is neutral evolution. Both should depend just on the mutation rate, as discussed above. This ratio should have the same value regardless of whether the sequences compared come from two populations of the same species or from different species.

Figure 3.24 describes the four kinds of observations that McDonald and Kreitman assembled. They counted the number of nonsynonymous and synonymous differences that separate different populations of the same species; these are called *polymorphisms*. They also counted the number of synonymous and nonsynonymous *fixed differences* that separate pairs of species; these are sites that are monomorphic within each species

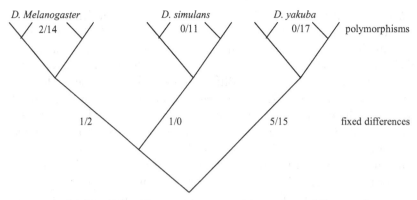

Figure 3.24 The number of nonsynonymous and synonymous differences that exist within and between three Drosophila species at the Adh locus. The within-species differences are called polymorphisms; the between-species differences are called fixed. Data from McDonald and Kreitman (1991); figure from Page and Holmes (1998: 267).

but vary between them. Here are the numbers of differences that McDonald and Kreitman found in the four categories:

	Fixed	Polymorphic
Synonymous	17	42
Nonsynonymous	7	2

If these sequences evolved neutrally, the ratio of synonymous to non-synonymous substitutions for the first column should be about the same as the ratio for the second. But they aren't close:

$$\frac{17}{7} \ll \frac{42}{2}.$$

There is an excess of nonsynonymous fixed differences (or a deficiency of polymorphic nonsynonymous substitutions). McDonald and Kreitman took this departure from the prediction of neutrality to be evidence for selection: Selection had reduced the within-species variation and amplified the between-species variation at nonsynonymous sites. They note that a population bottleneck could also explain the data but argue that the known history of *Drosophila* makes this alternative implausible.

The same epistemological questions arise in connection with the McDonald–Kreitman test that I raised about the relative rates test. The inference should not be thought of as an instance of probabilistic *modus tollens*. How, then, should it be conceived? McDonald and Kreitman used a *G* test, which is part of the Neyman–Pearson framework. The neutrality hypothesis once again is a null hypothesis, asserting that two ratios (or their expected values) are the same. One rejects the null and accepts the alternative to the null when the difference in the observed ratios deviates sufficiently from zero. A Bayesian treatment of this problem would need to answer the question of how much difference one would expect in the two ratios if selection were at work. This is the same type of unanswerable question I discussed in connection with the relative rate test. A third alternative is to reformulate the McDonald–Kreitman test in a model-selection framework. The selection model contains more adjustable parameters than the neutrality model; the former will, therefore, fit the data better. The question is whether the greater fit of the selection model suffices to compensate for its greater complexity.

Although the relative rate test is usually applied to three species or higher taxonomic groups while the McDonald–Kreitman test is usually applied to cases in which there is within- and between-species variation, the relationship between the two tests can be examined by seeing how they apply to the hypothetical example shown in Figure 3.25. Here *W* and *X* are local populations that belong to one species, and *Y* and *Z* are populations that belong to another. The relative rate test, applied to *(WX)Z* or to *W(YZ)*, checks whether the two in-group branches exhibit the same

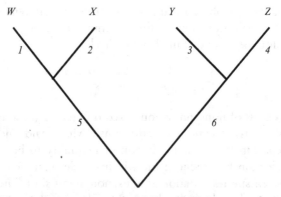

Figure 3.25 The relative rate test and the McDonald–Kreitman test focus on different events in this tree. The first tests neutrality's prediction that, in expectation, $s_1 + n_1 = s_2 + n_2$ and that $s_3 + n_3 = s_4 + n_4$; the second tests the neutral prediction that $(s_1 + s_2 + s_3 + s_4)/(n_1 + n_2 + n_3 + n_4) = (s_5 + s_6)/(n_5 + n_6)$, where s_i and n_i are, respectively, the number of synonymous and nonsynonymous changes on branch i.

total number of changes (the sum of synonymous and nonsynonymous changes). The test concerns just the time slice in which there is branching within the in-group. The McDonald–Kreitman test both separates synonymous from nonsynonymous changes and compares two different time slices: the polymorphisms that arose in branches 1–4 and the fixed differences that arose in branches 5 and 6. In deriving what the neutrality hypothesis predicts in both tests, I exploited the fact that *if* each site's mutation rate is constant through time, then the neutrality hypothesis has certain consequences (e.g., that there will be a molecular clock). This does not mean that the validity of the tests *depends* on each site's having a constant mutation rate. For example, if all mutation probabilities in all lineages changed uniformly with time, the relative rate test would still make sense. What this test requires is that the mutation rates in the two in-group branches will be about the same if there is neutral evolution; if there is neutral evolution but the in-group branches have very different average mutation probabilities, the data will probably lead to the erroneous conclusion that this is evidence against neutrality. A similar *caveat* applies to the McDonald–Kreitman test.

In the tests described, the selection hypothesis is rather unspecific; it predicts certain observational outcomes (that two distances will differ, that the tips in a tree will be over-dispersed, or that two ratios will differ), but it has little content beyond that. There obviously is more to the *process* of natural selection than these meager hypotheses capture; this is something that geneticists recognize when they note that *other* processes could in principle generate the outcomes that a test judges to be evidence favoring selection over drift. As more substantive molecular models of selection are developed, molecular tests of drift against selection will be sharpened (Kreitman 2000).

3.10 SELECTION VERSUS PHYLOGENETIC INERTIA[36]

I have focused so far on testing selection against drift. I now want to explore a second competitor to natural selection that biologists have considered. It is called *phylogenetic inertia*. To see what this idea involves, consider the following remark by Roger Lewin:

Why do most land vertebrates have four legs? The seemingly obvious answer is that this arrangement is the optimal design. This response would ignore, however, the

---

[36] The material in this section is drawn from Orzack and Sober (2001).

fact that the fish that were ancestral to terrestrial animals also have four limbs, or fins. Four limbs may be very suitable for locomotion on dry land, but the real reason that terrestrial animals have this arrangement is because their evolutionary predecessors possessed the same pattern. (Lewin 1980: 886)

There are two thoughts that this passage suggests. The first is a simple chronological point: Since the aquatic ancestors of land vertebrates already had four limbs, it is false that the trait initially became common in the lineage because of its utility for walking on dry land (Eaton 1960; Edwards 1989). This is no more controversial than the thought that cause must precede effect. However, once this point is granted, there is a second and more contentious thesis with which to reckon. This is the claim that the correct explanation for why land vertebrates are tetrapods consists in the fact that their ancestors had four limbs; it is incorrect to maintain that the trait remained in place because there was selection for the trait due to the fact that it facilitated walking on dry land. Selection for the ability to walk on dry land of course does not explain the *initial evolution* of the tetrapod morphology; the question is whether we also should reject the thesis that selection for walking was responsible for the trait's *subsequent maintenance*. The term "phylogenetic inertia" (Wilson 1975; Harvey and Pagel 1991) is sometimes used to refer to the explanation that Lewin favors, which might better be called *ancestral influence*.[37] The hypothesis of phylogenetic inertia and the hypothesis of stabilizing selection propose to explain the character state of a descendant in different ways; the former appeals to the lineage's ancestral state while the latter cites processes that the lineage subsequently experienced.

Why are these two possible explanations in conflict? Can't phylogenetic inertia *and* stabilizing selection both help explain why land vertebrates now have four limbs? This is not the position that Lewin takes in the quoted passage. Phylogenetic inertia is said to be "the real reason;" not only is selection not the *whole* story; it isn't even *part* of the story. Lewin seems to think that the hypothesis of phylogenetic inertia should be regarded as innocent until proven guilty; our default assumption should be that ancestral influence is the right explanation unless the data force us

---

[37] The word "inertia" misleadingly suggests that lineages have a tendency to continue evolving in a certain direction even after the initial "push" that got them started is no longer present. For example, if selection initially favors the evolution of longer fur in polar bears and evolution is "inertial," then fur length will continue to increase even if there ceases to be selection for longer fur. At the start of the twentieth century, the orthogenetic theory of evolution held that inertia in this sense explains why the Irish elk had such enormous horns (Gould 1977). This view of evolution is dubious, but inertia in this sense is not what "phylogenetic inertia" is now taken to mean.

to abandon that hypothesis.[38] If what we observe is consistent with the hypothesis of phylogenetic inertia and also with the hypothesis of stabilizing selection, we should prefer the former. Perhaps considerations of parsimony also suggest that we should prefer the one-factor inertia explanation over a pluralistic explanation that cites both inertia and selection.

Other evolutionary biologists have espoused other principles of default reasoning. George C. Williams (1966), in his influential book *Adaptation and Natural Selection*, asserts that adaptation is an "onerous concept" that should be embraced only if the data force one to do so. Ernst Mayr (1988: 150–1) takes the opposite stance – only after all possible selection explanations of a given trait have been explored and rejected can one tentatively conclude that the trait is a product of drift. Default principles also have been defended that give precedence to some types of natural selection over others. Williams maintains that the hypothesis of group selection is more onerous (by which he means *less parsimonious*) than the hypothesis of individual selection; our default assumption should be that when a trait evolves by natural selection, it evolves because it is advantageous to the individuals who possess it, not because it helps the groups in which it occurs.[39] Williams' principle, like Lewin's, says that we should accept one hypothesis rather than another even when the data fail to discriminate between them. These principles do not say that one should suspend judgment and remain agnostic.

The framework of model-selection theory (§1.7) throws light on these principles. Model-selection criteria such as AIC give weight to how parsimonious a model is (where parsimony is measured by how few adjustable parameters the model contains). This means that a model that postulates both individual *and* group selection will score worse than a model that postulates just individual selection (or one that postulates just group selection) *if* the two-factor model and the single-factor model fit the data about equally well. Notice the *if* in the preceding statement: model-selection theory does not permit one to ignore the data and to embrace the more parsimonious model just because it is more parsimonious. Notice also that model selection provides no basis for preferring a model that postulates just individual selection over one that postulates just group selection if the two models have the same number of adjustable parameters. And they do: individual selection is represented formally by the variance in fitness that exists within groups, and group selection is

---

[38] Ridley (1983) also recommends this policy.

[39] See Sober and Wilson (1998) for discussion of this and other arguments against group selection.

represented by the between-group variance in fitness. A single parameter does the representing in each case, so there is no difference in parsimony here.

Similar conclusions apply to the task of comparing phylogenetic inertia and stabilizing selection as possible explanations of why land vertebrates have four limbs. A model that postulates both phylogenetic inertia and stabilizing selection will score worse than a model that postulates just one of these possible causes *if* the two models are about equal in goodness of fit. But how do the two single-factor explanations that Lewin considers compare to each other? Why is a model that postulates phylogenetic inertia better than one that postulates stabilizing selection? Does the phylogenetic inertia model have fewer parameters? This question raises another that is more fundamental: What *are* the parameters that go into representing phylogenetic inertia? We also need to consider the fact that statisticians decline to use AIC and other model-selection criteria to evaluate a model unless there are considerably more items of data than there are adjustable parameters. But Lewin mentions just two observations: land vertebrates have four limbs, and their fish ancestors did too. Is this enough data? It might be replied that each species in these groups is a separate observation, and so the data set is in fact very large. However, these observations are not independent of each other, owing to the fact that the species are phylogenetically related (Felsenstein 1985).

In §3.5 I described how a simple Markov model can be used to represent the evolution of a dichotomous character subject to either drift or natural selection. This model also provides a natural representation of phylogenetic inertia. Consider a lineage that connects an ancestor A to a descendant D; suppose we observe that the descendant D is in state 1 ($D = 1$, for short). According to the Markov model, this outcome would have been more probable if the ancestor A had been in state 1 than if A had been in state 0. The Markov model represents the idea of ancestral influence because the *backwards inequality* is a consequence of this model (so long as ancestor and descendant are separated by a finite amount of time):

$$Pr(D = 1 \mid A = 1) > Pr(D = 1 \mid A = 0).$$

For any fixed pair of values for u and v (the two instantaneous probabilities of change) in the Markov model, the difference between these two probabilities is greatest when the time separating ancestor and descendant is small; the difference asymptotes to zero as the temporal separation grows larger. A current species is more influenced by its recent ancestors than it is by its ancient ancestors.

		Process at work in lineage	
		Selection for 1	Drift
	$A = 1$	p_1	p_2
State of the ancestor A	$A = 0$	p_3	p_4

Figure 3.26 Selection for character state 1 raises the probability that the descendant D will exhibit that character state ($p_1 > p_2$ and $p_3 > p_4$). According to the hypothesis of phylogenetic inertia, the probability that $D = 1$ increases if the ancestor was in character state 1 ($p_1 > p_3$ and $p_2 > p_4$). Cell entries are all of the form $Pr(D = 1 \mid -)$.

In §3.6, I described selection hypotheses by analogy with the impact of smoking on lung cancer. Smoking doesn't insure that you'll get lung cancer; in fact, smoking can be a positive causal factor in the production of lung cancer even if smoking doesn't make lung cancer more probable than not. Rather, the positive causal role played by smoking is that smoking *raises the probability* of lung cancer.[40] Similarly, there being selection in a lineage for character state 1 doesn't insure that the lineage will exhibit that trait value; and selection for character state 1 doesn't necessarily make the evolution of that character state more probable than not. Rather, selection for trait 1 *raises the probability* that the population will come to exhibit that trait. This fact about how selection ought to be understood is depicted in Figure 3.26. The same point holds with respect to the hypothesis of phylogenetic inertia. For phylogenetic inertia to be part of the explanation of why land vertebrates have four limbs, all that is required is that $A = 1$ raised the probability that $D = 1$. It isn't essential that the tetrapod morphology in the fish ancestors of land vertebrates guaranteed that land vertebrates would have four limbs; it isn't even required that the ancestral condition render that state of the descendant more probable than not. In summary, the hypothesis of natural selection entails the two horizontal inequalities shown in Figure 3.26; the hypothesis of phylogenetic inertia entails the two vertical inequalities.

The Markov model for a dichotomous trait treats inertia and selection a bit differently. There is nothing intrinsic to Markov modeling that requires that there be selection for trait 1; it is possible that $u > v$, but it also is possible that $u = v$ (drift) and that $u < v$ (selection for trait 0). However, as already noted, the Markov model entails the backwards inequality; inertia is therefore an inevitable consequence of Markov processes for dichotomous characters, but selection is not. This does not mean that hypotheses of phylogenetic inertia must be true; after all, the

---

[40] See below for a refinement of this idea that is needed if causation and correlation are to be distinct.

		Smoking	
		Yes	No
	Yes	q_1	q_2
Asbestos exposure	No	q_3	q_4

Figure 3.27 If smoking causally contributes to lung cancer, smoking should raise the probability of lung cancer for people who have the same degree of asbestos exposure ($q_1 > q_2$ and $q_3 > q_4$). If asbestos exposure causes lung cancer, asbestos should raise the probability of lung cancer for people who are alike with respect to how much they smoke ($q_1 > q_3$ and $q_2 > q_4$).[41] Cell entries are all of the form Pr(lung cancer | –).

Markov model might not be correct. The Markov model helps express what inertia means. But inertia needs to be tested just as much as selection does.[42]

The analogy with smoking and cancer provides a clue as to how the test of selection against inertia should be structured. Suppose we didn't already know that smoking and asbestos exposure each increase the risk of lung cancer, and we wanted to test these two causal hypotheses by examining the frequencies of lung cancer in different groups of people. The hypothesis that smoking is a cause predicts that smokers should get cancer more frequently than nonsmokers *who are otherwise the same*. The italicized rider is needed to guard against the possibility of misleading correlations. For example, it might turn out that smokers get cancer more frequently than nonsmokers, not because cigarette smoke is carcinogenic but because smokers tend to have some other property X far more frequently than nonsmokers do, and it is X that promotes lung cancer. This is why testing causal hypotheses by looking at frequency data requires *controlled comparisons* (Harvey and Pagel 1991: 37). To test whether smoking causes lung cancer, we must control for the presence of other possible causes. The same point holds for testing whether asbestos exposure causes cancer. Testing each of these causal hypotheses requires controlling for the other. The two causal claims are depicted in Figure 3.27.

---

[41] If smoking raises the probability of cancer among those who are not exposed to asbestos, but lowers it among those who are exposed, I'm inclined to say that smoking causes cancer in the one subpopulation but prevents it in the other. Can one also say that smoking causes cancer in the whole population (meaning that smoking is *sometimes* a positive causal factor in the population)? If this is acceptable, then smoking can be both a promoter and a preventer of cancer in the same population. I do not object to this mode of description. In any event, there is no need to take a stand on this question about "contextual unanimity" to see the point of controlled comparisons.

[42] Quantitative traits and traits with three or more characters need not obey a backwards inequality; as noted in §3.3, $Pr(D = x | A = x) > Pr(D = x | A = y)$, for all $x \neq y$, when a lineage is subject to drift, but the inequality isn't true when the lineage experiences directional selection that pushes towards an optimum other than x. Still, ancestral influence is built into the Ornstein–Uhlenbeck process in a way that selection is not; where a lineage begins *always* influences where the lineage ends, provided that the lineage has finite duration.

To test the smoking hypothesis, we need to see whether smokers get lung cancer more often than nonsmokers, when the two groups are alike with respect to asbestos exposure. And to test the asbestos hypothesis, we need to see whether people exposed to asbestos get lung cancer more often than people who were not exposed, when the two groups are alike with respect to whether they smoke. Thus construed, the smoking and the asbestos hypothesis are compatible with each other; each should be viewed as competing with a hypothesis that says that lung cancer is causally independent of the factor in question.

The selection and inertia hypotheses both make predictions about the frequencies of different kinds of events in different lineages. Let's consider this in connection with our running example about fur length in bears. Fur can be either "long" or "short" (I here set aside the qualms about dichotomous characters described in §3.5 for the sake of a simple example) and climate can be either "cold" or "warm." The selection hypothesis says that a cold climate produces selection favoring long fur and that a warm climate produces selection for short fur. The inertia hypothesis says that a descendant has a higher probability of having long fur if its ancestor had long fur than if its ancestor had short fur. Each of these hypotheses must be tested by controlling for the other. To test the inertia hypothesis, we must compare lineages that experience the same selective regime. And to test the selection hypothesis, we must compare lineages that started in the same ancestral state. These two protocols are depicted in Figure 3.28. The inertia hypothesis predicts that descendants should have long fur in lineages of type (1) more frequently than they

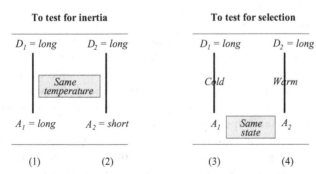

Figure 3.28 To test for phylogenetic inertia, lineages alike in their selective regimes must be compared. To test the selection hypothesis, lineages alike in their ancestral state must be compared.

have long fur in lineages of type (2); the selection hypothesis predicts that descendants should have long fur in lineages of type (3) more frequently than they have long fur in lineages of type (4).

I hope the analogy with smoking and asbestos exposure makes it clear why neither the inertia hypothesis nor the selection hypothesis should be regarded as innocent until proven guilty. Neither deserves to be treated as a default assumption. Rather, both are causal hypotheses, and both need to be tested. Frequency data may indicate that both inertia and selection have influenced the trait values of descendants, that only one of them has, or that neither of them has. Despite our *a priori* expectations, it may turn out that the frequency with which ancestors with long fur have descendants with long fur is no greater than the frequency with which ancestors with short fur have descendants with long fur. And our *a priori* expectation that fur length is an adaptive response to ambient temperature may fail to be borne out when we look at the frequencies of bears with long fur in different climates. The idea that inertia and stabilizing selection need to be tested on a level playing field also shows what is wrong with the following line of reasoning: "we know that inertia influenced the traits of descendants, so parsimony tells us to conclude that stabilizing selection did not." This is no more defensible than its mirror image: "we know that stabilizing selection influenced the traits of descendants, so parsimony tells us to conclude that inertia did not." Parsimony does not provide a justification for ignoring the data. Imagine oncologists reasoning this way about whether smoking and asbestos exposure cause lung cancer.

This way of framing the relationship of inertia and selection also shows why nothing much can be said about a data set consisting of a single lineage in which ancestor and descendant are in the same character state. This may be due to stabilizing selection, or to inertia, or to both, or to neither. This $n = 1$ data set needs to be augmented with information about other lineages. However, the situation is not improved if the new lineages considered are carbon copies of the first. Since causes are difference makers, causal hypotheses need to be tested by using frequency data in which there is variation (§3.6). A principle of default reasoning is no substitute for an impoverished data set.

Just as stasis in a lineage is no proof that inertia has played a role, so change in a lineage is no proof that it has not (Wake et al. 1983; Hansen 1997). To see this, consider a quantitative character, like fur length, and a set of bear lineages in which lineages that experienced the same temperature show a correlation between the fur length of descendants and

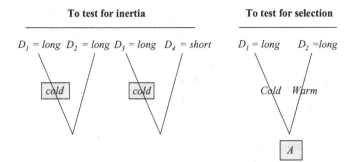

To test for inertia　　　　**To test for selection**

Figure 3.29　The fact that species have common ancestors permits phylogenetic inertia and selection to each be tested by means of controlled comparisons without estimating ancestral trait values.

the fur length of ancestors. Data of this sort justify an inference about expected values:

E(fur length of D | temperature x & $A = m$)

$> E$(fur length of D | temperature x & $A = n$), for all $m > n$.

Descendants can bear the imprint of their ancestors even if descendants manage to evolve away from ancestral trait values.

To test inertia and selection by applying the protocols depicted in Figure 3.28, the character states of ancestors must be known, a problem we considered in §3.3. The fact that species have common ancestors provides a solution to this problem, just as it did earlier in this chapter with respect to others. Figure 3.29 shows how each hypothesis can be tested while controlling for the other without any knowledge of the character states of ancestors. The fact of common ancestry makes this possible. Consider the test for selection shown in Figure 3.29. Since the descendants D_1 and D_2 have a common ancestor A, both lineages start with the same character state. The fact of common ancestry thus allows one to control for the possibility of phylogenetic inertia. If many such pairs of lineages are examined, and lineages that experience cold climates end up with long fur more often than sister lineages that experience warm climates, this is evidence for the selection hypothesis.[43] The same logic applies to testing an inertia hypothesis. Here we have four species. D_1 and D_3 experience the same selection pressure, but they differ in the character

[43] This is an instance of the sign test proposed by Burt (1988), discussed in §3.6.

states their sisters (D_2 and D_4, respectively) possess. If inertia influences the character states of D_1 and D_3, D_1 has a higher probability of exhibiting long fur than D_3 has. By looking at many such foursomes, one can ascertain whether species with long-furred sisters have long fur more often than species with short-furred sisters even though they experience the same selective regime. If there is ancestral influence, siblings should resemble each other.

Using the fact of common ancestry as a device for controlled testing of the inertia and selection hypotheses throws light on a method developed by Cheverud et al. (1985) in which one determines the degree to which a species' trait value is predicted by the trait values of its relatives. High predictability is said to reflect a strong effect of inertia. Residual differences are said to represent the effect of some other process, possibly natural selection. This procedure has much in common with the reasoning implicit in the quotation from Lewin; similarity of ancestor and descendant, or of close relatives, is assumed to indicate inertia, not selection. The method of controlled comparisons rejects the idea that the inertia hypothesis should be viewed as innocent until proven guilty. If relatives have similar trait values, it is impossible to say whether this is due to inertia until one knows how similar the hypothesis of natural selection predicts they should be.

The method of controlled comparisons described here also has implications concerning Felsenstein's (1985) proposal for how phylogenetic information should be taken into account in testing a selection hypothesis. Suppose you observe twenty bear species – ten live in warm climates and have short fur while the other ten live in cold climates and have long fur. These twenty observations might seem to provide strong evidence in favor of the hypothesis that fur length evolved as an adaptive response to ambient temperature. But suppose you then discover that the ten species with short fur are all close relatives and that the ten with long fur are too. Perhaps the first ten resemble each other because of phylogenetic inertia, not because of selection, and maybe the same is true of the second ten. Felsenstein's method of independent contrasts aims to control for the possibility of inertia and thus to provide a phylogenetically sensitive test of selection hypotheses. It is not the purpose of his method to test for phylogenetic inertia. Felsenstein derives his procedure under the assumption that the traits evolve by a Brownian motion process. However, if you want to test the hypothesis that selection causes lineages to evolve towards a stable optimum, the Brownian motion assumption is not appropriate. The assumptions used to test a selection hypothesis against others should be *independent* of which of those hypotheses is true (§2.12,

§3.1); the assumptions should not entail that the selection hypothesis is true, but neither should they entail that it is false.

3.11 THE CHRONOLOGICAL TEST

In the previous section, I identified a simple point only to set it to one side: If the lineage leading to land vertebrates had four limbs before vertebrates came up on dry land, then it is false that the trait initially evolved because it facilitated walking on dry land. It now is time to delve deeper. How do we know that one trait evolved in a lineage before another? If we know such things, then we have a method for testing selection hypotheses that does not require us to consider alternative explanations; the hypothesis that X caused Y is refuted if X occurred after Y. Here is a case in which it is possible to test a causal hypothesis without contrasting it with alternatives.

To know that X evolved before Y *in a lineage* requires more than knowing that X evolved before Y. By dating fossils we can see that some tetrapod fish fossils predate the earliest known fossils of terrestrial vertebrates. With a large enough sample and a big enough temporal gap, it is no great inferential leap to conclude that tetrapods existed before vertebrates appeared on dry land. The extra step comes with saying that tetrapod fish were *ancestors* of land vertebrates. We can't automatically assume that the fossils we observe are ancestors of present-day organisms; they may just be their close relatives (§3.3). It is more defensible to maintain that the fossils we observe provide evidence about the common ancestors they share with extant organisms. If fossil fish and land vertebrates have a common ancestor, the inference is that that common ancestor was a tetrapod. The same can be said of living fish and land vertebrates. The only difference is that fossil fish are temporally closer to that common ancestor than extant fish are, a point I discussed in connection with Figure 3.7.

When two descendants both have trait T, why conclude that their most recent common ancestor also had T? The usual answer is that this hypothesis is more parsimonious. If their most recent common ancestor had T, no change in character state had to occur in the two lineages leading from that ancestor to the two descendants. In contrast, if the ancestor lacked trait T, the trait would have had to originate separately and independently in the two lineages.[44] So, the hypothesis that the most

---

[44] This argument assumes that there is no horizontal gene transfer, an idea I'll discuss in §4.2.

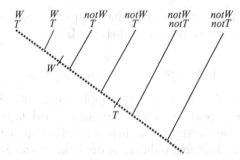

Figure 3.30 When the principle of parsimony is used to reconstruct the character
states of ancestors in this phylogenetic tree, the conclusion is that trait *T* and trait *W*
each evolved once, in the interior branches marked. It follows that trait *T* evolved
before trait *W* in the lineage drawn with a dotted line, which is the lineage leading to
present-day land vertebrates.

recent common ancestor had *T* is more parsimonious than the hypothesis
that the ancestor lacked *T*, where parsimony is measured by counting the
number of changes in character state that must have occurred in lineages
to produce the data we have. The inference that land vertebrates and
modern fish had a tetrapod common ancestor seems to be as compelling
and transparent as the inference that two word-for-word identical books
are copies of the same manuscript and were not written independently.

The chronological test of selection hypotheses relies on this kind of
reasoning. You begin with a phylogenetic tree that shows how extant
organisms are phylogenetically related to each other. Then you use the
principle of phylogenetic parsimony to assign character states to the common ancestors in the tree's interior. You then look at the relevant lineage and
see which traits evolved before which others. Figure 3.30 provides an
example. A parsimonious reconstruction of the character states of ancestors
entails that *W* (for *walking*) evolved in the lineage leading to land vertebrates
after *T* (*tetrapod limbs*) evolved. If so, *T* didn't evolve because it facilitated
W.[45] The chronological test is widely used.[46] For example, Lauder (1996:
75) argues that the presence of fingers in the human hand cannot be linked
to "any specific function that is unique to the human hand; fingers are an

---

[45] Understanding the chronological test requires a distinction between *ontogenetic* and *phylogenetic*
causation. In the lifetime of a land vertebrate (e.g., a human being), the organism develops four
limbs before it walks, but that does not rule out the possibility that selection caused the tetrapod
morphology to evolve because it facilitates walking. If the latter is true, there existed an ancestral
population in which the tetrapod trait became common because tetrapods walked better than
nontetrapods.

[46] Sterelny and Griffiths (1999) argue that it is the key to rendering adaptationist hypotheses testable.

ancient design feature of the vertebrate forelimb [...] and occur in many animals [e.g., salamanders and alligators] that do not have the manipulative abilities of the human hand." Of course, human beings *use* their fingers to manipulate objects. The point is that current utility can be different from the reason the trait first evolved (a thought discussed in §2.16 in connection with intelligent design). And not only *can* this be the case; a parsimonious reconstruction of ancestral trait values provides a reason for thinking that this *was* the case (see also Baum and Larson 1991).

We saw in §3.3 that the most parsimonious reconstruction of the quantitative character state of an ancestor A, given the character state of one of its descendants D, is the reconstruction of maximum likelihood when the character evolved by drift but not necessarily when the trait evolved by selection. However, if we shift from a quantitative character to a dichotomous character, the backwards inequality guarantees that the most parsimonious assignment of character state to A, given just the observation of the single descendant D, is also the reconstruction of maximum likelihood. It now is time to consider the reconstruction of character states in a *tree*, not just in a single *lineage*. This will lead to a reconceptualization of the question we need to pose; it isn't about the relationship of parsimony and *likelihood* but instead concerns the relationship of parsimony and *probability*.

In a causal chain that links a distal cause C_d to a more proximate cause C_p and then to an effect E, it often is plausible to assume that C_p *screens off* C_d from E, meaning that $Pr(E \mid C_p) = Pr(E \mid C_p \ \& \ C_d)$.[47] This is a standard assumption in the causal modeling literature; it also is standard in biological models of how ancestor/descendant chains of inheritance should be understood. However, if screening-off *does* hold in a phylogenetic tree, then a strict likelihood approach can assign character states to ancestors that are "shallow"; but once these are made, likelihood considerations cannot, in addition, be used to assign states to ancestors that are "deeper." Consider Figure 3.31. Different assignments of character states to A_1, A_2, and A_3 can have different likelihoods, but once those ancestors are assigned character states, the probability of the data is not further affected by assignments made to A_4 and A_5. For this reason, it is a mistake to think about reconstructing the character states of ancestors in a phylogenetic tree from a strict likelihood point of view. However, all is not lost. Assignments to A_4 and A_5 affect the probability of the character states of A_1, A_2, and A_3, and these in turn affect the probability of the character states of tip species. What we

---

[47] Screening off should not be expected if there is a second pathway from C_d to E that bypasses C_p.

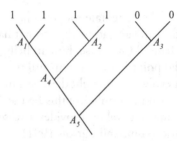

Figure 3.31 The probability of the data (the trait values of tip species) is affected by the character states assigned to ancestors A_1, A_2, and A_3. Once those assignments are made, the probability of the data is not further affected by assignments to A_4, and A_5, if there is screening-off.

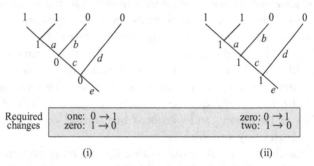

Required changes	one: $0 \rightarrow 1$ zero: $1 \rightarrow 0$	zero: $0 \rightarrow 1$ two: $1 \rightarrow 0$
	(i)	(ii)

Figure 3.32 Two reconstructions of ancestral character states. The character is dichotomous (its states are *0* and *1*); letters label branches. Reconstruction (i) requires fewer changes than reconstruction (ii); however, (i) and (ii) require different kinds of changes.

need to know is which assignments of character states to ancestors make the entire process that extends from the root of the tree to its tips most *probable*. We need to think of the problem from a Bayesian, and not a strictly likelihoodist, point of view (Farris 1986; Maddison 1991).

The relationship between a *parsimonious* reconstruction of ancestral character states and a *probable* reconstruction can be explored by considering the example depicted in Figure 3.32. Reconstruction (i) is more parsimonious than reconstruction (ii). Under what circumstances is (i) more probable than (ii)? Figure 3.33 describes, branch by branch, how the two reconstructions in Figure 3.32 disagree. With respect to branches *b* and *d*, the backwards inequality (§3.5, §3.10) guarantees that reconstruction (i) has a higher probability than reconstruction (ii); for branch *a*, the reverse is true. These comparisons hold, regardless of whether the trait evolves by selection or drift. What about branches *c* and *e*? If a drift

Branches	Reconstruction *i*	Probability ordering	Reconstruction *ii*
a	$0 \to 1$	<	$1 \to 1$
b	$0 \to 0$	>	$1 \to 0$
c	$0 \to 0$		$1 \to 1$
d	$0 \to 0$	>	$1 \to 0$
e	$\to 0$		$\to 1$

Figure 3.33 The two reconstructions of ancestral character states depicted in Figure 3.32 assign different events to branches *a–e*. The backwards inequality settles which reconstruction of branches *a*, *b*, and *d* has the higher probability.

process is at work, the two reconstructions have the same probability with respect to what they say about branch *c*. And if branch *e* has been evolving by drift for a long time, we can use the equilibrium value of $\frac{1}{2}$ as the probability that each reconstruction assigns to that branch. However, the assumption of drift, by itself, provides no verdict as to which reconstruction is more probable *overall*. The sticking point is the comparison of the single change from 0 to 1 postulated by the first reconstruction and the two changes in the opposite direction postulated by the second. This gap can be closed by an additional assumption: If the trait evolves by drift *and all branches have the same duration*, then reconstruction (i) is more probable than reconstruction (ii) (Goldman 1990); now parsimony and probability agree. With unequal branch durations, parsimony and probability can disagree, even when a drift process is at work.[48]

How do the two reconstructions compare when selection for trait 0 or selection for trait 1 is at work throughout the tree? Recall that the difference in probabilities described by the backwards inequality gets smaller as branches are given larger durations. So, if branches have very long durations, the difference in probability between what the two reconstructions say about branches *a*, *b*, and *d* will be negligible. However, with respect to branches *c* and *e*, reconstruction (*ii*) has the higher probability if there is selection for character state 1, and the magnitude of this difference *increases* as branches have larger durations. This means that the less parsimonious reconstruction (*ii*) will have the higher probability overall if branches all

---

[48] This conclusion is consistent with Maddison's demonstration (1991) that the most parsimonious reconstruction of ancestral character states for a quantitative character is the reconstruction of greatest probability when the trait evolves by drift (Brownian motion) and branches are given "equal weights." He notes that parsimony and probability will not coincide when there is drift but branches have unequal weights (Maddison 1991: 309–10). Maddison thinks of a branch's weight as reflecting the amount of change that has occurred on the branch in other characters; branches that have experienced lots of changes in other characters might have a higher probability of exhibiting a change in the character being reconstructed. My point about branch durations is another way of conceptualizing these weighting terms.

have large durations and there is strong selection for character state 1.[49] Whether parsimony is a guide to probability in the reconstruction of ancestral character states depends on the kind of process that governs trait evolution and on the amount of time there is in branches.

In thinking about which reconstruction in Figure 3.32 is overall more probable, I considered not just the *changes* in character state that the two reconstructions postulate but the instances of *stasis* as well. Considering only the former makes it tempting to think that reconstruction (*ii*) will be more probable when there is strong selection for state 0; after all, reconstruction (*ii*) requires two changes from state 1 to state 0 and none in the opposite direction, whereas reconstruction (*i*) requires a change from 0 to 1. In fact, the truth is just the reverse: It is strong selection for character state 1, not for 0, that makes reconstruction (*ii*) more probable when branches have long durations. Selection affects the probability of stasis just as it affects the probability of change. Reconstruction (*i*) has three lineages lingering in state 0, and the root of the tree begins in that state, whereas reconstruction (*ii*) has two branches remaining in state 1, which is also the state in which the tree begins.

To make clear how the above argument is Bayesian in character, consider the consequence of Bayes' theorem discussed in §1.3 that says that the ratio of posterior probabilities equals the ratio of likelihoods times the ratio of priors. Applied to the example under discussion, this means that

$$\frac{Pr(\text{Reconstruction } i \mid \text{data})}{Pr(\text{Reconstruction } ii \mid \text{data})}$$

$$= \frac{Pr(\text{data} \mid \text{Reconstruction } i)}{Pr(\text{data} \mid \text{Reconstruction } ii)} \times \frac{Pr(\text{Reconstruction } i)}{Pr(\text{Reconstruction } ii)}.$$

The data in our example are the character states of tip species. This means that the *likelihoods* of the two reconstructions reflect what the reconstructions say about shallow branches (*b* and *d*) while the *priors* reflect what the reconstructions say about deeper branches (*a*, *c*, and *e*):

$$\frac{Pr(\text{Reconstruction } i \mid \text{data})}{Pr(\text{Reconstruction } ii \mid \text{data})}$$

$$= \frac{Pr_b(0 \to 0)Pr_d(0 \to 0)}{Pr_b(1 \to 0)Pr_d(1 \to 0)} \times \frac{Pr_a(0 \to 1)Pr_c(0 \to 0)Pr_e(\text{root} = 0)}{Pr_a(1 \to 1)Pr_c(1 \to 1)Pr_e(\text{root} = 1)}.$$

---

[49] Symmetrically, reconstruction (*i*) is more probable if there is lots of time and there is selection for character state 0.

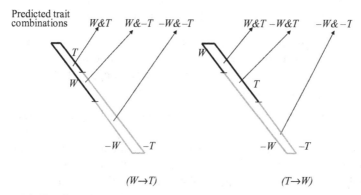

Figure 3.34 Two hypotheses about events in the lineage leading to land vertebrates that make different predictions about the trait combinations that land vertebrates and their relatives should exhibit; W = walking and T = the tetrapod morphology.

In conformity with the version of Bayesianism endorsed in Chapter 1, I take it that the priors and likelihoods in these expressions have objectively defensible values only given a model of character evolution.

What light does this throw on the chronological test? Is this test mistaken in claiming that the present distribution of walking and tetrapod morphology in vertebrates is evidence against the hypothesis that the morphology evolved to facilitate walking? No, it isn't this *conclusion* that is mistaken; rather, the flaw attaches to the *reasoning* that leads to that conclusion. The fact that different hypotheses about process disagree about which ancestral character state assignment is most probable is an objection to using parsimony in the chronological test. But perhaps the chronological test can be reconceived. Instead of using the data to construct a parsimonious assignment of character states to ancestors and pretending that this reconstruction provides independent testimony about the process hypotheses under test, we must let the hypotheses under test tell us what we should expect to see in the data.

Let's be more explicit about the hypotheses that are in contention concerning the evolution of walking and the evolution of four limbs in the vertebrate line. These are depicted in Figure 3.34. The hypothesis that fails the chronological test says the following:

($W \rightarrow T$) In the lineage leading to land vertebrates, first walking evolved. After walking was in place, there was selection for the tetrapod morphology, and the result was that four limbs eventually evolved.

A more sensible hypothesis postulates just the opposite causal ordering:

($T \rightarrow W$) In the lineage leading to land vertebrates, first the tetrapod morphology evolved. After the tetrapod morphology was in place, there was selection for walking, and the result was that walking eventually evolved.

Both hypotheses say that the lineage leading to land vertebrates begins with organisms that lack both T and W and that it ends with organisms that have both T and W. They disagree about the trait combinations that existed in between. The ($W \rightarrow T$) hypothesis says that in between there were organisms with the combination $W\&-T$; the ($T \rightarrow W$) hypothesis says that in between there were organisms with the combination $T\&-W$. If we observe organisms that stem from this lineage that are $T\&-W$, but none that are $W\&-T$, this favors ($T \rightarrow W$) over ($W \rightarrow T$). And this is precisely what we observe.[50] It isn't that the ($T \rightarrow W$) hypothesis says that it is *impossible* for a descendant of this lineage to have the combination $W\&-T$, nor does the ($W \rightarrow T$) hypothesis say it is *impossible* for a descendant to have the combination $-W\&T$. Rather, the two hypotheses confer different probabilities on each of these two possibilities:

$Pr[D$ has traits $T\&-W \mid D$ is a descendant of the line leading

  to land vertebrates & ($T \rightarrow W$)]

   $> Pr[D$ has traits $T\&-W \mid D$ is a descendant of the line leading

    to land vertebrates & ($W \rightarrow T$)].

$Pr[D$ has traits $-T\&W \mid D$ is a descendant of the line leading

  to land vertebrates & ($T \rightarrow W$)]

   $< Pr[D$ has traits $-T\&W \mid D$ is a descendant of the line leading

    to land vertebrates & ($W \rightarrow T$)].

The use of parsimony in the chronological test produces the right conclusion for the wrong reason. The present distribution of four limbs and walking among present-day vertebrates and their relatives provides strong evidence against the hypothesis that the tetrapod morphology evolved because it facilitated walking on dry land. The distribution of characters has this evidential meaning because it favors an alternative hypothesis.

---

[50] Although spiders walk without having four legs, their characteristics don't favor the ($W \rightarrow T$) hypothesis. Spiders are an *out*-group with respect to this problem; the hypotheses in question make predictions about the characters found in the *in*-group.

One of the most important distinctions in evolutionary biology is that between current utility and adaptation (§2.16). The idea, not surprisingly, traces back to Darwin; Williams (1966), Lewontin (1978), and Gould and Vrba (1982) have emphasized the point, and it has had a large and deserved influence. It is now widely agreed that to say that a trait is an *adaptation* for doing X in lineage L requires that the trait originally evolved in L because it helped organisms perform task X. As Lewontin notes, sea turtles use their front legs to dig nests in the sand, but this is not why front legs first emerged in the lineage, and so the legs are not adaptations for nest building. My criticism of the chronological test does not mean that there is no way to test hypotheses about adaptation so defined. Rather, my criticism concerns using *parsimony* to reconstruct the character states of ancestors to test a selection hypothesis. This criticism leaves it open that hypotheses about why a trait first evolved can be tested in some other way. But, in addition, we need to recognize that "adaptation" is not the only historical concept that is worth considering. If "adaptation" is defined in the way just described, we need another term for the concept of relevance here. Suppose, in a group of lineages, there was selection for longer fur in colder climates and for shorter fur in warmer. This is a claim about history, not about current utility, but it does not focus exclusively on the *first* emergence of long fur. Selection can cause a trait to evolve, but it also can maintain a trait in a lineage once it has evolved. It is wrong to think that historical questions about natural selection must focus exclusively on the former. The hypothesis that fur length evolved as an adaptive response to ambient temperature is a causal claim, like the hypothesis that smoking promotes lung cancer. To investigate the causal relation of smoking and lung cancer, you don't need to know whether lung cancer made its first appearance before or after the first appearance of smoking. This chronological information is not *necessary* (which is not to say that it would not be nice to have). The same holds for the question of whether fur length evolved as an adaptive response to the selection pressures created by cold and warmth. Whether long fur first appeared before or after the first appearance of cold climate is an interesting question, but there is another causal question that does not require one to ascertain what the chronological order was.

3.12 CONCLUDING COMMENTS

I began this chapter by focusing on the task of deciding whether the hypothesis of selection-plus-drift (SPD) or the hypothesis of pure-drift (PD), each conceptualized phenotypically, is better supported, given the trait value found in a single species. The reason I formulated the problem in this way is that it is simple, and so it is a good place to begin, and also because many biologists and popularizers of evolutionary biology invoke natural selection to explain single trait values and would be loath to think of drift as an alternative that is worth considering. If the trait is useful to the organisms that have it, the drift hypothesis gets dismissed out of hand. And if the trait is both useful and complex, the suggestion that drift should be considered may strike one as both pedantic and obscurantist. I have described the biological information that is needed if one wants to defend the claim that selection is more likely than drift. Appeal to intuition is not enough. Nor can probabilistic *modus tollens* (§1.4) be invoked as a reason for rejecting the drift hypothesis.

After exploring what information needs to be in place if the observation of polar bear fur length is to settle whether the SPD or the PD hypothesis has higher likelihood, I considered how the evidential situation shifts if the *explanandum* is reconfigured (§3.6). Instead of trying to explain why a single species has a given trait value, what happens if the fact to be explained is a correlation? Suppose we observe that bears living in colder climates tend to have longer fur than bears living in warmer climates. This change in *explanandum* is highly significant. It turns out that less biological information is needed for the likelihood comparison to go forward. One needn't know which fur length is optimal in which environment. And one needn't know the trait values of ancestors, or the curvature of the fitness function, the heritability of the trait, or the effective population size. The fact of common ancestry plays a central role in structuring the test of SPD against PD. Comparative data, and the testing of hypotheses that make different predictions about the patterns that should be found in such data, are the light and the way. Not that pure drift is the only alternative to natural selection that is worth considering. Phylogenetic inertia (aka ancestral influence) is an alternative alternative (§3.10), and there are others as well. Testing is contrastive, but it need not invariably revert to the same old contrasts.

We also had occasion to reflect on the use of phylogenetic parsimony as a tool for reconstructing the character states of ancestors. Biologists want to deploy "phylogenetically sensitive" tests of hypotheses about natural selection, and that is all for the good. But parsimony is not the tool it is often thought to be. Obviously, it is not an infallible guide to the character states of ancestors. I have argued for the further point that its authority depends on assumptions about the evolutionary process. In testing selection against drift, parsimony is not a neutral party; parsimony is partisan. It would be good if hypotheses about natural selection could be tested without relying on the use of parsimony to reconstruct ancestral character states. And they can be.

I began this chapter by thinking about how SPD and PD should be compared in a likelihood format, but model selection (§3.9) provides an important alternative approach. This framework illuminates the question of whether a pattern of variation across a number of species should be explained by a unified or a disunified model. For example, should the difference in longevities between human females and males be explained by the same factors that affect differences in longevity in other primates (§3.7)? Model-selection theory also provides a useful setting for testing hypotheses about molecular evolution. For example, are rates of change in a set of traits the same across lineages, or do they differ (§3.9)?

Hypotheses about natural selection describe the causal processes that occur in lineages, but biologists usually do not have snapshots of a lineage as it moves through time. This is the situation in which biologists find themselves when the trait of interest does not fossilize or if the fossil record is spotty. And even when a detailed fossil record is available, it is important to remember that the fossils we study may not be ancestors of present-day populations; they may merely be their relatives. In the second half of this chapter, I explored the question of how observing the current states of several lineages reveals what happened in those lineages in the past. The fact that present-day species share common ancestors plays a central role in answering this question – for example, in the various tests that population geneticists deploy when they use sequence data drawn from current populations to test selection against drift (§3.9). This raises the issue of why one should think that the different species alive today trace back to common ancestors. This issue is where the next chapter begins.

CHAPTER 4

Common ancestry

We saw in the last chapter that evolutionary theory places hypotheses about the causes of trait evolution within the framework of a phylogenetic tree. These hypotheses, whether they say that the trait of interest evolved by natural selection or by some other process, make claims about what happened in lineages, and different lineages stem from common ancestors. For example, different extant species have different kinds of eyes, and some have no eyes at all. The fact of common ancestry places a constraint on how the present distribution of trait values must be explained. If all these species have a common ancestor, the lineages descending from that common ancestor had to start with the same trait value. It follows that the task of explaining why vertebrates have camera eyes is essentially connected to the task of explaining why other groups have other kinds of eyes while still others have none at all.

Given how central the thesis of common ancestry is to evolutionary reasoning, one might expect there to be a vast literature in which the evidence for that claim is amassed. In fact, the question *is* discussed, but the literature on it is hardly vast. For most evolutionists, the similarities that different species share make it obvious that they have common ancestors, and there is no reason to puzzle further over the question. The kind of genealogical question that attracts far more attention in evolutionary biology concerns *how* various species are related to each other, not *whether* they are. Consider, for example, the question of how human beings, chimps, and gorillas are related. Two of the options are depicted in Figure 4.1. The first hypothesis, (*HC*)*G*, says that there is an ancestor shared by human beings and chimps that is not an ancestor of gorillas. The second, the *H*(*CG*) hypothesis, asserts that it is chimps and gorillas that are more closely related to each other than either is to human beings. Although these rival hypotheses disagree about the branching pattern, there is something on which they agree: *Go back far enough in time and you will find an ancestor common to all three.* This common ground is the

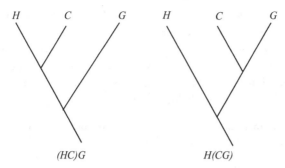

Figure 4.1 Two competing genealogical hypotheses about the phylogeny of human beings (*H*), chimpanzees (*C*), and gorillas (*G*). The hypotheses disagree about which two are more closely related to each other than either is to the third. They agree that there is an ancestor common to all three.

main subject of the present chapter. My goal is not to assemble evidence that justifies hypotheses of common ancestry but to understand the logic that dictates how such evidence should be interpreted. From Darwin down to the present, biologists have presented various arguments in favor of common ancestry. What rules should we apply to determine whether the arguments are strong or weak? I'll first take up the question of why, or under what circumstances, an observed similarity between species *X* and *Y* is evidence that they have a common ancestor. Then I'll turn to the question of which kinds of similarity provide stronger evidence for common ancestry and which provide only weaker evidence for that conclusion. After that, I'll address whether something besides the observed similarity of two or more species can provide evidence that bears on whether they are genealogically related; it is here that I'll consider intermediate fossils and biogeography. In the last section, I'll discuss the testing of more fine-grained genealogical hypotheses, such as (*HC*)*G* versus *H*(*CG*).

4.1 MODUS DARWIN

Similarity, ergo common ancestry. This form of argument occurs so often in Darwin's writings that it deserves to be called *modus Darwin*. The finches in the Galapagos Islands are similar; hence, they descended from a common ancestor. Human beings and monkeys are similar; hence, they descended from a common ancestor. The examples are plentiful, not just in Darwin's thought, but in evolutionary reasoning down to the present (Sober 1999a).

If two finch species have a common ancestor, and human beings and monkeys have a common ancestor, do those two common ancestors themselves have a common ancestor? How far does this knitting together of species proceed? In the last paragraph of *The Origin of Species*, Darwin gives a cautious answer to this question. In his famous exclamation that "there is grandeur in this view of life," he says that, in the beginning, life was "breathed into a few forms or into one" (Darwin [1859] 1964: 490). However, only a few pages before, Darwin takes a bolder position:

I believe that animals have descended from at most only four or five progenitors, and plants from an equal or lesser number. Analogy would lead me one step further, namely to the belief that all animals and plants have descended from some one prototype. But analogy may be a deceitful guide. Nevertheless all living things have much in common, in their chemical composition, their germinal vesicles, their cellular structure, and their laws of growth and reproduction. We see this even in so trifling a circumstance as that the same poison often similarly affects plants and animals; or that the poison secreted by the gall-fly produces monstrous growths on the wild rose or oak-tree. Therefore I should infer from analogy that probably all organic beings which have ever lived on this earth have descended from some one primordial form, into which life was first breathed.   (Darwin 1964: 484)

Notice that Darwin embraces the view that there is a single phylogenetic tree on the grounds that "all living things have much in common."[1] Perhaps if plants and animals were *less* similar, Darwin would have opted for the hypothesis that all animals are genealogically related and that all plants are too but that there is no ancestor that animals and plants have in common. This points to an obvious question about *modus Darwin*: How much similarity is needed for common ancestry to be a good inference? For example, human beings and chimps are about 98.5 percent similar at the level of their DNA sequences. If this is enough to justify the inference, what is the cut-off that 98.5 percent is said to exceed? Fifty percent? Twenty-five? Does this mean that if we find two species that are *less* similar than this cut-off that we should conclude that they *lack* a common ancestor?

If percentage similarity is the key to inferring common ancestry, how is it to be measured? Using percentage similarity in this way presupposes that there is a totality of *n* characteristics; once we have specified that totality, we then can contemplate how much two species must match on those *n* characteristics for the inference of common ancestry to make sense. To think about this total package of characteristics, we need to

---

[1] The *variorum* edition of *The Origin of Species* indicates that Darwin did not change his mind on this issue through successive editions. See Darwin (1959: 759).

understand how traits are to be separated from each other. For example, does having five digits on the left hand count as a different characteristic from having five digits on the right, or is there just one character here – that of having five digits on an appendage? Another question about using percentage similarity as our guide is that it assumes the principle of *one match one vote*; it assumes that all similarities have the same evidential significance – you just have to count them. This is clearly a dubious principle; surely some similarities are more telling than others.

The place to begin, I suggest, is with individual characteristics. When you think about the evidence that bears on the question of common ancestry, don't just think about percentage similarity; instead, think about the evidence trait by trait. If two species both have trait T, under what circumstances is this evidence that they share a common ancestor? Perhaps only some similarities count as evidence for common ancestry. If so, what distinguishes the similarities that count from the ones that do not? After considering when a similarity is evidence for common ancestry, we will investigate when one similarity carries more evidential weight than another. And, along the way, we will return to the question of how percentage similarity and other summary statistics are relevant.

The likelihood framework suggests two principles that bear on this inquiry, though these principles are not unique to that framework. The first is that we must think *contrastively*. If you are inclined to say that the 98.5 percent genetic match between humans and chimps is evidence that they share a common ancestor, ask yourself how similar you'd expect the two species to be if they had no common ancestor. If you can't answer this question, why do you think that the high degree of similarity is evidence for common ancestry? The same point is relevant when similarities are considered trait by trait; if the fact that species X and Y both have trait T is evidence for common ancestry, this will be because the matching favors the hypothesis that X and Y share a common ancestor over the hypothesis that they do not. The second principle we can glean from the likelihood approach is that we must *avoid selective attention*. If some similarities provide strong evidence for common ancestry, then surely there are some gross differences that would provide evidence against common ancestry. Even if creationists often focus exclusively on the *differences* that separate human beings from the rest of nature, evolutionists should not follow suit by focusing only on the *similarities*. The *principle of total evidence* (§1.3) requires that we consider *all* the data. If some similarities favor the common-ancestry hypothesis while some differences favor the hypothesis of separate ancestry, how should these similarities and differences be assembled into an overall assessment of the two hypotheses?

4.2 WHAT THE COMMON ANCESTRY HYPOTHESIS ASSERTS

In the passage from *The Origin of Species* just quoted, Darwin speculates about the genealogy of "all living things," by which he seems to mean animals and plants. Modern biology agrees that animals and plants trace back to a common ancestor, but this is no longer regarded as a speculative conjecture; there now is abundant evidence that strongly supports the common-ancestry hypothesis, evidence that Darwin did not possess. Another large shift has occurred since Darwin: We now know that there is far more to life than animals and plants. Plants and animals are parts of *Eukaryota* (meaning organisms with cell nuclei), but this group also includes fungi, algae, ciliates, and other groups. And besides the Eukaryotes, there are two other major lines, *Bacteria* and *Archaea* (Woese 1998). These last two comprise the Prokaryotes, "which embrace perhaps two-thirds of the biota and the first two-thirds of life's history" (Doolittle and Bapteste 2007). Plants and animals together comprise a twig on a branch, not the whole tree.

Before investigating the logic of the evidence behind claims of common ancestry, we need to get clear on what it means to assert or deny that X and Y have a common ancestor. X and Y could be two organisms (in the same or different species), or they could be two species;[2] in the latter case, we will say that two species have a common ancestor precisely when all the organisms in the first and all the organisms in the second have a common ancestor. In this sense, the common ancestry of species reduces to the common ancestry of the organisms in those species. But what kind of object is a "common ancestor?" If we think of an ancestor as a species, we need to say what a species is; we need to solve the notorious "species problem." One warning sign that this is not a path down which we should choose to tread is that the much-admired *biological species concept* (Mayr 2000) says that a species is a group of organisms that interbreed among themselves but which are reproductively isolated from other such groups. Understood in this way, a species must be made of sexual organisms. However, evolutionists agree that sexuality is a derived character; first there were asexual organisms. This means that the biological species concept is not the right choice if we wish to say that all life on Earth derives from a single *species*. Of course, there are other species concepts that might provide a satisfactory alternative, but there is a second reason why the ancestor postulated by the common

---

[2] X and Y could also be two genes, found in the same organism or in different organisms; I'll discuss the idea of gene genealogies later.

ancestry hypothesis should not be thought of as a species. How are the organisms in this supposed ur-species related to each other? If they trace back to a single ancestral organism, then we know that all current life forms have a single *organism*, not just a single *species*, as their original progenitor. Alternatively, if the organisms in this ur-species do not all trace back to a common ancestral organism, then it is appropriate to view this situation as a case of *multiple* ancestors. Whether there is a single ur-species is therefore not what is at issue.

So the common ancestry hypothesis says that all current life forms derive from a single *organism*, not a single *species*. But what is an organism? Well, it must be *alive*. Darwin and present-day Darwinians would not be satisfied if all life on Earth derived from the same large slab of rock whose nonliving materials produced numerous separate start-ups of life that never melded together but instead led separately to the several groups of organisms we now observe. In this case, we would say that animals and plants are "genealogically *un*related." They would have a common *origin* (the slab) but not a common *ancestor* (an organism). Another feature of ancestors is that they beget descendants. Your grandparents produced your parents, and your parents produced you; *ergo* your grandparents are among your ancestors. But what is this begetting relation? It is natural to think of reproduction in terms of genetic transmission. You received half your nuclear genes from your mother and half from your father, and they, in turn, received half of their genes from each of their parents. However, it does not follow that you received one-quarter of your genes from each of your four grandparents. Meiosis is a lottery, and one of your grandparents may have lost, meaning that you received 0 genes from him or her. As we consider ancestors of yours who are more and more remote, it becomes increasingly certain that some of them passed no genes to you. Ancestors have a shot at contributing genes to their descendants, but there is no guarantee that they succeed in doing so. Figure 4.2 provides a simple example of this point. Another reason not to define the relation of ancestor to descendant in terms of gene transmission is that the ancestors that existed when life first got started are thought to have predated the evolution of the genetic system; a self-replicating molecule might be an ancestor even if it has no genes to transmit.

Figure 4.2 also illustrates the fact that the genealogies of sexual organisms are reticulate; they are not strictly treelike. When a genealogy is strictly treelike, branches split but never join. Your family "tree" is not like this; you and your full sib have *two* most recent common ancestors (your mother and father), not *one*. Asexual organisms have treelike genealogies; each has

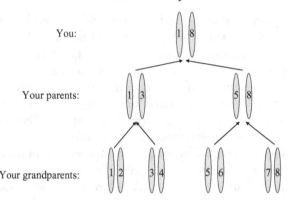

Figure 4.2 If you are a diploid organism with one chromosome pair, two of your four grandparents must have failed to make any genetic contribution to your genome.

one parent, not two. Still, even with reticulate genealogies, a distinction can be drawn between two organisms' having a common ancestor and their failing to do so. You and your half-sib share a parent. Darwin's thesis that all life traces back to a single common ancestor (i.e., an organism) does not require that the single genealogy be strictly treelike.

With these thoughts in mind, we now can clarify the Darwinian hypothesis that all current life traces back to a single common ancestor. Consider all the organisms that now are alive on Earth and all the fossils that now exist as well. Trace each of them back in time – to their recent ancestors and then to ancestors that are more ancient. As we trace lineages backwards in time, many coalesce, with two or more descendants eventually converging on a common ancestor. Is there a single ancestor that all these tracings eventually reach? The Darwinian hypothesis is *yes*: There was a last universal common ancestor (LUCA) of all the living and fossil forms that now exist. This hypothesis leaves it open that LUCA itself had ancestors; there is no claim that LUCA was the *first* organism on Earth, and no biologist thinks that it was. The hypothesis also is consistent with there having been thousands of start-ups of life from nonliving materials if all but one of them went extinct;[3] the hypothesis does not assert that all past and present organisms trace back to a single common ancestor. Nor does LUCA rule out there being many start-ups that all have descendants

---

[3] In the fifth edition of *The Origin of Species*, Darwin added the following remark to the passage I quoted at the beginning of the previous section: "No doubt it is possible, as Mr. G. H. Lewes has urged, that at the first commencement of life many different forms were evolved; but if so, we may conclude that only a few have left modified descendants" (Darwin 1959: 759).

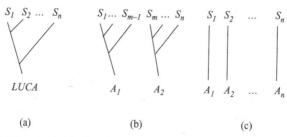

(a)   (b)   (c)

Figure 4.3 Hypothesis (a), that there was a LUCA, is denied by both (b) and (c), which disagree as to how much relatedness there is among the n organisms and fossils ($S_1 \ldots S_n$) that exist now.

alive today; LUCA can accommodate this possibility if there was a *bottleneck*. Here is a biologically implausible example that illustrates this conceptual possibility: Suppose there were numerous independent and simultaneous start-ups of sexual organisms from nonliving materials and that each parental pair subsequently produced just one offspring for a number of generations until the population dwindles to two individuals (one male, one female). These two individuals then have four offspring, and so do their descendants, thus yielding the many living things and fossils that exist today. The two individuals in this bottleneck are each LUCAs. Note, finally, that LUCA takes no stand on whether the start-ups that led to present-day organisms occurred *on Earth*; perhaps they occurred on another planet (*exogenesis*) and life arrived here on meteors.

LUCA can be placed at one end of the continuum depicted in Figure 4.3. For the moment, I'm ignoring the fact that sexual organisms have reticulate genealogies. Figures 4.3b and 4.3c should both be interpreted as saying that the ancestors at the roots of the genealogies they postulate do not themselves have common ancestors. Both therefore disagree with 4.3a; nonetheless, Figure 4.3b is "closer" to the LUCA idea than 4.3c is. The differences that separate these three examples can be captured by the following definition of a family of hypotheses, each of the form CA*i* (Sober and Steel 2002):

CA*i*: There exists a set A consisting of i organisms, and no set with fewer than i organisms, such that (1) no organism in A is ancestral to any other organism in A, (2) each current organism and each currently existing fossil (S_1, S_2, $\ldots$, S_n) has at least one ancestor in A, and (3) each organism in A is ancestral to at least one S_k.

Figure 4.3a, the LUCA hypothesis, asserts that $i = 1$; Figure 4.3b says that $i = 2$; and Figure 4.3c says that $i = n$. Figures 4.3b and 4.3c should make it clear that CAi does not mean that there were i universal common ancestors.

Sexual reproduction, we have seen, shows that the following principle is false: If A is an ancestor of D, then some of D's genes came from A. We now need to see that the converse of this principle is also wrong: It is false that if some of D's genes come from A, then A is an ancestor of D. This is incorrect because of *lateral gene transfer*. Disease vectors such as insects and bacteria can carry retroviruses from the organisms in one species to those in another, with nuclear genes coming along for the ride. You inherited the vast majority of your genes from your two parents, but perhaps a small number of them trace back via lateral transfer to a member of a different species (a worm, let us assume). This complication does not lead us to hesitate in saying that Mom and Dad are your ancestors, but the worm is not. Mom and Dad reproduced, and you were the result; the worm's acts of reproduction did not produce you. So, Mom and Dad are in your family tree, while the worm is not. An organism's genealogy is one thing, the genealogy of its genes is something else: The former traces organisms back to earlier organisms; the latter trace genes back to earlier genes.

Even when the genes in an organism trace back exclusively to genes in its ancestors (there being no lateral gene transfer), it needn't be true that *the same* ancestors are involved. An organism's genome is a composite entity whose parts can have different genealogies. Consider, for example, the mitochondria that all human beings have and the Y sex chromosomes that human males have. Human beings inherit their mitochondria from their mothers; biologists infer that there was a LUCA for those mitochondrial genes, who is sometimes called "mitochondrial Eve." The Y sex chromosomes that human males have are inherited paternally, and biologists infer that there was a LUCA for the different present-day copies of the human male sex chromosome, who is sometimes called "Y-chromosome Adam." Not only were this Eve and this Adam different organisms; biologists estimate that they lived many years apart – Adam and Eve never met.

Even though CA1 is the standard view in contemporary evolutionary biology, the field has no rock-bottom commitment to that hypothesis. The field would not go up in flames if CA2 or CA3 turned out to be true. For example, it is possible that life will someday be detected in extreme environments on Earth that have always been isolated from the rest of the biota, and that the living things found there will differ sufficiently from the rest of life that biologists will judge them to be the isolated descendants of a separate start-up. The same is true if biologists someday build an organism out of nonliving materials, an idea contemplated at the end of Chapter 2. In fact, there is no need to consider such hypothetical scenarios to see that the value of *i* in CA*i* is perfectly discussable within evolutionary biology. Consider Carl Woese's comment:

> The ancestor cannot have been a particular organism, a single organismal lineage. It was communal, a loosely knit, diverse conglomeration of primitive cells that evolved as a unit, and it eventually developed to a stage where it broke into several distinct communities, which in their turn became the three primary lines of descent. (Woese 1998: 6858)

As mentioned, Woese's three lines – bacteria, archaea, and eukaryotes – are now widely accepted; however, his idea that there was no LUCA is controversial. Notice that Woese's disagreement with CA1 depends on the ancestors described in that hypothesis being organisms or cells; a "loosely knit diverse conglomeration of primitive cells that evolved as a unit" is not a proper candidate for being *an* ancestor. Woese doubts CA1 because he thinks there was rampant lateral gene transfer in early life; Doolittle (2000) is a fellow doubter. They are more inclined to think that the version of CA3 depicted in Figure 4.4 is correct, though sometimes they express doubts as to whether it is possible to know whether CA1 or CA3 is right. According to the version of CA3 shown in Figure 4.4, there were repeated start-ups of life from nonlife and then massive lateral gene

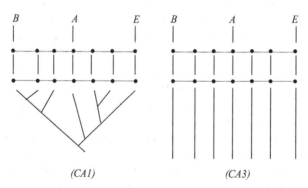

(CA1)   *(CA3)*

Figure 4.4 A CA1 and a CA3 genealogy for Bacteria (B), Archaea (A), and Eukaryotes (E), both of which involve rampant lateral gene transfer (represented by dotted lines) in early life.

transfer occurred. Out of this pattern, in which genes flowed to non-descendants as well as to descendants, there emerged the three lines of life that we now observe. A consequence of this pattern is that there can be universal features of present-day life that were not present in any universal common ancestor and that also did not evolve repeatedly.

As we make the value of i in CAi larger and larger, we approach the case in which each species alive now is said to be the result of a separate origination event. This huge value for i would be amazing and utterly shocking if it turned out to be correct. It flies in the face of the numerous similarities that knit species together, thus providing evidence for their common ancestry. The claim that no two species alive today have a common ancestor immediately calls to mind the doctrine of *special creation*, though creationists usually say that they are talking about the separate origins of the different "fundamental kinds" of organism, not about species in the biological sense of that term. Denying common ancestry and invoking intelligent design have been deeply enmeshed historically, but that should not obscure the fact that versions of CAi that assign big values to i do not logically entail a commitment to intelligent design. For this reason, the question of whether two or more species have a common ancestor should have an answer *within* evolutionary theory; it can and should be considered as a question separate from the creationist challenge *to* evolutionary theory.

The main subject of the next several sections is testing the common ancestry hypothesis for a set of organisms (or species) against the separate ancestry hypothesis. Notice that the word "ancestry" appears in both hypotheses; this means that we will *assume* that there are ancestors and *test* how they are related to the descendants we observe. In fact, when we consider two or more species, the question will be whether (or in what circumstances) the similarities they exhibit provide evidence that they stem from a common *cause*. It will not matter in our discussion whether the common cause of the similarities we observe, if such there be, is a common *ancestral organism*. True, it matters to biologists whether two organisms have genes in common because of inheritance from a common ancestor or because there was lateral gene transfer from an organism that was the ancestor of neither (or of only one); these possibilities are represented in Figure 4.5. It also matters whether the common cause that explains the fact that two species are similar is an ancestral organism or a slab of rock. However, in terms of the logic of the problem (as I see it), this is not the problem with which to begin. The first order of business is to determine whether there was a common *cause*. If the answer is *yes*, the second question

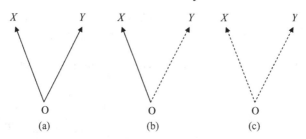

Figure 4.5 Three scenarios under which organisms X and Y share a trait because it was transmitted to them from an earlier organism O. (a) O is an ancestor of both X and Y; (b) O is an ancestor of X but not of Y; (c) O is an ancestor of neither. Solid lines represent ancestor/descendant relationships; dashed lines represent lateral gene transfer.

is whether that common cause was a common ancestor, an organism that was not a common ancestor, or no organism at all.[4] First things first.

We saw in Chapter 3 that the assumption of common ancestry plays a central role when a hypothesis that postulates natural selection is tested against hypotheses that postulate drift or phylogenetic inertia. These hypotheses usually concern a circumscribed set of organisms, not the whole of the living world. The question of why bears in cold climates have longer fur than bears in warm climates is typical in this regard. Biologists address this question by exploiting the fact that bears have a common ancestor; it doesn't really matter to that problem if *all* present-day organisms are genealogically related. The question about the value of i in CAi is important in terms of the big picture, but most biologists go about their business without needing the big picture. Still, the big question about all of life and the smaller question about a group of bears have the same logic: How do the similarities and differences we observe provide evidence as to whether the organisms of interest have a common ancestor?

4.3 A BAYESIAN DECOMPOSITION

When two species X and Y share a trait T, it is perfectly obvious that one cannot *deduce* from this that they have a common ancestor. Perhaps, then, we should seek to determine when the similarity renders the existence of a common ancestor *probable*. Bayes' theorem (§1.2) describes how that posterior prob-ability – the probability of the common ancestry (CA) hypothesis, conditional on the observed similarity – decomposes into three other quantities:

---

[4] The distinction drawn here resembles the one between inferring a tree topology and assigning character states to the interior nodes postulated by that tree topology. The former problem will be discussed in §4.8; the latter was discussed in §3.3 and §3.11.

(1) $Pr(\text{CA} \mid X \text{ and } Y \text{ share } T) = \dfrac{Pr(X \text{ and } Y \text{ share } T \mid \text{CA}) \, Pr(\text{CA})}{Pr(X \text{ and } Y \text{ share } T)}.$

Similarly, if we apply Bayes' theorem to the separate ancestry (SA) hypothesis, we obtain:

(2) $Pr(\text{SA} \mid X \text{ and } Y \text{ share } T) = \dfrac{Pr(X \text{ and } Y \text{ share } T \mid \text{SA}) \, Pr(\text{SA})}{Pr(X \text{ and } Y \text{ share } T)}.$

Equations (1) and (2) together yield the following:

(3) $Pr(CA \mid X \text{ and } Y \text{ share } T) > Pr(SA \mid X \text{ and } Y \text{ share } T)$ if and only if
$Pr(X \text{ and } Y \text{ share } T \mid CA) \, Pr(CA) > Pr(X \text{ and } Y \text{ share } T \mid SA) Pr(SA).$

Equation (3) tells you how to determine which hypothesis has the higher posterior probability. This depends on the prior probabilities and on how probable each hypothesis says the observed similarity is (i.e., on the likelihoods of the hypotheses).

Let's begin with the priors. Without considering the observed similarities and differences that characterize two species, how probable is it that they trace back to a common ancestor? Is this more probable than that they have separate ancestry? We want our answer to this question to be *objective*. It is not epistemologically relevant how strong someone's prior prejudices are on the subject. To find prior probabilities that have some authority, we need an empirically well-grounded theory that addresses the following questions. How often should we expect start-ups (the origination of living things from nonliving materials) to have occurred? And once a start-up occurs, what is the probability that it will have descendants among the species and fossils that currently exist on earth? What reasons could such a theory offer for thinking that we should expect species X and Y to have a common ancestor even before we examine their similarities and differences? One possibility is that the theory might show that start-ups are so vastly improbable that there probably was just one of them in the whole time since the Earth began. Oparin (1953) developed a second reason for expecting all present-day life to be part of a single genealogical tree. Perhaps the first start-up was not terribly improbable, but once it occurred, it destroyed the conditions needed for further such events to occur. Darwin (1887: III, 18) speculates about this possibility in one of his letters:

It is often said that all the conditions for the first production of a living organism are now present, which could ever have been present. But if (and oh! What a big if!) we could conceive in some warm little pond with all sorts of ammonia and phosphoric salts, light, heat, electricity, etc. present, that a proteine compound was formed, ready to undergo still more complex changes, at the present day such matter would be instantly devoured or absorbed, which would not have been the case before living creatures were formed.

A third possibility, related to the second, is that several start-ups probably occurred, but that it was very improbable that more than one of them produced descendants or fossils that now exist. This might have been because organisms in different start-ups competed so intensely that all present-day species and fossils trace back to just one of them.

As far as I know, these possibilities are now open biological questions. There is at present no detailed and well-confirmed biological theory that assigns a prior probability to the hypothesis that all present-day organisms and fossils trace back to a single common ancestor. However, if these prior probabilities are obscure, the same will be true of the posterior probabilities. This can be seen by inspecting proposition 3. It follows that if *modus Darwin* licenses the conclusion that there *probably* is a single tree of life, it is questionable whether this form of argument really makes sense. It does not follow, however, that scientists have no basis for the CA1 hypothesis. Rather, what we need to consider is the possibility that *modus Darwin* is a species of likelihood inference. Even if observed similarities do not render the common ancestry hypothesis more *probable* than not, it is possible that these observations *strongly favor* the one hypothesis over the other. The fact that it is hard to see how *modus Darwin* is a probability inference should not send alarm bells ringing among evolutionists. As emphasized in Chapter 1, it often happens in science that likelihood arguments stand on their own, even when probability arguments are not to be had. In §1.2, I briefly discussed how Eddington tested Newtonian mechanics against the general theory of relativity by observing an eclipse. The test provided evidence that favored one theory over the other; this does not require that the theories be assigned prior probabilities.

4.4 A SINGLE CHARACTER: SPECIES MATCHING AND SPECIES MISMATCHING

I now turn to the question of why the observation of a similarity – that species *X* and *Y* both have trait *T* – should be thought to favor the common ancestry hypothesis over the separate ancestry hypothesis. I'll begin by considering a dichotomous trait, but then I'll examine discrete *n*-state traits and continuous traits. I hope that the welter of detail that I'll present will not obscure how simple and intuitive the argument from similarity to common ancestry is.

When each student in a philosophy class is required to submit an essay on the meaning of life, and the essays that Smith and Jones submit are word-for-word identical, it is *possible* that they wrote their essays independently, and it also is *possible* that the students plagiarized from a common source, say a document on the Internet (Salmon 1984). Though the separate-origin and the common-origin hypotheses are both logically consistent with the observations, the similarity of the two essays is evidence that favors the hypothesis of common origination. I hope it is clear how this interpretation of the evidence can be represented in the likelihood framework. The matching of the essays is extremely improbable if they were written independently but would be much less surprising if they were obtained from a common source. The same line of reasoning is at work when we reason from the similarity of two species to their common ancestry. Our present task is to flesh out this likelihood argument in more detail.

Modus Darwin *for dichotomous characters*

Reichenbach's (1956) work on the *principle of the common cause* provides a sufficient condition for the common-ancestry hypothesis to have higher likelihood than the separate ancestry hypothesis,[5] relative to the observation that species X and Y both occupy the same character state. To start with the simplest case, I assume that the similarity in question concerns a dichotomous variable, whose two states are 0 and 1; the observation is that $X = i$ and $Y = i$ (where $i = 0$ or $i = 1$). There are nine assumptions that need to be stated. In each case, I'll state the idea in English and then I'll formalize it in terms of one or two numbered propositions that are expressed in terms of the variables used in Figure 4.6.

The first two assumptions assert that a descendant's probability of having a trait depends on the trait of its ancestor not on how many other descendants that ancestor has:

(1) $\qquad Pr(X = i \mid Z = j) = Pr(X = i \mid Z_1 = j)$, for $i, j = 0, 1$.

---

[5] I criticized Reichenbach's (1956) principle in §3.8; it says, recall, that when two event types are correlated, either the one causes the other, or the other causes the one, or the two trace back to a common cause. Although this principle is too strong, there is something of value in Reichenbach's argument. He proved that a common-cause model that is set up in a certain way *deductively entails* that the joint effects will be correlated. I will use this result in the present chapter to address a problem that differs from the one that Reichenbach considered. The problem here involves a comparative question (does the common-cause hypothesis have higher likelihood than the separate-cause hypotheses?), and the *explanandum* I consider is the similarity of two event *tokens*, not the correlation of two event *types* (Sober 1988).

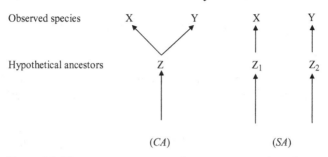

Figure 4.6 The common-ancestry and separate-ancestry hypotheses.

(2) $Pr(Y = i \mid Z = j) = Pr(Y = i \mid Z_2 = j)$, for $i, j = 0, 1$.

The next assumption is that an ancestor's prior probability of being in a given character state does not depend on how many descendants it has, and if it has just one, its probability of being in a given state does not depend on which descendant it has:

(3) $Pr(Z = i) = Pr(Z_1 = i) = Pr(Z_2 = i)$, for $i = 0, 1$.

Assumptions 4 and 5 say that *descendants and their ancestors are positively correlated.*[6]

(4) $Pr(X = i \mid Z = i) > Pr(X = i \mid Z = j)$ (for $i \neq j$).

(5) $Pr(Y = i \mid Z = i) > Pr(Y = i \mid Z = j)$ (for $i \neq j$).

This restates the backwards inequality that is a consequence of the Markov model presented in §3.5. Assumption 6 asserts that *two lineages stemming from a common ancestor evolve independently of each other:*

(6) $Pr(X = i$ and $Y = j \mid Z = k)$
       $= Pr(X = i \mid Z = k)Pr(Y = j \mid Z = k)$, for $i, j, k = 0, 1$.

The seventh assumption says that the same independence relation holds for the descendants of the two ancestors postulated by the *separate-ancestry* hypothesis:

---

[6] Notice that assumptions (4) and (5) do not assert that the process leading from Z to X and the process leading from Z to Y are probabilistically equivalent. Different processes can occur in the two lineages.

(7) $Pr(X = i \text{ and } Y = j | Z_1 = k \ \& \ Z_2 = l)$

$= Pr(X = i | Z_1 = k) Pr(Y = j | Z_2 = l), \text{ for } i, j, k, l = 0, 1.$

Assumption 8 says that the two ancestors postulated by the separate-ancestry hypothesis have character states that are probabilistically independent of each other:

(8) $Pr(Z_1 = i \ \& \ Z_2 = j) = Pr(Z_1 = i) \, Pr(Z_2 = j), \text{ for } i, j = 0, 1.$

And the ninth assumption speaks for itself:

(9)   All probabilities have values strictly between 0 and 1.

This exclusion of 0s and 1s as possible probability values is a substantive postulate; it is not a consequence of the axioms of probability.

These nine assumptions entail that the common-ancestry hypothesis (CA) has higher likelihood than the separate ancestry hypothesis (SA), relative to the observation that $X = 1$ and $Y = 1$. To see why, let

$Pr(X = 1 \mid Z = 1) = x \quad Pr(X = 1 \mid Z = 0) = a \quad Pr(Z = 1) = p$

$Pr(Y = 1 \mid Z = 1) = y \quad Pr(Y = 1 \mid Z = 0) = b.$

Assumptions 4 and 5 now can be stated as $x > a$ and $y > b$. It follows that

$$Pr(X = 1 \text{ and } Y = 1 \mid CA) > Pr(X = 1 \text{ and } Y = 1 \mid SA)$$

precisely when

$$pxy + (1 - p)ab > [px + (1 - p) \, a][py + (1 - p) \, b],$$

which simplifies to

(10) $(x - a)(y - b) > 0.$

The same conclusion would follow if the observations were $X = 0$ and $Y = 0$. Assumptions 1–9 therefore suffice to show that an observed *matching* favors the CA hypothesis over the SA hypothesis. By parity of reasoning, a *mismatch* must favor SA over CA. This is because the probabilities of the different possible observations, given a single hypothesis, must sum to unity:

$$Pr(X = 1 \ \& \ Y = 1 \mid CA) + Pr(X = 0 \ \& \ X = 0 \mid CA)$$
$$+ Pr(X = 1 \ \& \ Y = 0 \mid CA) + Pr(X = 0 \ \& \ Y = 1 \mid CA) = 1$$
$$Pr(X = 1 \ \& \ Y = 1 \mid SA) + Pr(X = 0 \ \& \ X = 0 \mid SA)$$
$$+ Pr(X = 1 \ \& \ Y = 0 \mid SA) + Pr(X = 0 \ \& \ Y = 1 \mid SA) = 1.$$

For dichotomous characters, matches favor CA over SA and *therefore* mismatches favor SA over CA.

If one of the nine conditions just described fails to be true for two species and a trait they share, does it follow that this similarity is *not* evidence favoring CA over SA? The answer depends on the assumption in question. For example, consider assumption 6, which says that the descendants of a common ancestor evolve independently of each other. Suppose two lineages evolve in continuing physical contact with each other so that the characteristics present in one lineage influence which traits evolve in the other. If so, assumption 6 is false. However, this does not show that similarity fails to provide evidence for CA in this circumstance, only that the argument just given fails to apply. The situation with respect to assumption 7, which says that ancestors have their character states independently of each other, is a bit different.

If there is a *perfect* correlation between the separate ancestors postulated by the separate-ancestry hypothesis, CA and SA will be evidentially indistinguishable. A similar judgment attaches to assumption 9, which says that none of the probabilities involved have values of 0 or 1. Suppose we violate this assumption by stipulating that the ancestors postulated by the common-ancestry and the separate-ancestry hypotheses were in character state 0 (or in character state 1) with probability 1. It then follows that the two hypotheses have *identical* likelihoods; the observed similarity of X and Y does not favor CA over SA, nor does it have the opposite evidential significance. Within the likelihood framework, it is essential to think of the evolutionary process probabilistically if we are to see how similarity can be evidence for CA (Sober 1988).

What about assumption 3: that an ancestor's probability of having a given trait is independent of whether that ancestor gives birth to two descendants or just to one? This is plausible for many traits but not for traits that affect the probability of speciation events. The literature on *species selection* provides a number of hypothetical examples of such traits (Sober 1984: 355–68; Coyne and Orr 2004: 442–5). For example, Stanley (1979) describes a hypothetical clade of grasshoppers; some species have wings while others are wingless. Wingless grasshoppers have higher probabilities of producing peripheral isolates and thus of giving birth to new daughter species by allopatric speciation. Characteristics of this sort can be such that $Pr(Z = \text{wingless}) \neq Pr(Z_1 = \text{wingless})$, thus violating assumption 3. But once again, the failure of this assumption in Stanley's example does not mean that winglessness fails to be evidence of CA; rather, the proper conclusion is that the argument given above does not apply.[7] Assumption 3 also requires that the two ancestors postulated by the separate-ancestry hypothesis have the same prior probability of exhibiting a given character state; cancel that and the simple algebra that leads to proposition 10 falls by the way.

A different situation arises in connection with assumptions 4 and 5, which assert that each lineage is such that ancestor and descendant are positively correlated. Notice how these premises serve to establish the conclusion, proposition 10. The conclusion would still be true if there were a *negative* correlation in both lineages (i.e., $x < a$ and $y < b$). The idea that ancestor and descendant are positively correlated is standard in evolutionary biology; this is pretty much what Darwin meant by his strong law of heredity: that like tends to beget like. Still, for traits in which there is no influence of parents on offspring – in which offspring traits are *independent* of the traits of parents – an observed similarity will fail to discriminate between CA and SA. And if there were a positive correlation in one lineage and a negative correlation in the other, the similarity of X and Y would favor the SA hypothesis over that of CA.

The idea of *competitive exclusion* provides a possible scenario in which similarity favors SA over CA. Suppose that two species that stem from a common ancestor will probably live in the same locale and will therefore have a high probability of diverging in their character states; if X exhibits character state 1, Y will probably exhibit character state 0. Suppose further that if the two species originate separately, they probably will live in separate locales and therefore will evolve independently of each other. In this situation, the observation that X and Y are in the same state favors SA over CA. This scenario involves a violation of assumption 6 – that descendants stemming from a common ancestor evolve independently.

Before working through the details of premises 1–9, you may have been inclined to think that it is *obvious* that similarity is evidence for

---

[7] Here's another possible example, this time concerning organismic, not species, genealogies: In a crowded environment, an organism may have a higher probability of having a second offspring if the first offspring is small.

common ancestry. Perhaps you also were disposed to ask the following rhetorical question – who, besides a philosopher, would bother belaboring the obvious? This dismissive complaint contains a grain of truth – it *is* a favorite pastime of philosophers to ask what justification there is behind the obvious. I hope the examination of premises 1–9 makes it clear why there is no *intrinsic* reason why similarity must count as evidence for common ancestry. *Whether this is so depends on nontrivial empirical matters of fact.* Propositions 1–9 are not consequences of the axioms of probability. Neither are they *necessary* conditions for common ancestry to have a higher likelihood than separate ancestry, and, for this reason, it would be wrong to regard them as *assumptions* that *modus Darwin* requires; however, these propositions *suffice* for similarity to be evidence for common ancestry, and they have broad applicability.

Homology

My focus has been on how a similarity (or a dissimilarity) that characterizes a pair of species provides evidence that discriminates between the common-ancestry and the separate-ancestry hypotheses. Isn't this to ignore the fundamental biological point that it is *homologies* that provide evidence for common ancestry? There is a large literature on how the concept of homology should be understood, but the question at hand in fact has a simple answer. Homologies are usually taken to be similarities that are present because of inheritance from a common ancestor; the wings of sparrows and robins are homologies in this sense. A *homoplasy*, in contrast, is a similarity that is not due to inheritance from a common ancestor but instead arose because independent origination events occurred in separate lineages; the wings of birds and bats are an example. So defined, the concept of homology already has built into it the claim of common ancestry. If our goal is to *test* the common-ancestry hypothesis against the separate ancestry hypothesis by looking at data, then it would beg the question to say that our data consist of "homologies" in this sense.[8] What counts as an observation in this problem must be knowable without one's already needing to have an opinion as to which of the competing hypotheses is true (§2.14). This is why similarities are the right place to begin.

---

[8] Sober (1988) argues that if synapomorphies are to be evidence for one phylogenetic tree over another, then the concept of a synapomorphy should not be defined to mean that the trait is a homology.

Multistate and continuous characters

Conditions 1–9 suffice for similarities to favor CA over SA and for dissimilarities to have the opposite evidential significance, when the trait in question is dichotomous. Does the argument extend to the case of discrete characters that come in more than two states?

There is a logical difference between dichotomous and n-state characters (where $n > 2$). For a dichotomous trait, if matches are evidence *for* CA, it must be true that mismatches are evidence *against*. The situation with respect to an n-state discrete character (where $n > 2$) is more complicated. If two species are in the same character state, that still is evidence favoring the CA hypothesis. However, if they differ in character state, their difference may or may not be evidence against. It all depends on the nature of the difference. In some cases, their different character states can actually be evidence *for* CA. To see this possibility, consider the number of chromosome pairs a diploid species might possess. If two species have exactly the same number of chromosomes, this is evidence that they have a common ancestor, by the argument of the previous section. But suppose that one species has twenty-three pairs and the other has twenty-four, as is the case for human beings and chimpanzees, respectively. What is the evidential significance of this difference?

The way to think about this question is not to subtract 23 from 24 and focus on the fact that the difference is a small number. Rather, we need to think about this question from the point of view of the law of likelihood (§1.3). Are the observed trait values more probable under the common-ancestry or the separate-ancestry hypothesis? The answer depends on the processes that govern the evolution of chromosome number. Consider, for example, the two *transformation series* depicted in Figure 4.7. The transformation series for a character gives the probabilities of different types of change from one character state to another. Figure 4.7a describes an n-state character in which all changes have the same probability (u); it is no harder for a lineage to change from T_2 to T_{10} than it is for it to change from T_9 to T_{10}.[9] A different arrangement is depicted in Figure 4.7b; here there is an ordering of the n states and the probability of changing from one state to an adjacent state is u; no direct jumping to a nonadjacent state is possible. With this transformation series, the probability of a lineage's changing from T_2 to T_6 is smaller than its probability of changing from T_5

---

[9] Figure 4.7a depicts the Jukes and Cantor (1969) model of nucleotide evolution when $n = 4$; this will be discussed in §4.8.

(a) $\quad T_i \; \underset{\longleftarrow}{\overset{u \;\longrightarrow}{\boxed{}}} \; T_j$

(b) $\quad T_i \; \underset{u \;\longrightarrow}{\overset{\longleftarrow}{\boxed{}}} \; T_{i+1}$

Figure 4.7 Two possible transformation series for a trait T that has n states (T_1, T_2, ..., T_n). Each describes the probabilities of changes in character state. In (a), all changes have the same probability; in (b), there is an ordering of character states and the only changes that are possible are changes to adjacent states.

to T_6. In the transformation series shown in Figure 4.7a, every mismatch between the species X and Y has the same evidential significance. Since exact identity of character state is evidence favoring CA over SA, every difference between the two species is evidence favoring SA over CA. The transformation series depicted in Figure 4.7b is different. In this case, two species exhibiting traits T_2 and T_{10} differ more than two species that exhibit traits T_9 and T_{10}. As before, an exact match in character state favors CA over SA. A very large difference in character state between the two species is evidence favoring SA over CA. A more modest difference might have either evidential meaning, depending on further details.

With the distinction between these two transformation series in mind, let's return to the example of chromosome number. If every change in chromosome number had the same probability as every other, then the fact that chimps have twenty-four chromosomes and humans have twenty-three would be evidence favoring SA over CA. However, if the transformation series is the one shown in Figure 4.7b, the near identity of chromosome number of chimps and humans might be evidence favoring CA. In contrast, the fact that a subspecies of the fern *Ophioglossum reticulatum* has 720 pairs (Stace 2000) whereas females in the ant *Myrmecia pilosula* have a single pair while the males are haploid (Gould 1991) is evidence favoring SA over CA if we assume the kind of transformation series depicted in Figure 4.7b and if this difference in chromosome number is the largest one possible. This does not entail that the two species have no common ancestor, only that their chromosome number favors that conclusion (if the assumption about the transformation series is true); other traits that they share might favor CA over SA.

It is intuitive that the laws of motion given in Figure 4.7a entail that every difference between species X and Y favors SA over CA. But what explains the fact that large differences favor SA over CA while modest differences can have the opposite significance when the transformation series is the one given in Figure 4.7b? Suppose our observation is that $X = 4$

and $Y = 5$ where the character in question has ten states, and let's further suppose that the ancestors depicted in Figure 4.6 have a non-negligible chance of producing descendants with those character states only if the ancestors are within two units of their descendants. This is tantamount to saying that u^3 is so small that it can safely be ignored in comparing likelihoods. If the character we are considering has ten character states, then the ancestor postulated by the common-ancestry hypothesis can produce these descendants if and only if it is in one of states 3–6, while the two ancestors postulated by the separate-ancestry hypothesis can produce these descendants precisely when the first is in states 2–6 and the second is in states 3–7. If the ten character states an ancestor might occupy have equal probability,[10] then the probability that the common ancestor is in the range 3–6 is 0.4, while the probability that each of the separate ancestors is in range of its respective descendant is $(0.5)(0.5) = 0.25$. So CA has the higher likelihood with respect to the observation that $X = 4$ and $Y = 5$. But now suppose there is a wider gap between X and Y; for example, suppose that $X = 4$ and $Y = 8$. Then the single ancestor postulated by the common-ancestry hypothesis must be in state 6 if it is to have a chance of producing these two very different descendants, whereas the two ancestors postulated by the separate-ancestry hypothesis can yield these descendants if the first is in one of the states 2–6 while the second is in one of the states 6–10. With equal probabilities on ancestral character states, the ancestor postulated by the common-ancestry hypothesis has a 0.1 chance of being in a state that can generate the data, while the two ancestors postulated by the separate-ancestry hypothesis again have a $(0.5)(0.5) = 0.25$ chance of doing so. Now it is SA that makes the observations more probable. I hope this informal argument makes it intuitive why a small difference between X and Y can favor CA over SA whereas a larger difference will have the opposite evidential significance, if the transformation series is the one given in Figure 4.7b.

The two models of evolution shown in Figure 4.7 do not exhaust the possible transformation series that might be true of an n-state character. A more complex and realistic model might allow there to be a different probability for every type of change. And even when larger changes usually have smaller probabilities than more modest changes, there can be exceptions. Consider, for example, the process of *polyploidy*, wherein chromosome number doubles or triples or quadruples. In modeling the evidential

---

[10] This prior probability for ancestral character states is the equilibrium distribution for the process depicted in Figure 4.7b; it is not an additional postulate.

significance of chromosome number, we'd have to take this process into account as well as the process of adding or deleting a single chromosome pair. Maybe it is easier to go from twenty to sixty chromosomes than it is to go from twenty to fifty-nine. This means that if two species have twenty and sixty pairs respectively, this might be evidence favoring CA over SA, whereas the opposite conclusion might be correct if they exhibit twenty and fifty-nine. Once again it is a mistake to focus on how much or how little two species differ in their character state; what is fundamental is the processes at work in the evolution of their trait values.

These complications should not obscure the essential points. Exact identity of character state is *always* evidence favoring CA over SA so long as assumptions 1–9 from the previous section hold true. And regardless of whether the character is dichotomous or multistate, *some* differences in character state *must* count as evidence favoring SA over CA. Whether *all* differences in character state constitute evidence favoring SA over CA depends on the transformation series; this is true of the rules represented in Figure 4.7a, but it isn't true of others. Notice that assumptions 1–9 say nothing about the transformation series; this means that interpreting differences in character state for *n*-state characters requires additional assumptions beyond those that suffice for interpreting differences in dichotomous traits.

The transformation series represented in Figure 4.7b has no bias; evolving from one state to another has the same probability as evolving in the opposite direction. This transformation series is therefore appropriate for modeling a drift process but not for modeling how selection leads a lineage to evolve towards a single optimum. In any event, as the number of character states in this transformation series is increased, and *u* is made small, one approaches a model of the evolution of a continuous character subject to drift. By viewing a continuous trait as the limit of an *n*-state trait (where *n* is made large), you can see how the conclusions stated in the previous paragraph apply to continuous characters.

We have just seen that *n*-state characters differ epistemologically from dichotomous characters. But which are *better* to use in testing CA against SA? No sweeping generalization can be expected in answer, but there is a special case of this question that the principle of total evidence addresses. Returning to the example of human and chimp chromosome number, let's define a dichotomous character *W* (for "weak") by saying that a species has *W* precisely when it has twenty-four or more chromosome pairs. Human beings lack W but chimps have *W*, so, by the argument of the previous section, their difference with respect to this dichotomous character is evidence against the hypothesis that they have a common ancestor. But now let's define a logically stronger feature S that comes in n states; S_1 means that a species has one chromosome pair; S_2 means it has two, and so on. The n-state character S is logically stronger than the dichotomous character W because the trait value a species has for S logically entails its value for W, but not conversely. Suppose the laws governing the evolution of chromosome number have the consequence that humans having twenty-three and chimps having twenty-four chromosome pairs is evidence in favor of their having a common ancestor. If so, the dichotomous character state W and the n-state character S point in opposite directions. Which should we take more seriously? The principle of total evidence says that we should use the strong description (S), not the weak one (W).

The reader will have noticed a dismaying arbitrariness that enters into the definition in this example of the dichotomous trait W. I defined the cutoff between W and $notW$ as twenty-four, with the consequence that humans and chimps differ with respect to W. But I could just as easily have defined a dichotomous trait W^* that applies to a species when it has more than fifteen chromosome pairs, with the result that humans and chimps both have W^*. Differing with respect to W favors SA over CA, but sharing W^* has the opposite significance. So which dichotomous trait, W or W^*, should we use to evaluate the common-ancestry and separate-ancestry hypotheses? This arbitrariness does not arise if we move to the count property S. However, the price of using this more informative description of the two species is that we face a new epistemological problem; we need a substantive theory of how the n-state character evolves before we can say what the evidential significance is of the observed difference between the two species.

The paradox of the heap is a philosophical staple that traces back to ancient Greece; there is no precise number of pebbles that separates a heap from a nonheap, and no precise number of hairs that separates the bald from the not bald. These familiar facts do not and should not deter us from using those concepts, since, after all, there are plenty of clear cases of heaps and bald people. The present problem is different; it is evidential, not taxonomic. When a dichotomous character is imposed on an underlying reality of quantitative difference, the dichotomy can make intuitive sense (as when we say that some people are bald while others are not) or it can seem utterly arbitrary (as when I invented the character W). But whether the dichotomy is familiar or not, other dichotomies laid on the same underlying quantitative reality are logically possible, and so the question arises of which we should use to describe the evidence. Shifting from a dichotomous to a multistate or continuous character cuts this

Gordian knot, and it has the additional advantage of being sanctioned by the principle of total evidence.[11]

These points are relevant to discussion of how the "universality" (or "near universality") of the genetic code is said to bear on the question of common ancestry. Back when it was thought that all organisms now alive use exactly the same genetic code, evolutionary biologists argued that this universal code is compelling evidence that all present life forms trace back to a single common ancestor (see, for example, Crick 1968). However, when biologists started discovering exceptions to this "universal" code in the late 1970s, this did not lead them to retreat from their endorsement of the CA1 hypothesis (Cavalcanti and Landweber 2004). Is this an example of biologists wanting to have it both ways? If a universal code would be evidence *for* CA, doesn't it follow that nonuniversality must be evidence *against*? Answering this question requires that we be careful to distinguish logically stronger from logically weaker descriptions of the evidence, just as we did for the case of chromosome number. If there are *n* possible codes, an exact match between the codes used by two or more organisms is evidence that they share a common ancestor, given assumptions 1–9. And if we use a logically weak description of two organisms that use different codes and say only that they differ (without saying how they differ), then the conclusion must be drawn that this weak description of the evidence favors SA over CA. However, once we describe *how* the two organisms differ (rather than just saying *that* they differ), the different trait values of the two organisms do not automatically lead to the conclusion that SA is more likely than CA. We have to look at the details.

The standard code is used in the nuclear genomes of all multicellular plants and animals, and this is evidence that they spring from a common ancestor. But what about the nonstandard codes (there is more than one) found in some prokaryotes, in some fungi and algae, and in the mitochondria of eukaryotes? Are these differences in code evidence against CA1? This follows no more than it follows that humans' having twenty-three chromosome pairs and chimps' having twenty-four is evidence that they lack a common ancestor. The question is whether the different codes we observe differ "enough," and this question requires a substantive theory of code evolution to answer.[12]

---

[11] The question of whether a dichotomous or a continuous character should be used to describe the character state of a species arose in §3.6 in connection with the problem of testing selection against drift.

[12] For a useful summary of the different codes now known to exist and a survey of ongoing work on theories of code evolution, see Knight et al. (2001).

Phenotypic and genetic characters

The discussion in the previous section of the relationship between a logically weaker dichotomous character W and the logically stronger multistate characters S bears on the relationship of phenotypic and genetic characters. If a genetic character G entails a phenotypic character P, but not conversely, then the former is logically stronger, and the principle of total evidence says that it is the genetic character, not the phenotype, that should be used in inferences about common ancestry. Suppose species X and Y both have phenotype P, but it then turns out that they manifest this commonality because they use two very different gene complexes, G_1 and G_2. The shared phenotype P favors CA over SA, but it may turn out that the fact that X has G_1 and Y has G_2 is evidence favoring SA over CA. According to the principle of total evidence, the genetic difference trumps the phenotypic similarity.[13] This point throws light on Zuckerkandl and Pauling's (1965) contention that phylogenetic trees inferred from genetic data are better than those inferred from phenotypic data because genes are "causally prior" to phenotypes. A similar pattern applies to some pairs of phenotypic traits. Sharks and whales both have fins, but a closer examination of what is inside those fins shows that there are morphological differences. Sharing fins is evidence for CA, but the difference in morphology may be evidence against. Which of these traits deserves more weight? If the detailed morphology of each organism guarantees that it has fins, we have an answer. Logically stronger traits take precedence over traits that are logically weaker.

Sequence data without alignment

The usual technique for using sequence data in phylogenetic inference problems is first to align the sequences, which means sliding one sequence along the other and stopping when the number of matching sites is greatest. I won't discuss here the use of this procedure in problems where the goal is to determine which tree is best supported by the data, where all the trees considered assert that the tip taxa trace back to a common ancestor; this is the kind of problem depicted in Figure 4.1. In inference

---

[13] Does considering the possibility that genotype entails phenotype involve flirting with some discredited notion of genetic determinism? Does it contradict the truism that an organism's phenotype is the joint effect of its genes and its environment? No; all that is required is that the organisms under study, *in the environment they actually inhabit*, obey the principle "same gene, *ergo* same phenotype" but not the converse. This leaves it open that a change in environment might produce a change in phenotype.

problems of this sort, the common-ancestry hypothesis is *assumed*; it is not *tested*. My present concern is how sequence data can be used to test CA against SA. At first glance, alignment seems not to make sense in this problem. Since matching at a site is evidence for CA, aligning the sequences so as to maximize matching seems to load the dice in favor of the common-ancestry hypothesis. But the problem is deeper. If two sequences have a common ancestor, it makes sense to say that a site in one sequence "corresponds" to a site in the other; this correspondence means that the two sites derive from a site in their common ancestor. But if there was no such common ancestor, what would alignment even mean? If we want to *test* the separate-ancestry hypothesis rather than just *assume* from the outset that it is false, we need to rethink the question of how sequence data can be used.

To begin, imagine that two sequences, each 2,000 sites long, are drawn from species X and Y and that each sequence consists of 500 repetitions of the nucleotides G–A–T–C in that order. The two sequences therefore occupy the same state of a character that has $4^{2,000}$ possible states. By the argument given earlier, this matching counts as evidence of CA. There is no need to align the sites to say this. The same point applies when the sequences (each n sites long) drawn from the two species do not match perfectly. They then will occupy different states of a single character that has 4^n possible states. Whether this difference between the two species favors CA or SA depends on the rules of evolution that govern how this complex character evolves. The situation is precisely the same as the one considered earlier in connection with number of chromosome pairs. The question is simply whether the observed mismatch has a higher probability of arising under the common-ancestry or the separate-ancestry hypothesis. To answer this question, all that is needed is the two *un*aligned sequences and a reasonable model of the process of sequence evolution. An inference that begins with aligned sequences is valid to the extent that it mimics the verdicts of the procedure that uses unaligned sequences. When this is true, aligning sequences is not loading the dice.

Two inference problems about common ancestors

There are two questions one might ask about common ancestors:

- Does the observation that two species (X and Y) are in the same character state favor the CA hypothesis over the SA hypothesis?

- Does the observation that two species (X and Y) are in the same character state favor the hypothesis that their common ancestor occupied the same character state?

The first question concerns *testing* the CA hypothesis; the second *assumes* that there was a common ancestor and asks how one might test hypotheses about that ancestor's characteristics. Corresponding to these two questions are two possible answers, each taking the form of a likelihood inequality:

- $Pr(X=i$ and $Y=i \mid CA) > Pr(X=i$ and $Y=i \mid SA)$.

- $Pr(X=i$ and $Y=i \mid CA$ & the most recent common ancestor of X and Y was in state $i) > Pr(X=i$ and $Y=i \mid CA$ & the most recent common ancestor of X and Y was in state j$)$, for all i $\neq$ j.

Propositions 1–9 entail *both* these inequalities if the trait in question is dichotomous. However, for continuous traits, those nine propositions entail the first but not the second. For continuous characters, inferring the *existence* of a common ancestor is one thing, inferring its *characteristics* another. Propositions 1–9 are neutral on whether lineages experienced drift or selection. This neutrality does not prevent those propositions from answering the first question, but it does prevent them from answering the second if the trait is continuous.[14]

These points bear on the questions explored in §4.2. There are three ways in which a similarity observed to unite organisms X and Y might trace back to a common cause:

- The common cause is a common ancestor.
- The common cause is an organism though not a common ancestor.
- The common cause is not an organism.

The second of these possibilities might involve lateral gene transfer; the third would be true if X and Y originated from the same slab of rock but lacked a common ancestor. When the human beings and chimps alive now share a characteristic (and propositions 1–9 are true), this is evidence that there was a common cause. But what characteristics did that common cause have? Biologists do not doubt that it was an *organism*, but they go much farther – it was an animal, a vertebrate, a mammal, and a primate.

---

[14] Not that dichotomous characters are *always* a safe haven. Recall that estimating ancestral character states in a *tree* depends on the process at work in lineages even when the trait is dichotomous (§3.11).

The same point holds when we consider organisms that are more distantly related. Human beings and daffodils are both Eukaryotes, and biologists take it to be very clear that their most recent common ancestor was a Eukaryote. However, for a Prokaryote and a Eukaryote, their commonalities point to an organismic common cause, but the possibility of rampant lateral gene transfer may make it difficult to say whether that organism was an ancestor of both.

Matching and mismatching summarized

The modest assumptions embodied in propositions 1–9 entail, for any dichotomous character, that sameness of character state favors CA over SA while a difference favors SA over CA. This result is represented by the two horizontal inequalities in Figure 4.8: $1 - e_1 > e_2$, which of course is equivalent to $1 - e_2 > e_1$. Half of this result applies to multistate characters; exact matches favor CA, but if the two species differ, further information is needed to say whether the different character states they occupy favor CA or SA.

These epistemological conclusions should not obscure a point of logic. The hypothesis that two species share a common ancestor is logically consistent with their being as dissimilar as you please. It is perfectly possible for natural selection to adapt each lineage to its own special environment; given enough time and enough divergent selection pressure, the two descendants will evolve away from the suite of characters present in their common ancestor. And just as descendants of a common ancestor can be dissimilar, it also is possible for two species that lack a common ancestor to be rendered similar by powerful selection pressures. The result is wholesale convergence. The four logical possibilities are represented in Figure 4.7. It is important to keep this logical point separate from the epistemological one. The CA hypothesis is *logically consistent* with similarity and difference, and the same is true of the SA hypothesis. However, it

		Genealogical hypotheses	
		CA	*SA*
Possible	**X and Y match**	*1–e₁*	*e₂*
observations	**X and Y differ**	*e₁*	*1–e₂*

Figure 4.8 When *X* and *Y* are scored for whether they match on a dichotomous trait, there are two possible observations. Since there are two possible genealogical hypotheses, there are four possibilities to consider. Probabilities of the form *Pr*(Observation | Genealogical hypothesis) are given in the four cells.

would be a mistake to conclude, by invoking a Popperian falsifiability criterion (§2.8), that neither hypothesis is testable. Evidence can *discriminate* between the two hypotheses even though the evidence cannot *deductively entail* that either hypothesis is false. The law of likelihood does the needed work; it does what the criterion of falsification cannot do. Assumptions 1–9, coupled with the law of likelihood, allow the data to speak.

This likelihood analysis illustrates how misleading it can be to assess the evidence for CA by asking what the hypotheses of common and SA "predict." Prediction can of course be probabilistic (as when a weather forecaster predicts rain tomorrow based on the fact that the probability of rain is 0.98); that is not the source of the stumbling block. But suppose you believe the following simple principle concerning what the word "predict" means: *If a hypothesis predicts that O is true, then the hypothesis does not predict that notO is true.* If this is right, and if the hypothesis confers a nonzero probability on both *O* and on *notO*, which of these outcomes does it predict? Presumably, the answer is that it predicts the outcome that it says is more probable. This means that if the CA hypothesis predicts that species X and Y will match with respect to the dichotomous trait T, then CA must confer on that outcome a probability that is greater than 0.5. No such claim is entailed by the Reichenbachian argument developed here. In terms of Figure 4.8, the main result so far is the identification of a sufficient condition for the "horizontal inequality" $(1 - e_1) > e_2$, and so for $(1 - e_2) > e_1$. These horizontal relationships do not settle the vertical questions of whether $(1 - e_1) > e_1$ or whether $(1 - e_2) < e_2$. If you want to know what CA and SA "predict," you need to know whether $e_1 < 0.5$ and whether $e_2 < 0.5$. These vertical questions are a distraction if you are using the law of likelihood to think about the interpretation of evidence. From that point of view, the right advice is . . . *get horizontal.*[15]

4.5   MORE THAN ONE CHARACTER

Overall similarity in a data set

We so far have considered the evidential significance of a single character. The next step is to ascertain the evidential meaning of a data set that

---

[15] The question of whether these vertical inequalities are true is the question of whether pairs of species are apt to generate evidence that is misleading with respect to whether they have or lack a common ancestor. I will address a version of this query in §4.8.

describes both the similarities *and* the dissimilarities that characterize species X and Y.

For a set of dichotomous characters, each similarity "votes" for CA over SA, and each dissimilarity "votes" for SA over CA. Should we therefore adopt a principle of majority rule, according to which CA is overall more likely precisely when the similarities outnumber the dissimilarities? This rule can be wrong if the similarities and the dissimilarities are governed by different evolutionary processes.[16] But even if the traits evolve independently and by the same process, the principle of majority rule can still be wrong. Here's an argument that helps explain why this is so. If a dichotomous character has trait values of 0 and 1, then there are four possible observations you can make when you score species X and Y; call them 11, 10, 01, and 00. Let's lump the first and fourth possible observations together and say that X and Y "match" when they are both in state 1 or both in state 0; when X and Y differ, we have a "mismatch." If the likelihood ratio of CA to SA with respect to a match has the value c/s, then the likelihood ratio of SA to CA with respect to a mismatch will have the value $(1 - s)/(1 - c)$. By the argument from §4.4, both these likelihood ratios are greater than unity – matching favors CA and mismatching favors SA. But which ratio is greater? The majority-rule principle says that they are equal in value. To see how this can be wrong, suppose that $c = 0.001$ and $s = 0.00001$; this means that the first ratio has value of 100 while the second has a value only slightly greater than 1. If all the traits in a data set have these values for c and s, then one similarity speaks more strongly in favor of CA than ninety-five differences speak in favor of SA. This is why the rule of "one trait, one vote" is not inevitably true. More generally, if two traits evolve according to the same probabilistic rules, and X and Y match with respect to one and mismatch with respect to the other, the two characters together favor CA over SA precisely when

$$\frac{c}{s} > \frac{(1 - s)}{(1 - c)}$$

which (given that $c > s$) simplifies to $c + s < 1$. The match outweighs the mismatch precisely when $Pr(\text{match} \mid \text{CA}) + Pr(\text{match} \mid \text{SA}) < 1$. The example values of $c = 0.001$ and $s = 0.00001$ make this inequality

---

[16] By "wrong" I don't simply mean that the rule can say that one hypothesis is better supported than the other by a data set when, in fact, the "better" hypothesis is false and the "worse" hypothesis is true; that can happen in any nondeductive inference. What I mean is that "majority rule" can err in its evaluation of which hypothesis has the higher likelihood.

true; according to these values, the common-ancestry hypothesis says that a matching is improbable, but the separate-ancestry hypothesis says that the matching is even more improbable. For characters of this sort, a match *strongly* favors CA over SA while a mismatch favors SA over CA, but only very *weakly*. When $c = 0.001$ and $s = 0.00001$, it may sound wrong to say that either hypothesis "predicts" that the two species will match, but that does not matter.   .

This example shows that a data set can favor CA over SA even when the two species X and Y *mis*match on most of their characters. But suppose they match on most. Is that enough for CA to have the higher likelihood? Again the answer is *no*. Everything depends on the processes at work that produced the data. If each dichotomous trait is subject to drift and lineages have been evolving for a long time, the expected percentage of similarity under the separate-ancestry hypothesis is 50 percent. On the other hand, if there is strong selection for state 1 of each of the many dichotomous characters considered, so that the probability of a descendant's exhibiting state 1 for a given character is, say, 0.99, then the expected similarity under the separate-ancestry hypothesis is $(0.99)^2 + (0.01)^2 = 98$ percent. A data set in which species X and Y are 98 percent similar strongly favors CA over SA if the lineages have been drifting for a long time, but these data can fail to discriminate between CA and SA if there is strong selection for the observed traits.

Much is made in the popular press of the fact that the DNA sequences of chimps and humans are about 98 percent similar. Is this similarity compelling evidence that we and chimps share a common ancestor? Since each site in a sequence is characterized by one of four nucleotides, we have here a set of four-state, not dichotomous, characters, but the point about the processes generating the data is the same. The expected percentage similarity for humans and chimps, according to the separate-ancestry hypothesis, would be 25 percent, if each site evolved independently by the process of random genetic drift (and the lineages had been evolving for a long time). In this case, the observed similarity of 98 percent would strongly favor CA over SA. However, if there was strong selection in each lineage for the traits that one observes, the expected degree of similarity would be about the same, regardless of whether the common-ancestry or the separate-ancestry hypothesis is true.

Both of the following thoughts are therefore naive: "humans and chimps must share a common ancestor because they are so similar" and "humans and mushrooms must have arisen independently because they are so different." Within a probabilistic framework, there is no "must" in either case. But, in addition, the transition from degree of similarity to a claim about CA versus SA in both examples must be mediated by information about the processes generating the traits in question.

Comparing kinds of similarity

We have seen that sameness of character state always favors the common-ancestry (CA) hypothesis over the separate ancestry (SA) hypothesis, if the assumptions described in §4.4 hold. This is true whether the trait is dichotomous or multistate. In this respect, *all* similarities are on the same page. But which types of similarity provide strong evidence for CA over SA and which provide evidence that, though positive, is relatively weak? In *The Origin of Species*, Darwin ([1859] 1964: 427) answers this question in a way that makes a great deal of sense:

adaptive characters, although of the utmost importance to the welfare of the being, are almost valueless to the systematist. For animals belonging to two most distinct lines of descent, may readily become adapted to similar conditions, and thus assume a close external resemblance; but such resemblances will not reveal – will rather tend to conceal their blood-relationship to their proper lines of descent.

This is why the torpedo shape found in sharks and dolphins does not provide strong evidence for their having a common ancestor; natural selection favors this shape in large aquatic predators, so we'd expect it to be present in modern sharks and dolphins regardless of whether they have a common ancestor. In contrast, the gill slits found in human embryos and in many fish are evidence of common ancestry precisely because they have no adaptive utility in human embryos. The term "vestigial" carries the double meaning that Darwin intended; vestigial traits are useless to the organism and they are vestiges of a bygone age. By recognizing that they are useless, we see that they provide substantial information about the past.

Darwin's principle – that selectively advantageous traits are "almost valueless" as evidence of CA – can be represented in terms of the law of likelihood. As a first approximation (to be refined shortly), the claim is that

$$\frac{Pr(X \text{ and } Y \text{ have trait } T \mid \text{there was selection for } T \& CA)}{Pr(X \text{ and } Y \text{ have trait } T \mid \text{there was selection for } T \& SA)} \approx 1.$$

His complementary claim – that similarities due to neutral evolution provide substantial evidence of CA – also has a likelihood representation. Here is a first pass at stating it:

$$\frac{Pr(X \text{ and } Y \text{ have trait } T \mid T \text{ evolved by neutral evolution \& } CA)}{Pr(X \text{ and } Y \text{ have trait } T \mid T \text{ evolved by neutral evolution \& } SA)} \gg 1.$$

The task at hand is to investigate how the process governing the evolution of a similarity affects that similarity's evidential significance. In addition to the adaptive and neutral similarities just mentioned, we will consider similarities that involve disadvantageous traits and adaptive similarities that arise when there is a certain kind of frequency-dependent selection. The point of this exercise is to investigate in a more rigorous and systematic way the prima-facie plausible generalization that Darwin formulated.

Three processes

The three types of evolutionary process I want to discuss are represented in Figure 4.9. In Figure 4.9a, the fitnesses of the two traits A and B do not depend on their frequencies in the population. Because A is fitter than B at all frequencies, natural selection can be expected to lead A to evolve to fixation.[17] In Figure 4.9b, the traits A and B have identical fitnesses, and so the evolution of these traits in a lineage will be a random walk. In Figure 4.9c, selection is frequency dependent; the majority trait is the one that selection favors, so a lineage can be expected to evolve to either 100 percent A or to 100 percent B, depending on the trait's starting frequency.[18]

We know from the previous section that the observation that two species X and Y are in the same character state favors the common-ancestry over the separate-ancestry hypothesis, regardless of what processes are at work in

---

[17] The models considered here are phenotypic and, as is usual for such models, I will assume that "like begets like." As mentioned in Chapter 3, it is well known that there are genetic arrangements (e.g., heterozygote superiority) in which the fitter trait cannot evolve to fixation. I will ignore these genetic complications in what follows in order to isolate what I think are some fundamental epistemological points.

[18] A genetic example of frequency-dependent selection for the majority trait is provided by a one-locus two-allele model in which the genotypic fitnesses are $w_{AA} = w_{aa} > w_{Aa}$. In this case of heterozygote inferiority, the genotypic fitnesses are frequency independent, but the allelic fitnesses are frequency dependent, with the majority allele being the fitter.

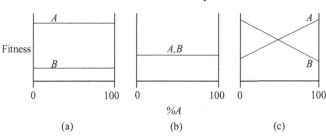

Figure 4.9 Three processes: (a) frequency independent selection for trait A; (b) drift; (c) frequency dependent selection for the majority trait.

lineages (as long as assumptions 1–9 are correct). The "favoring" there described involves a difference in likelihoods; the assumptions entail that $Pr(X$ and Y have $T \mid$ CA$) > Pr(X$ and Y have $T \mid$ SA$)$. The present question is how the *magnitude* of this favoring is affected by which process is at work. In what follows, I'll measure strength of evidence in terms of the quantitative measure that the law of likelihood recommends: the likelihood *ratio* (§1.3). Our questions will therefore be: Which of the following three likelihood ratios is the biggest? Which is second? And which comes in third?

(FIS-ADV) $\dfrac{Pr(X \text{ and } Y \text{ have } T \mid \text{ CA \& Frequency independent selection for } T)}{Pr(X \text{ and } Y \text{ have } T \mid \text{ SA \& Frequency independent selection for } T)}$

(D) $\dfrac{Pr(X \text{ and } Y \text{ have } T \mid \text{ CA \& Drift})}{Pr(X \text{ and } Y \text{ have } T \mid \text{ SA \& Drift})}$

(FDS) $\dfrac{Pr(X \text{ and } Y \text{ have } T \mid \text{ CA \& Frequency dependent selection for the majority trait})}{Pr(X \text{ and } Y \text{ have } T \mid \text{ SA \& Frequency dependent selection for the majority trait})}$

There is a fourth possibility. Suppose there is frequency-independent selection, but the trait that evolves is not the one selection *favors*; instead, what evolves is the trait that is selected *against*. The similarity exhibited by the descendants X and Y will then involve a *dis*advantageous trait. How does the evidence that this kind of similarity provides compare with the evidence provided by the other three? To answer this question, we need to consider a fourth likelihood ratio:

(FIS-DIS)
$$\frac{Pr(X \text{ and } Y \text{ have } T \mid \text{CA \& Frequency independent selection against } T)}{Pr(X \text{ and } Y \text{ have } T \mid \text{SA \& Frequency independent selection against } T)}$$

Models for a dichotomous trait

To compare the impact of the three processes described in Figure 4.9 on the evidential significance of the observed similarity of species X and Y, I want to shift from describing X and Y in terms of trait frequencies to describing each in terms of a single dichotomous variable, whose two states are 0 and 1. I discussed the usual Markov model representation of this process in §3.5. The model begins with the thought that within a very small period of time (an "instant"), there is a small probability (u) of a lineage's changing from state 0 to 1, and a possibly different probability (v) of changing from 1 to 0. A lineage's probability of ending in state j, if it begins in state i ($i, j = 0,1$) and there have been t units of time in between, $Pr_t(i \rightarrow j)$, is a function of these instantaneous probabilities and the amount of time.

How can this simple model be used to represent the three different processes we wish to investigate? I will deploy the same representations that were used in Chapter 3. Drift will be represented in this Markov format by the constraint that $u = v$. With drift, there is no bias in a lineage's tendency to change; it has the same chance of going from 0 to 1 that it has of going from 1 to 0. Frequency independent selection for character state 1 means that $u \gg v > 0$. And frequency-dependent selection for the majority trait means that $u \gg v > 0$ when the ancestor is in state 1 and $v \gg u > 0$ when the ancestor is in state 0. By never setting u or v equal to zero, I am assuming that all three of these processes experience a mutational input. Drift does not have 0 percent A and 100 percent A as absorbing states, and what frequency-independent selection has as its expected outcome (when there is lots of time for the lineage to evolve to its equilibrium) is not 100 percent A, but something a bit less, say 99 percent. When $t = \infty$, $Pr_t(i \rightarrow j) = Pr_t(j \rightarrow j)$ if the process is one of drift or frequency independent selection. However, this equality fails to hold if there is frequency-dependent selection for the majority trait. In that case, $Pr_t(i \to j) \approx 0$ and $Pr_t(j \to j) \approx 1$ when $t = \infty$; this is because, with frequency dependence, a lineage beginning in state 1 will experience strong selection for state 1, while a lineage beginning in state 0 will experience strong selection for state 0. In cases (a) and (b) depicted in Figure 4.9, the process that occurs in a lineage is independent of the lineage's initial state; this is not true in (c).

In general, the likelihood ratio of CA to SA, where the data are that $X = 1$ and $Y = 1$, will have the form:

$$(G) \quad \frac{Pr(X = 1 \,\&\, Y = 1 \,|\, CA)}{Pr(X = 1 \,\&\, Y = 1 \,|\, SA)} = \frac{(1 - p)Pr_t(0 \to 1)^2 + (p)Pr_t(1 \to 1)^2}{[(1 - p)Pr_t(0 \to 1) + (p)Pr_t(1 \to 1)]^2}.$$

Here p is the probability that the relevant ancestor has character state 1. I will assume in what follows that the processes we are considering don't suddenly start in the lineages leading from the ancestors (Z, Z_1, Z_2) to the descendants X and Y but are present in all the lineages depicted in Figure 4.3. This means that ancestors in the drift process have a probability of about $\frac{1}{2}$ of being in state 0, while the ancestors in the frequency-independent selection process have a probability of about 0.99 of being in state 1. These priors for ancestors are the equilibrium values for the processes considered. For the case of frequency dependence, the process itself does not provide a prior for the ancestral condition, and I will not introduce a postulate that fixes its value.

The generic formula (G) for the likelihood ratio will take different specific forms, depending on the kind of process at work in lineages. For drift, the formula becomes:

$$(D) \quad \frac{\left(\frac{1}{2}\right)a^2 + \left(\frac{1}{2}\right)(1 - a)^2}{\left[\left(\frac{1}{2}\right)a + \left(\frac{1}{2}\right)(1 - a)\right]^2}.$$

In a drift process, $p = \frac{1}{2}$ and $a = Pr_t(0 \to 1) = Pr_t(1 \to 0)$; $a = 0$ if $t = 0$, and $a = \frac{1}{2}$ if $t = \infty$. At t_0, the drift ratio D has a value of 2; at t_∞, D = 1. When the two species X and Y match with respect to a neutral dichotomous character, the matching always favors CA over SA, but how much evidence this matching provides declines as the amount of time in the lineages increases.

When there is frequency-independent selection favoring character state 1, and the species we observe both exhibit that advantageous trait, the generic formula for the likelihood ratio takes the following form:

(FIS-ADV) $$\frac{(0.01)b^2 + (0.99)(1)}{[(0.01)b + (0.99)(1)]^2}.$$

Here b represents $Pr_t(0 \rightarrow 1)$ when there is frequency-independent selection for trait 1; $b = 0$ when $t = 0$ and $b \approx 1$ when $t = \infty$. The quantity $Pr_t(1 \rightarrow 1)$ is here assigned a value of unity; if there is (strong) selection for trait 1, a lineage that begins in that state cannot escape it. This is a slight simplification; strictly speaking, $Pr_t(1 \rightarrow 1) = 1$ at $t = 0$ and monotonically declines in value until $Pr_t(1 \rightarrow 1) = 0.9999$ when $t = \infty$. When two species match with respect to an advantageous dichotomous character, the FIS-ADV ratio has a value of 1.01 at t_0; at t_∞, the ratio has a value of unity. Here again, the amount of evidence that a matching provides declines as the amount of time in the lineage increases.

These values for the FIS-ADV ratio provide a vindication (within this simple model) of Darwin's statement that adaptive similarities are "almost valueless" in terms of their ability to indicate genealogical relationships. It is interesting that Darwin said "almost" valueless, rather than "completely." Royall (1997: 89) suggests defining the idea that there is "strong" evidence favoring one hypothesis over another as a likelihood ratio of at least 8. This convention entails that an adaptive similarity (with respect to a dichotomous trait) that evolves by frequency-independent selection never represents "strong" evidence for CA. It is worth remembering, however, that this claim concerns the evidential significance of a *single* similarity. Two or more similarities, if they are independent of each other (conditional on each of the phylogenetic hypotheses considered) will have an evidential significance characterized by the *product* of the likelihood ratios for each. Even if *each* of several characters fails to provide strong evidence, they *collectively* may provide very strong evidence indeed.[19]

---

[19] Just as increasing the number of *similarities* can strengthen the evidence favoring CA over SA, so increasing the number of similar *species* can have the same effect. The problem explored here, of *two* species and *one* similarity, is thus the weakest case, in each of two dimensions.

The third process we need to consider is frequency-dependent selection for the majority trait; here the relevant likelihood ratio is:

(FDS)
$$\frac{(1-p)c^2 + (p)(1-c)^2}{[(1-p)c + (p)(1-c)]^2}.$$

Here $c = Pr_t(0 \rightarrow 1)$ when there is selection for trait 0 and $Pr_t(1 \rightarrow 1) = (1 - c)$ when there is a selection for state 1. The parameter c has a very small value if the frequency-dependent selection process is very strong, regardless of how much time there is in lineages, so the frequency-dependent selection ratio has a value that is approximately equal to $\frac{1}{p}$, regardless of how much time there is in lineages.

The last process in our menagerie involves a disadvantageous trait that evolves in spite of the fact that there is frequency-independent selection against it. As before, our observation is that $X = 1$ and $Y = 1$, but we now suppose that selection favored state 0 over state 1. The equilibrium value for this process, which provides a valid prior for the states of the ancestors, is $Pr(Z = 0) = 0.99$. Using the generic formula for the likelihood ratio of the common-ancestry and separate-ancestry hypotheses, we obtain the following likelihood ratio for the observation of a disadvantageous similarity:

(FIS-DIS)
$$\frac{(0.99)(0) + (0.01)d^2}{[(0.99)(0) + (0.01)d]^2}.$$

Here $d = Pr_t(1 \rightarrow 1)$ when there is selection for state 0; $d = 1$ at t_0 and $d \approx 0$ at t_∞. I'm here assuming that a lineage has no chance of evolving from 0 to 1 if there is selection for state 0. The FIS-DIS ratio is approximately 100 regardless of the duration of lineages.

The results of this four-way competition are summarized in Figure 4.10. Notice first that all four likelihood ratios are greater than unity as long as time is finite; in all four cases, similarity provides evidence favoring CA over SA. Among these four ratios, why does FIS-ADV come in last? The key is that the process of frequency-independent selection for trait 1 makes it very probable that X and Y will both be in state 1 *regardless of whether CA or SA is true*. The contrast with FIS-DIS is telling; here we observe that X and Y are in state 1 even though there was frequency-independent selection for state 0. With this process at work, it is highly probable that ancestors were in state 0 and highly improbable that ancestors in state 0 would give rise to descendants that are in state 1. So

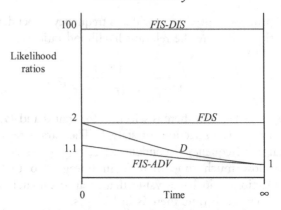

Figure 4.10 Four likelihood ratios, two of which depend on the amount of time between ancestor and descendants. For illustrative purpose, I assume that $Pr(Z = 1 \mid \text{FDS}) = \frac{1}{2}$.

when we observe that X and Y are in state 1, we have to decide whether one very improbable event occurred, or two. The same reasoning illuminates why similarities generated by drift provide more substantial evidence for common ancestry than adaptive similarities do. Under the drift hypothesis, the probability that an ancestor is in state 1 is only $\frac{1}{2}$. The fact that the ordering of the three ratios is FIS-DIS > D > FIS-ADV is explained by the fact that $Pr(Z = 1 \mid \text{FIS-DIS}) < Pr(Z = 1 \mid \text{Drift}) < Pr(Z = 1 \mid \text{FIS-ADV})$. The fourth process also conforms to this pattern. Even though frequency-dependent selection differs structurally from the other three processes examined here,[20] it remains true that the weight of evidence provided by frequency-dependent selection goes up as p declines, where $p = Pr(Z = 1 \mid \text{FDS})$. Here is the general pattern: *The more improbable it is that an ancestor has trait T, the stronger the evidence that X and Y's having T provides for their common ancestry.*

The ratio FIS-DIS always exceeds the D and FIS-ADV ratios; disadvantageous similarities always provide stronger evidence for common ancestry than do neutral or advantageous similarities (when all these characters are dichotomous – see the next section). The gill slits that human fetuses possess are useless, as is the lanugo, which is a covering of

---

[20] Frequency-dependent selection for the majority trait violates a rule obeyed by frequency-independent selection (whether the similarity of X and Y is advantageous or disadvantageous) and drift, namely that $Pr_t(0 \rightarrow 1)$ and $Pr_t(1 \rightarrow 1)$ approach each other as t increases; for this reason, the frequency-dependent selection process does not provide an equilibrium value for the probability of the state of ancestors, as already noted.

hair grown five months after conception that is usually shed right before birth. The appendix is different; it is worse than useless. Gill slits and the lanugo provide evidence for common ancestry, but the appendix provides better evidence still. The only other similarity in our menagerie that can touch disadvantageous similarities is a similarity due to frequency-dependent selection; whether it does so depends on p, the probability that each ancestor in that frequency-dependent process had of being in state 1.[21]

The four processes described here have something else in common. Increasing the duration of the lineages connecting ancestors to descendants cannot increase the amount of information that descendant states provide on the question of common versus separate ancestry. In the case of drift and the evolution of advantageous traits under frequency independent selection, the information declines with time; in the case of frequency-dependent selection for the majority trait and the case of disadvantageous similarities, the information content stays about the same. This nonincrease in information is just what one would expect in a Markov process. A causal chain extending from E at t_1 to F at t_2 to G at t_3 is Markovian precisely when the intermediate stage of the process *screens off* earlier and later stages from each other: $Pr(G$ at $t_3 | F$ at $t_2) = Pr(G$ at $t_3 | F$ at $t_2 \text{ \& } E$ at $t_1)$. The *information processing inequality* assures us that the $I(G$ at t_3, F at $t_2) \geq I(G$ at t_3, E at $t_1)$, where $I(\alpha, \beta)$ is the mutual information between α and β; as a lineage's duration is increased, the mutual information between its end state and its beginning state cannot increase (Sober and Steel 2002).

N-*state characters*

How would the likelihood ratio for the common-ancestry and separate-ancestry hypotheses generated by the observation that species X and Y are in the same character state be affected if the trait considered had more than two states? As noted in §4.4, switching from a dichotomous to a multistate character does not change the fact that X and Y's matching counts as evidence favoring CA; the likelihood ratio of CA to SA is still greater than unity. However, a change in the number of character states affects the *value* of the likelihood ratio for one of the processes considered but not for the others.

---

[21] The approach taken here can be used to investigate the question of when a *complex* adaptive similarity provides stronger evidence for CA than a *simple* adaptive similarity does. I leave this as an exercise for the reader.

Suppose that the trait in question has n states and that the transformation series shown in Figure 4.4a is correct. This transformation series represents a drift process in which all changes have the same probability. If $X = 1$ and $Y = 1$ and the character in question has n states, the likelihood ratio takes the form:

$$(D_n) \qquad \frac{[\frac{(n-1)}{n}]a^2 + [\frac{1}{n}][1 - (n-1)a]^2}{\{[\frac{(n-1)}{n}]a + [\frac{1}{n}][1 - (n-1)a]\}^2}.$$

Here $a = Pr_t(i \rightarrow j)$, for each $i \neq j$. Notice that D_n has a value of n at $t = 0$ and a value of unity at $t = \infty$. For all finite times, $D_n > D$ when $n > 2$. When two descendants are in the same state of an n-state character, and the character evolved by drift, their matching provides stronger evidence for common ancestry the larger n is. Once again, the similarities that provide the strongest evidence for common ancestry are the ones generated by processes in which it is highly improbable that the ancestors in question have the trait observed in their descendants.

The likelihood ratios for similarities generated by the other processes I have discussed are not affected by shifting from dichotomous to multi-state characters. Consider the ratio for the frequency-dependent process. This has a value of $\frac{1}{p}$, where, recall, $p = Pr(Z$ is in state 1). Subdividing Z's being in the alternative state 0 into several alternatives does not affect the value one should assign to p, illicit invocations of the principle of indifference notwithstanding. Similar remarks apply to the adaptive similarities generated by frequency-independent selection (FIS-ADV). If character state 1 is optimal, it does not matter whether there is one alternative state or many; the equilibrium value for the probability an ancestor has of being in state 1 will still be close to unity. Similar remarks apply to disadvantageous similarities (FIS-DIS).

Correlation in a data set

Propositions 1–9 (§4.4) entail a *qualitative* result for *each* dichotomous character – that a match favors CA over SA and a mismatch favors SA over CA. Those propositions do not answer the *quantitative* question of how much a match favors CA or a mismatch favors SA. Nor do they say which hypothesis is more likely, relative to *all* the data, if the data set includes both matches and mismatches. To answer these further questions, I ventured beyond propositions 1–9 and considered in more detail the evolutionary processes that can generate different characters.

Biologists may well wonder if there is a way to sidestep the requirement that we have information about the evolutionary process beyond what propositions 1–9 provide. Do we really need to know which characters evolve by drift and which by selection? Do we really need to have a ballpark estimate, for each character, of the probability that an ancestor occupied this or that character state? The concept of correlation holds out the promise that the comparison of CA and SA can get by with less (Forster 1986).

To see why, let's go back to Reichenbach's example of the acting troupe (§3.8). As before, we keep track of the sick days for two actors (X and Y) over several years, but suppose now that each gets sick on half the days and that the two get sick together one day in four. These observations entail that there is no association in the data between X's getting sick and Y's getting sick, in the sense that

$$f(X \text{ and } Y \text{ get sick}) = f(X \text{ gets sick})f(Y \text{ gets sick}).$$

Using propositions 1–9, we can construct a model that says that there is a common cause of the two actor's sick days and another model that says that the two actors' sick days are causally independent. The second of these confers a higher probability on this fact about frequencies than the first. The same sort of reasoning can be applied to the data shown in Figure 4.11a. Species X and Y are each scored for their character states (0 or 1) on a set of dichotomous characters with the result that each species has character state 1 for half of the characters considered while the frequency of characters on which they both occupy state 1 is 0.25. There is no association in the data between X's being in state 1 and Y's being in state 1:

$$f(X = 1 \text{ and } Y = 1) = f(X = 1)f(Y = 1).$$

We then can construct a common-ancestry model and an separate-ancestry model with the result that this fact about frequencies is more probable according to the second model than it is according to the first. I will call this analysis of the data *the correlational approach*.

In the correlational analysis of the actors' sick days, we don't consider sick days one by one, puzzling over the causes of sickness and wellness one day at a time. Rather, we lump the sick days together for each actor and compute the frequency with which each actor gets sick and the frequency with which they get sick together. This suggests that in testing CA against SA, we don't need to consider traits one by one, puzzling over the

		25%	25%	25%	25%
(a)	X	1	1	0	0
	Y	1	0	1	0

		25%	12.5%	37.5%	25%
(b)	X	1	1	0	0
	Y	1	0	1	0

Figure 4.11 Two character distribution for the two species *X* and *Y*. In (a), there is zero association; in (b), there is a positive association. In both, the overall similarity of *X* and *Y* is 50 percent.

evolutionary processes that give rise to each. Rather, we can lump traits together and compute the frequencies with which each species has character state 1 and the frequency with which they match on this character state. The trait-by-trait likelihood analysis requires assumptions additional to propositions 1–9 if we wish to assess the evidential meaning of a data set like that given in Figure 4.11a that includes both matches and mismatches. But if all we need to explain is the fact that the traits of the two species are not associated in the data, propositions 1–9 seem to suffice.

Although propositions 1–9 do not settle whether CA or SA has the higher likelihood for data sets that contain both matches and mismatches, they do suffice when the characters all exhibit the same pattern. In such cases, the trait-by-trait likelihood analysis can conflict with the correlational approach. For example, suppose that species *X* is in state 1 for all the characters scored and that the same is true of species *Y*. There is no association in the data between the two species (since 100 percent = 100 percent × 100 percent) and so the correlational approach concludes that SA is better supported than CA. However, if propositions 1–9 are true, each character favors CA over SA, and so the entire data set must do so as well. The principle of total evidence (§1.3) cuts through this disagreement. It says that we should use the logically stronger description of the data rather than a description that is logically weaker. Describing the characteristics of each species trait by trait is logically stronger than simply saying how much association there is in the data. The situation here is reminiscent of the example about toast and eggs discussed in connection with Figures 1.6 and 1.7.[22]

---

[22] The other way in which all the traits in a data set can exhibit the same pattern is for *X* and *Y* to always be in the opposite character state. With 100 percent mismatches, propositions 1–9 entail

In Reichenbach's example, an actor's being sick on one day is viewed as the same type of event as his or her being sick on another. The similar events that occur on different days are assembled to give rise to a quantity we call "the actor's frequency of sick days." This seems entirely straightforward. However, the parallel conceptualization in the context of our question about common ancestry reveals a nontrivial issue. Figure 4.11a describes the frequencies with which a set of characters exhibits the four possible character distributions. There is no association in this data set, since $0.25 = 0.5 \times 0.5$. But suppose we recode some of the characters that are represented in Figure 4.11a as $X = 1$ and $Y = 0$ and describe them instead as $X = 0$ and $Y = 1$, thus obtaining the distribution described in 4.11b. Now there is a positive association in the data, since $0.25 > 0.375 \times 0.625$. We need to choose between these two ways of describing the data if we wish to use the correlational approach. Is there an association in the data, or not? To answer this question, we need to decide whether a 1 on character T_i is "the same type of event" as a 1 on character T_j, or if a 1 on T_i is the same type of event as a 0 on T_j.

This problem finds a solution in the fact that we want to use the frequencies found in the data to estimate the values of probabilities, perhaps by using the method of maximum likelihood estimation (§1.2, §1.5). When we look at the frequencies of days on which the two actors are sick, both singly and together, and infer from this whether the sick days are correlated, we are using a probability model according to which each actor has a constant but unknown probability of getting sick on each day. This is why an actor's being sick on one day counts as "the same type of event" as his or her being sick on another: Both events provide evidence for estimating the value of a single probabilistic parameter. Symmetrically, if we are prepared to use a model in which each species has a constant but unknown probability of evolving state 1 on each character, this settles which of the two codings of the data shown in Figure 4.11 we should take to be correct. It follows that implementing the correlational approach to testing CA against SA requires assumptions beyond propositions 1–9, just as is true of the trait-by-trait likelihood approach. The additional assumptions describe how characters in the data are *related to each other*,

that separate ancestry is better supported than CA. In contrast, the correlational approach regards both positive and negative associations as evidence for a common cause. Despite appearances, I do not think this example illustrates a fundamental conflict between the two approaches, since a natural reaction on the part of friends of trait-by-trait analysis is to consider a different model of the CA hypothesis – one that rejects proposition 6, which says that descendants of a common ancestor evolve independently.

whereas propositions 1–9 describe what is true of each character *taken on its own*.

This problem does not disappear, nor is its solution any different, if we consider sequence data. It may seem arbitrary to say that a "1" on the wing/no-wing morphological character is "the same" as a "1" on the carnivore/herbivore character, but not at all arbitrary to say that a *G* at one site in a sequence is "the same" as a *G* at another. However, the meaning of the two judgments is the same: You are saying that the probability of wings is the same as the probability of carnivory and you are saying that a *G* at the one site has the same probability as a *G* at the other. Linking distinct token events together as instances of the same type of event makes sense in the interpretation of evidence only when they are evidence for the value of the same probabilistic quantity.

Although I have taken pains to highlight an assumption made by the correlational approach, this does not mean that the approach is always wrong. The correlational approach and a trait-by-trait likelihood analysis each have their place, depending on the biological information we have available. If we know that the traits described in the data all evolve by the same process but don't know what that process is, we can use the correlational approach. However, if we think that the traits evolve by different rules and have views about which processes governed the evolution of which traits, we should perform a trait-by-trait likelihood analysis. Both approaches use the law of likelihood (§1.3); they differ over which description of the data they use to evaluate competing hypotheses. Lurking in the background is a third approach which is not likelihoodist at all. This is to think about the common-ancestry and the separate-ancestry hypotheses each in conjunction with different possible process models (where each process model contains adjustable parameters) and then to determine (perhaps using AIC) how varying the process model affects model selection scores. We will consider this strategy in §4.7.

4.6 CONCLUDING COMMENTS ON THE EVIDENTIAL
SIGNIFICANCE OF SIMILARITY

Modus Darwin, the inference of common ancestry from observed similarity, is instantly appealing. A moment's reflection elicits the thought that it has the structure of a likelihood inference. Our confidence that two organisms or two species share parents has the same kind of justification as our confidence that two languages or two texts have a common ancestor. The observed similarity would constitute a great coincidence if the two objects arose independently whereas the similarity would be less surprising if there were a common cause. The first part of this chapter spells out a framework for justifying the likelihood inequality that is at the heart of this argument. To understand why *modus Darwin* makes sense it is essential to understand when it does not. Similarities sometimes *fail* to discriminate between the common-ancestry and separate-ancestry hypotheses. There are even cases in which similarities are evidence *against* CA, not *for* it.

If assumptions 1–9 are true of a trait's evolution, then the resulting similarity favors CA over SA even when the similarity is adaptive, the result of frequency-independent selection. If drift or frequency-independent selection are the processes at work, then assumptions 4 and 5, which say that the traits of ancestors and descendants are positively correlated, demand that the time between descendants and ancestors be finite; however, if the process is one of frequency-dependent selection for the majority trait, those assumptions can hold true even when time is infinite. It may seem absurdly counterfactual to ask what is true when time is infinite, but there is a point to this question. If information asymptotes to zero, it will be extremely attenuated for finite lengths of time that are "big enough," and this suggests that questions about CA may be unanswerable when they concern *deep time*. This fear – that some *temps* may be *temps perdu* – is justified when evolution is governed by some processes but not when it is governed by others. Notice that the likelihood ratios represented in Figure 4.11 approach unity as time increases for only two of the four processes considered.

In §4.5, I turned to a second question that presupposes an answer to the first. If two similarities both provide evidence in favor of CA, how are we to determine which provides the stronger evidence? And if a similarity provides evidence *pro* and a dissimilarity provides evidence *con*, when does the *pro* outweigh the *con*? These questions led us to consider different processes that can produce similarities. Besides similarities produced by drift (D) and by frequency-dependent selection (FDS) for the majority trait, I considered two kinds of similarity that can evolve in the process of frequency-independent selection. In the first (FIS-ADV), the resulting similarity is advantageous and it evolves because of selection; in the second (FIS-DIS), the similarity is disadvantageous and it evolves in spite of the fact that there is selection against it. With finite time between ancestors and descendants, and when selection is understood as a frequency-independent process, the kinds of similarity that descendants might exhibit can be ordered according to how strongly they favor CA over SA:

Deleterious similarties > neutral similarties > adaptive similarties.

The fourth type of process considered, frequency-dependent selection for the majority trait, can be placed in this chain of inequalities once one says how probable it is that an ancestor exhibit state 1 if the trait evolves by this process. If this is sufficiently improbable, this type of similarity can provide stronger evidence than all the rest.

This result concerning frequency-dependent selection may seem a mere curiosity, but it is arguable that selection for the majority trait is a pervasive phenomenon. Consider the example of chromosome number discussed in §4.4. Even if changing chromosome number does not reduce the survival probabilities of an organism, it may well reduce its reproductive prospects. For example, a cross between a diploid and a tetraploid will yield a triploid offspring, which is often sterile. When this is so, a mating individual gains a reproductive advantage by having the same chromosome number that the majority of the population exhibits. Another example is the visual cues that flowers provide their pollinators. If bees specialize on flowers based on visual cues, any trait that affects how a flower looks is likely to be subject to frequency-dependent selection. It is better to look like your conspecifics if you want bees to visit.

Although the frequency-dependent selection model was formulated to describe frequency-dependent selection for the majority trait, its actual content applies to any situation in which there are multiple stable con-figurations each maintained by selection once it is achieved. This can be true even when all trait combinations have frequency *in*dependent fitness values. For example, consider two dichotomous traits, A and B, where the four trait combinations have constant fitnesses that are related as follows: $w_{A\&B} = w_{-A\&-B} > w_{A\&-B} = w_{-A\&B}$. This has the consequence that A is fitter than $-A$ when B is universal, whereas the reverse is true when $-B$ is universal. The fitness of A does not depend on *its own* frequency, but on the frequency of B. The same is true of the fitness of B: It depends on the frequency of A. The trait combination $A\&B$ will be maintained by sta-bilizing selection, and the same is true of the combination $-A\&-B$. This further widens the scope of application for the frequency-dependent selection model.

Consider a pair of organisms that use the same genetic code. As noted earlier, what is now called the "standard" code is not universal. From the fact that several codes are in place, we can see that all of them are functional, but this does not mean that codes evolved by drift. Rather, once a code evolves, stabilizing selection tends to keep it in place. There is first a frequency-dependent effect. An organism in a sexual species is apt to be less reproductively successful if its code differs from that used by the organisms with which it mates. And there is a powerful frequency-independent effect as well. Consider an asexual organism that has a functioning genetic code, but whose offspring has a code that differs in even one detail from that of its parent; it is highly probable that the offspring will be nonviable. This is because changing the meaning of a codon entails that the codon produces the "wrong" amino acid every time it is used. Changing the code has the same upshot as a massive number of point mutations. Functioning codes, once in place, constitute adaptive peaks. This point does not depend on all peaks having the same height; the argument is consistent with the hypothesis that the current code is optimal, as Freeland et al. (2000) have argued.

The fact that similarities that evolve by frequency-dependent selection can provide strong evidence for common ancestry shows that an important qualification is needed in Darwin's claim that adaptive similarities are "nearly valueless" in the evidence they provide about genealogy. He was right for cases in which there is a single adaptive equilibrium but wrong when there is more than one. When there are multiple peaks, traits that evolve by natural selection can be *more* informative than traits that evolve by drift.

I have concentrated so far on comparing the common-ancestry and the separate-ancestry hypotheses when each is formulated as a claim about just two species. The arguments generalize when there are more than two. If the fact that species X and Y use the same code favors CA over SA, then the fact that X, Y, and Z share the same code favors CA over SA even more. Matching favors CA over SA more strongly, the more matching objects there are. If the two actors in Reichenbach's theater company get sick on a given day, this provides some evidence for a common cause, but when the whole troupe falls ill, that is stronger evidence still. And if n independent and reliable witnesses all report that proposition p is true, their unanimity carries more weight the larger n is (§1.3).

The two questions explored here – why similarity rather than dissimilarity favors CA over SA, and which kinds of similarity provide the strongest evidence for CA – have different answers. Assumptions 1–9 suffice for an observed similarity to favor CA over SA; it does not matter how probable or improbable it was that the ancestors postulated by the two hypotheses came to occupy the character state we see in their descendants, so long as those probabilities are strictly between 0 and 1. However, the key to answering the second question is that the observation that $X = 1$ and $Y = 1$ produces stronger evidence favoring CA over SA the *lower* the probability is that the ancestors postulated by the two hypotheses were in state 1.

This last result provides a reminder of how important the contrastive framework is for evaluating evidence. It seems to offend against common sense to say that E is stronger evidence for the common-ancestry hypothesis the *lower* the value is of $Pr(E\,|\,\text{CA})$. This seems tantamount to saying that the evidence better supports a hypothesis the more miraculous the evidence would be if the hypothesis were true. Have we entered a Lewis Carroll world in which down is up? No, the point is that, in the models we have examined, the ratio $Pr(E\,|\,\text{CA})/Pr(E\,|\,\text{SA})$ goes up as $Pr(E\,|\,\text{CA})$ goes down. An easy way to see this point is to imagine that $Pr(1 \rightarrow 1) = 1$, $Pr(0 \rightarrow 1) = 0$, and let $Pr(Z = 1) = p$, where Z, recall, is an ancestor of the observed species X and Y. Then the likelihood of CA is p and the likelihood of SA is p^2, so the likelihood ratio of CA to SA is $\frac{1}{p}$. Now it is obvious how the evidence for CA gets stronger as p gets smaller. When the likelihoods of the two hypotheses are linked in this way, it is a point in favor of the common-ancestry hypothesis that it says that the evidence is very improbable.

4.7 EVIDENCE OTHER THAN SIMILARITY

What else, besides the similarities and differences of species X and Y, could be evidence that bears on the question of whether they have a common ancestor? In what follows, I examine three lines of argument. The first holds that when different data sets agree on which phylogenetic tree is best supported, that this consensus is evidence that the taxa share a common ancestor. The second appeals to the existence of fossils that are intermediate in character between X and Y. The third concerns how X and Y figure in a larger biogeographical distribution.

Using a tree-construction method to test the common-ancestry hypothesis[23]

Penny et al. (1982) propose a method for testing the common ancestry hypothesis that differs substantially from the approach described so far. They look at different sets of data for the same taxa and then use

---

[23] Here I draw material from Sober and Steel (2002).

phylogenetic parsimony to determine which phylogenetic tree is best supported by each data set. It turns out that the trees singled out as best for the different data sets are very similar. Penny et al. take this to be evidence in favor of the common-ancestry hypothesis. Although the authors use phylogenetic parsimony as their method of tree construction, the logic of their argument could be applied to other tree-construction methods – for example, to a method based on maximum likelihood.

I pointed out at the start of the present chapter in discussing Figure 4.1 that methods of tree construction, parsimony included, assume that the species under study all share a common ancestor. At the risk of introducing a bit of redundancy into our notation, we can think of parsimony and these other methods as attempting to discriminate among the following alternatives: $CA \& T_1$, $CA \& T_2$, ..., $CA \& T_n$, where the T_is are the different trees under consideration. This is equivalent to the more familiar representation of the alternatives as T_1, T_2, ..., T_n, since each tree entails the CA hypothesis. We can begin considering the logic of Penny et al.'s test by seeing how it applies to the special case in which the different data sets are in complete agreement as to which tree is best: If each of the m data sets favors $CA \& T_1$ over $CA \& T_2$, ..., $CA \& T_n$, does this unanimity favor CA over SA?

Before addressing this question, I want to comment on one that is slightly different. If each data set favors $CA \& T_1$ over $CA \& T_2$, ..., $CA \& T_n$, do the data also favor CA over SA? This is a slightly different question, since now we're asking whether *the data* favor CA over SA, not whether the fact that *different data sets agree as to which tree is best* has that evidential significance. In any event, there is a general feature of likelihood comparisons that is relevant here: *It is possible for data to discriminate among a set of hypotheses without saying anything about a proposition that is common to all the alternatives considered.* A simple example from cards illustrates this point. I am the (trustworthy) dealer in a card game and tell you that the card I am about to turn up is red. This information favors the hypothesis that the next card will be the Jack of Hearts over the hypothesis that it will be the Jack of Spades, but it does not favor the hypothesis that it will be a Jack rather than some other rank. The same pattern occurs when the competing models all use the same idealization. Suppose you are testing various evolutionary hypotheses against each other that all assume random mating or infinite population size. A number of data sets may all agree that one of those models is better than the rest, but this says nothing as to whether the data sets favor random mating over some alternative, or favor infinite population size over the hypothesis that the population is finite (Orzack and Sober 1993).[24] These examples suggest a general point about the method that Penny et al. propose: The fact that different data sets agree as to which *tree* is best supported leaves open whether the data provide evidence for a proposition on which all the trees agree – that the species considered have a common ancestor. This point is a close cousin to an idea that will be familiar to philosophers who have thought about confirmation theory; Hempel's (1965b) *special consequence condition* is wrong. If E confirms H and H entails X, it does not follow that E confirms X.

Now let's look at the specifics. Penny et al. find the best tree for each of the m data sets and then compute how similar those m best trees are. They also compute how similar m randomly constructed trees would be, on average. Penny et al. reason that if the actual degree of similarity for the m best trees constructed for the different data sets greatly exceeds the average similarity of m randomly constructed trees, that this is evidence for CA. How should this problem be organized within a likelihood framework? The observation on which Penny et al. focus, I take it, is the degree of similarity of the best trees constructed for the m data sets. The question is whether this observation is rendered more probable by the CA hypothesis or by the SA hypothesis:

> $Pr(s$ is the degree of similarity of the best trees for the different
>
> data sets $|\,CA) > Pr(s$ is the degree of similarity of the best trees
>
> for the different data sets $|\,SA)$.

Penny et al. do not calculate either of the quantities mentioned in this inequality. The probabilistic quantity that they do calculate has nothing to do with the data; it just describes the expected degree of similarity for several trees drawn at random from the set of possible trees.

It may seem that the first likelihood in the above inequality must have a high value, but, in fact, whether this is so depends on what rules of evolution are followed by the traits in the different data sets. For example, if the different data sets evolve on the same tree but follow different rules, those rules can make it highly probable that the data sets will *disagree* as to which tree is best. On the other hand, if all traits evolve by the same process, then one expects them to agree as to which tree is best.

---

[24] Yet another example is provided by the discussion of the McDonald–Kreitman (1991) test in §3.9; the fact that their observations favor selection over drift does not entail that they also favor the *Drosophila* phylogeny they assumed over some alternative.

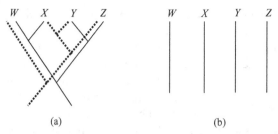

Figure 4.12 Two alternatives to the hypotheses that all the traits of the taxa *W*, *X*, *Y*, and *Z* stem from a single common ancestor. In (a) different traits have different trees. In (b), all traits have the same non-tree.

Penny et al. offer no reason to think that the different data sets all follow the same rules of evolution. Another possibility is that different traits evolve on the same tree by different processes and the characters grouped in a data set are obtained by random sampling from the total set of characters. Here, again, one expects the different data sets to agree as to which tree is best. However, Penny et al. did not construct their data sets by randomly sampling from the set of characters.

Does the second likelihood in the above inequality have a low value? To answer, we first need to distinguish two ways in which the separate-ancestry hypothesis could be true. These are shown in Figure 4.12. The separate-ancestry hypothesis might say that different data sets have different trees,[25] or it might say that all data sets evolved on the same nontreelike genealogy. The former seems to be what Penny et al. have in mind, given that they construct *m* trees at random. However, the question is then what the different data sets produced by the different trees in Figure 4.12a will look like and how similar the trees constructed from those data sets will be. It doesn't matter how similar trees will be if one ignores the data and constructs them by sampling from the space of all possible trees. With respect to the second interpretation of what the separate-ancestry hypothesis might say – namely, the one depicted in Figure 4.12b – the point is that it is not inevitable that different data generated by this non-treelike genealogy will lead parsimony to construct trees that are very different from each other. Again, that depends on the processes at work in branches. For example, suppose that all dichotomous characters evolve on the branches in Figure 4.12b according to the same rules, which make it highly probable, for each character, that $W = 1$, $X = 1$, $Y = 0$, and $Z = 0$. If

---

[25] The discussion in §4.2 of gene genealogies is relevant here.

you draw several data sets, it is very probable that they will agree that (XX) (YZ) is the best unrooted tree. For a second example, suppose that a trait evolves by the same process on each branch but that different traits evolve by different processes; in this case, different data sets will tend to agree on which tree is best as long as the data sets are assembled at random.

Penny et al. take agreement among data sets as to which tree is best to be evidence favoring CA over SA. This is very different from claiming that the data taken as a whole favor CA over SA. It may well be true that

$$Pr(\text{all } n \text{ data sets} \mid CA) > Pr(\text{all } n \text{ data sets} \mid SA).$$

If *each* data set favors CA over SA, and the data sets are independent of each other, conditional on CA and conditional on SA, then the total evidence provides stronger evidence than any one data set does on its own (§1.3). It is agreement among data sets on the fact that CA is more likely than SA that would be significant, not agreement as to which *tree* is most likely. This is a conclusion sanctioned by the principle of total evidence, which instructs us to use a logically stronger, rather than a logically weaker, description of the data if the hypotheses considered make predictions about both (§1.3).

Intermediate fossils

The sufficient condition identified in §4.4 for sameness of character state to be evidence for common ancestry does not depend on whether X and Y are both extant organisms, or both are fossils, or they are one of each. In this sense, fossils introduce no novelty into the problem of inferring common ancestry. However, there is another way in which fossils do introduce a new dimension. Suppose we observe that there is a fossil whose trait value is intermediate between those of X and Y. How does the discovery of a fossil intermediate affect the question of whether X and Y have a common ancestor?

Creationists frequently claim that the absence of intermediate fossil forms is evidence against evolutionary theory. Evolutionists reply by pointing to the numerous intermediate fossils that have been discovered; these link dinosaurs with birds, land tetrapods with fish, reptiles with mammals, and land mammals with whales. Of course, if you discover an I that is intermediate between species X and Y, the question can still be raised as to where the forms are that fall between X and I and between I and Y. There always will be "gaps"; they just get narrower. Biologists do not interpret these gaps, whether they are narrow or wide, as evidence

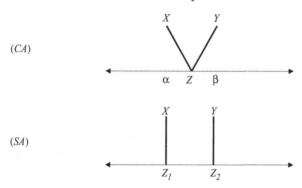

Figure 4.13 If the evolutionary process is gradual, the CA hypothesis predicts the existence of ancestors that had intermediate forms, regardless of the character state of the common ancestor Z. What the SA hypothesis predicts depends on the states of the postulated ancestors Z_1 and Z_2.

against CA. Gaps are simply chalked up to "the imperfection of the fossil record" – fossils often don't get formed, and even when they do, it is easy enough for them to be destroyed or for biologists to fail to find them. Is this reaction on the part of evolutionists an instance of their wanting to have their cake and eat it too? If finding intermediate forms is evidence *for* CA, isn't failing to find them evidence *against*?

Figure 4.13 depicts what the common-ancestry and the separate-ancestry hypotheses say about the existence of intermediate forms. We observe the character states of the extant species X and Y. If evolutionary change proceeds gradually, there *must* be intermediate forms if X and Y have a common ancestor (Z). Slide Z along the scale that represents its possible trait values; no matter what character state Z occupies, the lineage leading from Z to X or the lineage leading from Z to Y must contain organisms whose trait values fall between the values α and β that attach to X and Y, respectively. This is not true for the separate-ancestry hypothesis. If the lineage passing through Z_1 on its way to X never has a trait value that is greater than α and the lineage passing through Z_2 on its way to Y never has a trait value that is less than β, then there will be no ancestors in the lineages leading to X and Y that have trait values that fall between α and β. Historically, the SA hypothesis has been associated with the claim of *evolutionary stasis* – the thesis that ancestors were, in the main, just like their descendants. However, the separate-ancestry hypothesis is not logically committed to stasis. We can and should separate the separate-ancestry hypothesis from the assumption that lineages do not change trait values.

	CA	SA
There existed an intermediate.	1	*q*
There did not.	0	*(1–q)*

Figure 4.14 Either X and Y have a common ancestor (CA) or they do not (SA). Cells represent probabilities of the form $Pr(\pm\text{intermediate} \mid \pm\text{CA})$. Gradualism is assumed.

The upshot is that CA and SA provide different answers to the question of whether intermediate forms once existed; given the assumption of gradualism, CA answers that they *must have existed*, while SA's reply is that they *may have*. It is only a short step to the following likelihood inequality:

$Pr(\text{an organism intermediate between } X \text{ and } Y \text{ existed} \mid \text{CA})$

$> Pr(\text{an organism intermediate between } X \text{ and } Y \text{ existed} \mid \text{SA})$.

Given gradualism, the first of these likelihoods has a value of unity. If the separate-ancestry hypothesis allows that there is some chance that the lineages leading to X and Y never strayed into the "intermediate zone" between α and β, then the second likelihood has a value less than unity. If we add to the SA hypothesis the stronger assumption of evolutionary stasis, the second likelihood has a value of zero. These points are summarized in Figure 4.14. Notice that entries in each column must sum to unity. If we use the likelihood *ratio* to represent how strong the evidence is that favors one hypothesis over the other, we obtain an asymmetry. If there is an intermediate form, this favors CA over SA, and the strength of the favoring is represented by the ratio $\frac{1}{q}$. This is greater than unity if $q < 1$. On the other hand, if there is no intermediate, this *infinitely* favors SA over CA, since $(1 - q)/0 = \infty$ (assuming that $q < 1$). The nonexistence of an intermediate form would have a more profound evidential impact than the existence of an intermediate.

Although the assumption of gradualism plays a role in these likelihood comparisons, it is important to remember that gradualism is not plausible for some traits. Consider the example discussed in §4.4 of chromosome number. There is no iron law of evolution that says that a lineage that evolves from twenty-four pairs of chromosomes to forty-eight must evolve from twenty-four to twenty-five to twenty-six to … forty-seven to forty-eight. Polyploidy (the doubling or tripling of chromosome number) is a known process.[26] Still, gradualism is usually assumed when the evolution

---

[26] Developmental genetics provides numerous examples (e.g. hox genes) in which small genetic changes induce discontinuous phenotypic changes; see Carroll (2005) for an introduction.

	CA	*SA*
We have observed an intermediate.	*a*	*qa*
We have not.	*1–a*	*1–qa*

Figure 4.15 Either *X* and *Y* have a common ancestor (CA) or they do not (SA). Cells represent the probability that we have observed an intermediate, or that we have not, conditional on CA and conditional on SA.

of a continuous character is under discussion. And discussion of "intermediate" forms usually involves continuous characters.

We now can turn to the accusation that evolutionists play a game of "heads I win, tails you lose" when they appeal to the imperfection of the fossil record to excuse the fact that no fossil that is intermediate between *X* and *Y* has yet been observed. The key is to not confuse the *existence* of intermediates with our *observing* such intermediates. As we have seen, the CA hypothesis is committed to the first of these so long as gradualism is correct. But the CA hypothesis does not guarantee that we will have *observed* those intermediate forms. That depends on how often they fossilize, on how long those fossils last, and on how much fossil hunting there has been. The probability of *our having observed* an intermediate form, and of our having failed to do so, conditional on each of the two hypotheses, is represented in Figure 4.15. As before, $q = Pr$(there exists a fossil intermediate | SA). But now let us introduce a new quantity:

$a = Pr$(we observe a fossil intermediate | CA & there exists a fossil
 intermediate)

$= Pr$(we observe a fossil intermediate | SA & there exists a fossil
 intermediate).

I'm assuming here that the probability of observing an intermediate, if one exists, is the same, regardless of whether CA or SA is true. Notice that the likelihood ratio of CA to SA, given that we have observed an intermediate fossil, is $\frac{1}{q}$; in this respect Figures 4.14 and 4.15 agree. However, when we have *not* observed a fossil intermediate, the likelihood ratio of SA to CA takes on the value

$$\frac{Pr(\text{we have not observed a fossil intermediate} \mid SA)}{Pr(\text{we have not observed a fossil intermediate} \mid CA)} = \frac{1-qa}{1-a}.$$

This ratio is greater than unity if $a > 0$ and $q < 1$. As long as there is *some* chance that we'll observe a fossil intermediate if one exists, and there is

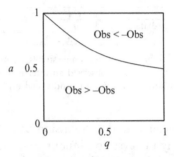

Figure 4.16 Observing an intermediate fossil favors CA over SA, and failing to so observe favors SA over CA, if $a > 0$ and $q < 1$. Here $a = Pr$(we have observed an intermediate fossil|an intermediate fossil exists) and $q = Pr$(an intermediate fossil exists | SA). Observing an intermediate favors CA more strongly than failing to so observe favors SA precisely when $1/(1 + q) > a$. This condition divides parameter space into two regions.

some chance that intermediate fossils will not exist if the separate-ancestry hypothesis is true, the failure to observe an intermediate favors SA over CA.

So far we have a symmetry: Observing an intermediate fossil favors CA over SA, and failing to observe one has the opposite significance (provided that a and q are constrained as just described). However, there may yet be an asymmetry – the fact that each of the two likelihood ratios is greater than unity does not settle which of them is bigger, or by how much. We can address this additional question by noting that observing an intermediate fossil favors CA over SA more strongly than failing to do so favors SA over CA precisely when the two likelihood ratios are related as follows:

$$\frac{1}{q} > \frac{1 - qa}{1 - a}.$$

Rearranging a little yields

$$\frac{1}{1 + q} > a.$$

This criterion says that each possible value of q puts a constraint on how large a is allowed to be, as shown in Figure 4.16. If q is fairly small, practically any value for a will satisfy this inequality; if $a < \frac{1}{2}$, the inequality is true no matter what value q has. And if q and a are *both* small, the first likelihood ratio will *greatly* exceed the second. If there is a small probability of our having observed a fossil intermediate when it exists, and if fossil intermediates have a small probability of existing when

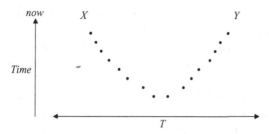

Figure 4.17 These dated fossils form an intermediate series between the extant species *X* and *Y*. Horizontal location indicates a taxon's score on the quantitative character *T*.

the separate-ancestry hypothesis is true, then observing an intermediate favors CA over SA far more profoundly than failing to so observe favors SA over CA.

There is an old motto that scientists often repeat: *Absence of evidence is not evidence of absence.* What is true in the present context is that if you don't even look for fossil intermediates, you will certainly fail to find them, and this will be true whether the common-ancestry or the separate-ancestry hypothesis is correct. Failing to find an intermediate in this circumstance provides zero evidence concerning the competition between the common-ancestry and separate-ancestry hypotheses. But this special case aside, the motto embodies an exaggeration. Suppose you *look* for intermediate fossils and fail to find them. This outcome isn't *equally* probable under the two hypotheses if $0 < q, a < 1$. Entries in each column must sum to unity in Figure 4.15 just as they must in Figure 4.14. What is true, without exaggeration, is that for many values of the relevant parameters, not observing a fossil intermediate provides *negligible* evidence favoring SA over CA – absence of evidence is *almost* valueless.[27]

If discovering a fossil intermediate provides evidence that favors the hypothesis that *X* and *Y* have a common ancestor, then discovering an intermediate fossil *series* provides even stronger evidence for the same conclusion. Figure 4.17 depicts an example; not only are there a number of intermediates, but the dating of these fossils has the consequence that as one moves back in time, the fossils get closer together in terms of their score on the quantitative character *T*. What more stunning evidence could there be that the extant species *X* and *Y* stem from a common ancestor?

---

[27] Notice the echo of Darwin's carefully qualified comment that adaptive similarities are *almost* valueless as evidence of CA (§4.5).

The human eye finds it irresistible to connect the dots. Suppose you visit a beach and see two people standing at the water's edge; you also see footprints that lead back from each of them and converge at the parking lot. The thought comes to mind that the people and the footprints are connected. Perhaps the people got where they are by walking from the parking lot; they now stand at different places on the beach, but they trace back to the same location. Similar reasoning is at work in the inference of common ancestry from an intermediate fossil series. I noted in §3.3 that we can't assume that the fossils we observe are *ancestors* of present-day organisms; they might just be *relatives*. This point also applies to Figure 4.17; when we observe an intermediate fossil series, we should not assume that earlier fossils are the ancestors of later ones. The same is true of the footprints on the beach. In a line of footprints that lead from the parking lot to the water's edge, it is false that earlier footprints cause later ones; rather, they are effects of a common cause, a person walking. A path of footprints leads us to infer that there was a person moving through space who left traces; an intermediate fossil series leads us to infer that there was a lineage moving through time that also left traces. In neither case is it necessary or particularly plausible to conclude that earlier traces caused later ones.[28]

Biogeography

Darwin cites three types of evidence in *The Origin of Species* to support the hypothesis that species alive today share common ancestors. The first is that they possess telling similarities; for example, they share vestigial organs that are useless or worse than useless. The second is the existence of intermediate fossils. The third involves facts about biogeographical distribution.

On the face of it, the spatial proximity of two species and their phenotypic similarity seem to be facts of different type. When biologists think that a phenotypic similarity is evidence for CA, they usually assume that the trait is influenced by genes; this assumption seems to be

---

[28] Reichenbach (1956) introduced the term "pseudo-process" to describe situations of this sort. Salmon (1984) gives the example of a circle of light that moves across the ceiling of the Astrodome. Earlier circles do not cause later ones; rather, they are effects of a common cause, the rotating searchlight that sits on the floor of the Astrodome. The Weismannian conception of heredity provides another example; it says that the phenotypic resemblance of parents and offspring is the result of a common cause. Parental genes cause the parents' phenotype as well as the offspring's genotype; the cause of an offspring's having blue eyes is not that its parents had blue eyes. For other examples and for discussion of the epistemology of pseudo-processes, see Shapiro and Sober (2007).

important because genes are passed from ancestors to descendants. However, there apparently are no genes for geographical location; organisms change location by moving, not by mutating. This thought makes it puzzling why geographical proximity should provide evidence for CA. The puzzle can be solved by considering the discussion in §4.4 of why phenotypic similarity is evidence for CA. Any trait T that satisfies conditions 1–9 will have the following property: if species X and Y both have T (or if both lack T), this observation favors CA over SA, in the sense that the first hypothesis will have the higher likelihood. Conditions 1–9 do not mention genes. All that is required with respect to the issue of "heritability" is that ancestor and descendant be positively correlated:

$$Pr(\text{Descendant has } T \mid \text{Ancestor has } T)$$
$$> Pr(\text{Descendant has } T \mid \text{Ancestor lacks } T).$$

This inequality may be true because of genetic transmission, but that is just one mechanism for securing the correlation. Geographical locale in many cases satisfies this inequality. If a descendant is in the geographical region R, which hypothesis makes that outcome more probable – that its ancestor was there already, or that its ancestor was somewhere else? If the former (and if the other eight conditions listed are satisfied), then the fact that species X and Y both live in region R provides evidence favoring CA over SA by the same logic that allows a phenotypic similarity to do so.

Once we assimilate spatial proximity to phenotypic similarity, it is no surprise that a problem that surfaced in our discussion of the latter also pops up in connection with the former. Are a bird species in the Galapagos Islands and a bird species on the west coast of South America in "the same" geographical region? Well, if we draw a circle that encompasses them both, the answer is *yes*, and we then can conclude that their occupying the same locale is evidence in favor of CA. But we could draw a smaller circle instead, one that circumscribes the Galapagos and leaves South America outside, and now the verdict is that the two species do not inhabit "the same" region. Since they fail to occupy the same locale, we conclude that this difference favors SA over CA. As before, it is arbitrary to impose a dichotomous character on an underlying quantitative reality. The solution is to leave dichotomous characters behind and deal explicitly with the fact that spatial proximity is a matter of degree. But now we face a new question: Are the two species close enough spatially for this fact to favor CA over SA, or are they sufficiently separated that the opposite is true? How close is close enough?

Let's explore this question by using a random walk model of dispersal in the context of the representation in Figure 4.6 of the common-ancestry and the separate-ancestry hypotheses. For a simple example, consider a line on which there are ten equally spaced points, numbered 1 through 10. These points might be thought of as islands in an archipelago. The organisms in question disperse from one island to an adjacent island with a fixed probability μ per unit time; for them to go to an island that is not adjacent, they must pass through the islands in between. Organisms have no chance of going beyond either of the two endpoints. This is the spatial analog of the transformation series depicted in Figure 4.7b. The equilibrium values for this process of random dispersal provide priors for the states of the ancestors postulated by the common-ancestry and the separate-ancestry hypotheses; each ancestor has a 1 in 10 chance of being at any given spatial location. We can calculate the expected spatial separation between the separate ancestors Z_1 and Z_2 postulated by the separate-ancestry hypothesis. With ten locations, the expectation under the separate-ancestry hypothesis is that X and Y will be a bit more than three islands away from each other. If X and Y are more spatially proximate than this, then CA has the higher likelihood; if not, not. Neutral evolution within an ordered n-state character is formally just like random dispersal across an n-island archipelago.

In the previous section on fossils, I pointed out that fossils introduce a new kind of evidence when they constitute *third terms*. When a phenotypic similarity that unites X and Y favors CA over SA, it doesn't matter whether the two species are extant or fossilized. However, the discovery of a fossil intermediate between X and Y provides a new kind of information about X and Y. The same is true for spatial proximity. That X and Y are spatially proximate is just a kind of similarity that they share. But suppose there exists a third species that is spatially intermediate between X and Y and that it has an intermediate phenotype. This is a new kind of datum. Spatial proximity and phenotypic similarity are now correlated. Biogeography describes the spatial distributions of species that exist at the same time (now); paleobiology describes the temporal distribution of extant and fossilized species that exist in the same broad geographical locale. An intermediate fossil series is evidence that X and Y have a common ancestor, and if X and Y are on islands in an archipelago, with intermediate forms occupying the intervening islands, this too is evidence that X and Y have a common ancestor. Fossils are to time what biogeography is to space.

This connection between biogeographical data and fossils is something that Darwin contemplated. In *The Origin of Species* ([1859]1964: 409),

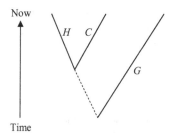

Figure 4.18 *H*, *C* and *G* are each temporally extended lineages; time slices drawn at random from *H* and from *C* can be expected to be temporally more proximate to each other that time slices drawn at random from *H* and from *G* (or from *C* and from *G*).

he says that "there is a striking parallelism in the laws of life throughout time and space: the laws governing the succession of forms in past times being nearly the same with those governing at the present time the differences in different areas." Darwin then enumerates several similarities that unite the way a lineage will change through time if it stays in the same place and the way a group of related species at the same time will vary spatially. He ends with a thesis about the spatial and temporal consequences of propinquity of descent: "the more nearly any two forms are related in blood, the nearer they will generally stand to each other in time and space" ([1859]1964: 410). Darwin's idea might be put like this:

(ST) If *X*, *Y*, and *Z*, are extant organisms or fossils and the true genealogical grouping is *(XY)Z*, then it is probable that *X* and *Y* will be geographically more proximate to each other, and temporally closer to each other as well, than either is to *Z*.

I'll call this Darwin's *space–time principle*. I use the term "probable" to make precise what I think Darwin intended by the qualifier "generally." And although Darwin talks about "two forms," I have put his point in terms of three.

The temporal part of the ST principle is on firm ground. Consider the branching diagram of humans, chimps, and gorillas in Figure 4.18. The three groups depicted are not just present-day populations; they are lineages that persist through time. Since the human and chimp lineages split off from each other more recently than the human–chimp clade split from the gorilla line, a human and a chimp fossil can be expected to have dates that are closer together than the dates of a human and a gorilla fossil. Notice that there is a part of the tree in Figure 4.18 that is a broken line;

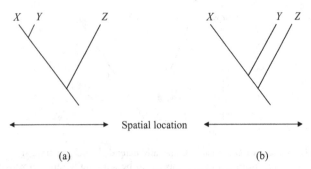

Figure 4.19   The genealogy of *X*, *Y*, and *Z* is (*XY*)*Z*. In (a), spatially more proximate
species are more closely related; in (b), this is not true.

this does not indicate any doubts about the existence of this lineage but
just that it calls for special comment. How should the organisms in this
line be described? Were they gorillas? If nót, what should we call them?
I won't address this question; let's just call them "*X*." Darwin's (ST)
principle says that two fossils drawn at random from *H* and *C* will
probably have dates that are closer together than two that are drawn at
random from *H* + *C* + *X* on the one hand and *G* on the other. The
temporal part of the ST principle is correct.

   The spatial part of the ST principle is less secure; it is true in some
circumstances, but not in others. Consider Figure 4.19, which depicts
three species (*X*, *Y*, and *Z*) that exist now. Their true genealogical
grouping is (*XY*)*Z*. What does this genealogy predict about which of them
will be spatially closer together and which more distant? In scenario (a),
geographical location mirrors propinquity of descent, but in (b) it does
not. When Darwin says that closeness of blood relationship is "generally"
associated with spatial proximity, he is saying that (a) is the common
pattern and (b) the rare one. This would be correct if the dispersal of
organisms were usually an unbiased process, like random drift. Under this
process, the expected spatial separation between two species is greater the
more time there has been since their most recent common ancestor. *X* and
Y in Figure 4.19a are spatially more proximate than either is to *Z* because
the lines leading to *X* and to *Y* split relatively recently, whereas the
lineages leading to *X* and *Y* on the one hand and to *Z* on the other split
longer ago. The aberrant pattern in Figure 4.19b arises when dispersal is
nonrandom, causing the lineages leading to *Y* and *Z* to remain spatially
proximate, with the lineage leading to X moving away from both. Thus,
the isomorphism between space and time is not perfect. The spatial part

of the ST principle is true under a model of random dispersal but can fail to be true when dispersal is nonrandom. The temporal part of the principle requires no such qualification. The concept of propinquity of descent has a closer conceptual connection with time than it has with space.

The ST principle stated above describes what is *probable* – meaning, I take it, what is more probable than not. It would be better to formulate the idea in terms of a likelihood inequality. The claim would then be that

$Pr[X$ and Y are closer together spatially (temporally)

than either is to $Z \mid (XY)Z]$

$> Pr[X$ and Y are closer together spatially (temporally).

than either is to $Z \mid X(YZ)]$.

There is no requirement that the first probability in this inequality has a value greater than 0.5. The spatial inequality is true if dispersal is random; the temporal inequality is true without this qualification.

Darwin's biogeographical inferences sometimes go beyond the conclusion that spatial proximity is an indicator of propinquity of descent. Consider the following argument:

[W]hy should the species which are supposed to have been created in the Galapagos Archipelago, and nowhere else, bear so plainly the stamp of affinity to those created in America? There is nothing in the conditions of life, in the geological nature of the islands, in their heights or climate, or in the proportions in which the several classes are associated together, which resembles closely the conditions of the South American coast: in fact there is considerable dissimilarity in all these respects. On the other hand, there is a considerable degree of resemblance in the volcanic nature of the soil, in climate, height, and size of the islands between the Galapagos and Cape de Verde Archipelagos: but what an entire and absolute difference in their inhabitants! The inhabitants of the Cape de Verde Islands are related to those of Africa, like those of the Galapagos to America. I believe this grand fact can receive no sort of explanation on the ordinary view of independent creation; whereas on the view here maintained, it is obvious that the Galapagos Islands would be likely to receive colonists, whether by occasional means of transport or by formerly continuous land, from America; and the Cape de Verde Islands from Africa; – the principle of inheritance still betraying their original birthplace. (Darwin [1859]1964: 398–9)

To analyze this line of reasoning, we need to sort out the hypotheses that are in contention. Darwin contrasts CA with independent "creation," but as noted earlier in this chapter, the hypothesis of separate *origination* has nothing logically to do with the idea of *intelligent design*.

In addition, Darwin in this passage associates the hypothesis of separate origination with the conjecture that the organisms in a locale have the features they do because these features "fit" them to their environment.[29] This hypothesis (which is consistent both with intelligent design and with natural selection) leads to the erroneous prediction that the organisms on the Galapagos should resemble those on Cape Verde more than the Galapagos organisms resemble those on the west coast of South America. Since the hypothesis of organism–environment fit is at variance with the observations, Darwin concludes that this counts against the hypothesis of independent origination. But just as the separate-ancestry hypothesis is distinct from the idea of intelligent design, so too does it differ from the hypothesis of organism–environment fit.

In the passage quoted, Darwin is not arguing that Galapagos tortoises and iguanas have a common ancestor based on the fact that they happen to live side by side. Not that he denied that they share a common ancestor, but this is not what he is here concluding. Still less is Darwin arguing for the false proposition that the organisms on the Galapagos are more closely related to each other than any of them is to an organism that lives on the west coast of South America. Rather, Darwin is considering the two biotas; the fact that there are similarities binding members of the first to members of the second *and* the fact that the two geographical areas are spatially proximate together provide strong evidence in favor of CA. The first step in Darwin's argument is this:

There are n pairs of species – X_1 and Y_1, X_2 and Y_2, ... , and X_n and Y_n. The species in each X–Y pair are similar; in addition, the Xs live in one locale, the Ys live in another, and the first locale is geographically close to the second. This complex of facts provides evidence that each of the X–Y pairs traces back to a common ancestor.

We have already examined how spatial proximity and phenotypic similarity can both be evidence for common ancestry. These two types of similarity support the claim that for each $X_i - Y_i$ pair, there is a common ancestor Z_i. But there is more to Darwin's argument than the conclusion that there are

[29] Darwin is here reacting against the biogeographical theory of Buffon – that "the earth [the conditions that exist in a locale] makes the plants [there], and the earth and the plants make the animals" and also against the theory of his teacher Lyell, who denied common ancestry and held that the characteristics that species in the same genus or family share with each other are determined providentially so that species fit the ecological conditions that obtained at their place of origin (Hodge 1987: 240).

n such common ancestors. There is a stronger conclusion: that there was a single geographical locale (an "original birthplace") from which the two present-day biotas arose.[30] Here is an analogy: Consider two villages – one in Argentina, the other in Italy. We find that individuals now in the first village have different names, many of them rare; and individuals now in the second village have a very similar array of last names. Sharing a rare last name is evidence of CA,[31] but there is something more that can be said here. The pair-wise matching is evidence that the n common ancestors lived in the same place. Why?

The relevant point concerns the nonindependence of dispersal events. If one bird is blown from South America to the Galapagos by a storm, this raises the probability that another bird on the continent ends up on the islands as well. And if a log floating from the west coast of South America carries one insect, this raises the probability that other insects come along for the ride. The "agents" (or "vectors") of dispersal – the physical processes that transport organisms – are common causes, often impinging on multiple organisms in the same locale and taking them to the same destination. Dispersal can be random, and still this failure of independence obtains. In terms of our model of the ten points on a line, the randomness of dispersal means that

$$Pr(Z_i \text{ ends at location 4} \mid Z_i \text{ start at location 5})$$
$$= Pr(Z_i \text{ ends at location 6} \mid Z_i \text{ start at location 5}).$$

Nonindependence means that

$$Pr(Z_1 \text{ and } Z_2 \text{ end at 4} \mid Z_1 \text{ and } Z_2 \text{ start at 5})$$
$$> Pr(Z_1 \text{ ends at 4} \mid Z_1 \text{ starts at 5})Pr(Z_2 \text{ ends at 4} \mid Z_2 \text{ starts at 5}).$$

Let us add to this the assumption of independence when organisms start in different places:

$$Pr(Z_1 \text{ and } Z_2 \text{ end at 4} \mid Z_1 \text{ starts at 3 and } Z_2 \text{ start at 5})$$
$$= Pr(Z_1 \text{ ends at 4} \mid Z_1 \text{ starts at 3})Pr(Z_2 \text{ ends at 4} \mid Z_2 \text{ starts at 5}).$$

---

[30] If each X_iY_i pair has a most recent common ancestor Z_i, that ancestor must have had some spatial location or other; this does not deductively entail that there was a single place where all these Z_i ancestors resided. To think otherwise is to commit the *birthday fallacy* (§2.2). The inference of a center of origin is a further, nondeductive, step.

[31] That sharing a rare name is stronger evidence for CA than sharing a common one is obvious; Sober (1988) proves the point about strength of evidence within the context of a simple mathematical model; it is the Smith–Quackdoodle theorem.

And we have already touched on the analog of "heritability" with respect to spatial location:

$Pr(Z_1$ and Z_2 end in the same place | Z_1 and Z_2 start in the same place)

> $Pr(Z_1$ and Z_2 end in the same place | Z_1 and Z_2 start in different places).

Given such assumptions about dispersal, there is a likelihood argument concerning centers of origin that builds on the inference about CA. If each $X_i - Y_i$ pair exhibits phenotypic similarity and geographical proximity, this is evidence for the existence of a common ancestor Z_i, one for each pair. If the different X_is live in one place and the Y_is in another, this biogeographical distribution is more probable if the n common ancestors all lived in the same place than it would be if they had distinct locations.

Biogeographical data describe spatial variation; fossil data provide a temporal dimension. Some data sets provide both. The acheulean stone tools used by premodern humans are found in Africa, Asia, and Europe. The oldest such tools are found in Africa, and areas that are spatially closer to Africa tend to have older tools than sites that are farther away. It is possible that acheulean tools arose independently in different locations, but the spatiotemporal correlation just described throws doubt on this hypothesis. Far more likely is the hypothesis that tool users spread from Africa, bringing their innovations with them. Or perhaps the ideas were transmitted from one more-or-less sedentary group to another. The CA hypothesis makes it unsurprising that distance from Africa and recency of finds are associated. As this example shows, inferring common ancestry makes sense in the context of cultural evolution mediated by teaching and learning just as it does in the arena of evolution mediated by genetic transmission.

4.8   PHYLOGENETIC INFERENCE: THE CONTEST BETWEEN LIKELIHOOD AND CLADISTIC PARSIMONY[32]

This chapter has focused, so far, on the question of how observations can provide evidence for (or against) the hypothesis that two or more species have a common ancestor, not on the question of how evidence can be used to discriminate between more fine-grained hypotheses about propinquity of descent. In terms of Figure 4.1, the question of interest has been how evidence bears on whether humans, chimps, and gorillas all

---

[32] This section draws on material from Sober (2004a).

trace back to a common ancestor, not whether the data favor $(HC)G$ over $H(CG)$. It now is time to take up this second type of question.

In the kind of "classic" phylogenetic inference problem I want to discuss, the observed taxa are assumed to be the tips of a bifurcating tree, and the goal is to infer just the "topology" of the tree, not the amount of time between branching events or the amount of evolution that has taken place on branches, or the character states of interior vertices.[33] Two of the main methods that biologists now use to solve such problems are *maximum likelihood* (ML) and *maximum parsimony* (MP); distance methods constitute a third approach, which I won't examine (not that they aren't interesting). ML seeks to find the tree topology that confers the highest probability on the observed characteristics of tip species. MP seeks to find the tree topology that requires the fewest changes in character state to produce the characteristics of those tip species. Besides saying what the "best" tree is for a given data set, both methods also provide an *ordering* of trees, from best to worst. The two methods sometimes disagree about this ordering – most vividly, when they disagree about which tree is best supported by the evidence. For this reason, biologists have had to think about the methodological conflict between ML and MP; they can't set it aside as a merely philosophical dispute of dubious relevance to scientists in the trenches.

The main criticism that has been lodged against ML is that it requires the adoption of a model of the evolutionary process that one has scant reason to think is true. ML requires a process model because hypotheses that specify a tree topology (and nothing more) do not, by themselves, confer probabilities on the observations. Here we face yet another instance of the Duhem–Quine thesis, which was a *leitmotif* in Chapters 2 and 3. This thesis asserts that theories in science typically do not make predictions about observables all by themselves but need to be supplemented by *auxiliary propositions* if they are to do so. As before, we need to give this thesis a probabilistic twist. From a likelihood point of view, it isn't essential that hypotheses about the topology of a phylogenetic tree *deductively entail* observational claims about the characteristics of species.[34] What is required is that they *confer probabilities* on those observations. The problem is that, all by themselves, they do not. In the

---

[33] The task of reconstructing the character states of the ancestors in a tree that is presumed to be true was discussed in §3.3 and §3.11 in connection with testing selection hypotheses.

[34] In Sober (1988: Chapter 4), I discuss and criticize some attempts to justify phylogenetic parsimony in terms of Popperian ideas about falsification (§2.8).

language of statistics, these genealogical hypotheses are composite, not simple.

The main objection that has been made against MP is that parsimony implicitly assumes this or that dubious proposition about the evolutionary process. The force of this objection is somewhat unclear, since it is controversial which propositions the method in fact assumes. Does MP assume that evolution proceeds parsimoniously? That is, if a lineage starts with one character state and ends with another, is one obliged to assume that the lineage got there via a trajectory that involved the smallest possible number of evolutionary changes? This allegation has been strenuously denied by proponents of parsimony (e.g., Farris 1983), some of whom maintain that parsimony assumes only that there has been descent with modification.[35]

Which is better – using a method that explicitly makes unrealistic assumptions or a method whose assumptions are unclear? I will argue that this unhappy dilemma misrepresents the dialectical situation twice over. Although ML has usually been implemented in the way described, where the analysis is carried out by stating a single process model and assuming that it is true, there is every reason to shift to a model-selection framework (§1.7) in which multiple process models can be taken into account. This means that a statistical approach to phylogenetic inference is not stopped dead by the objection against ML that I just described. With respect to the criticism of MP, something substantive *is* known about what parsimony assumes, though the issue of parsimony's presuppositions has often been misunderstood.

The debate about ML and MP may seem to be settled by the type of data one wishes to analyze, the thought being that aligned sequences require ML and phenotypes require MP. To be sure, ML is often applied to sequences and rarely to phenotypes (see Lewis 2001 for an exception) while MP is often applied to morphological data and with increasing reluctance to sequences. However, this is a sociological fact, not a logical inevitability. In what follows I'll try to show that the questions that need to be answered when ML is applied to sequence data also are central to the task of applying ML to phenotypes. Symmetrically, MP can be applied to sequence data just as it can be applied to morphology. In addition, ML and MP are sometimes *equivalent* (more on this below), so it is hard to see how MP can be tied essentially to one type of data and ML to another.

---

[35] For discussion of Farris's argument, see Sober 1988.

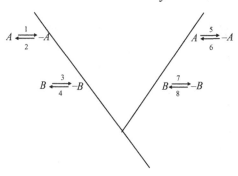

Figure 4.20   Each of the dichotomous traits A and B can experience two changes and each change can occur on each of the two branches. There are eight parameters $(p_1, \ldots, p_8)$ – one per change, per trait, per branch.

Although ML methods are most familiar in the context of analyzing sequence data, I want to start discussing that methodology in the context of models of phenotypic evolution. To get a feeling for the different process models that might be used, consider two dichotomous traits that evolve on the two branches of the phylogenetic tree depicted in Figure 4.20. If we assign a separate parameter to characterize the probability of each change that might occur in each trait on each branch, there will be eight parameters. We can reduce the number of parameters by introducing *constraints*; these constraints require various parameters to have the same value. Here are three examples:

- A constraint on changes within traits within branches: $p_1 = p_2, p_3 = p_4, p_5 = p_6, p_7 = p_8$.
- A constraint on changes across traits within branches: $p_1 = p_3, p_2 = p_4, p_5 = p_7, p_6 = p_8$.
- A constraint on changes within traits across branches: $p_1 = p_5, p_2 = p_6, p_3 = p_7, p_4 = p_8$.

A very simple model can be constructed by imposing all three of these constraints; I'll call this the *yes–yes–yes* model. This model contains a single parameter; it rules out biased processes such as natural selection, since it says that a change from A to $-A$ has the same probability as a change from $-A$ to A. At the opposite extreme is the *?–?–?* model; this is the eight-parameter model just mentioned. It does not *deny* the equalities expressed in the constraints just described; rather, this model simply declines to assert that they are true (this is why I use three question marks

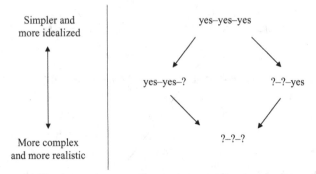

Figure 4.21 Models are more complex the larger the number of adjustable parameters they contain. Arrows represent deductive implication; "$M_1 \rightarrow M_2$" means that if M_1 is true, M_2 must be true.

rather than three "no"s to represent this model). This model is compatible with drift or selection, and with homogeneity and heterogeneity between branches and between different traits on the same branch. In between the one parameter *yes–yes–yes* and the eight-parameter *?–?–?*, there are six intermediate models. For example, the *yes–yes–?* model rules out natural selection, but it allows that the two branches might experience different rates of neutral evolution. And the *?–?–yes* model allows that selection is possible, but requires that a given character experience the same process across branches (be it biased or unbiased). These different models are related to each other by the relation of logical implication, as shown in Figure 4.21. The most constrained model is a special case of all the less constrained models. Removing constraints produces a logically weaker model.[36] Notice that the two intermediate models described in the figure, *yes–yes–?* and *?–?–yes*, are not related to each other by the entailment relation; neither is a special case of the other.

Although this taxonomy of process models applies to dichotomous phenotypic traits, it easily generalizes to sequence data. Each site in a sequence has one of four possible states (*G*, *A*, *T*, and *C*). Consider two aligned sequences drawn from different branches of a phylogenetic tree, as shown in Figure 4.22. The models usually used in phylogenetic inference

---

[36] Even with just two characters on two branches, further complications might be introduced. For example, the eight models described all assume that traits on the same branch evolve *independently*; models that allow for correlated changes within branches would introduce additional adjustable parameters.

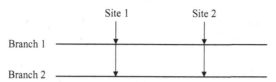

Figure 4.22 Two sites in two aligned sequences that come from different branches of a phylogenetic tree.

for molecular characters are a small subset of the possibilities. Virtually all are *time reversible*; it is assumed that a change from one state to another in a site on a branch has the same probability as a change in the opposite direction (Swofford et al. 1996: 433). This excludes selection. And a change at one site on a branch is assumed to have the same probability as the same change at a different site on the same branch. However, branches are allowed to differ; even if a model says that all changes have the same probability *per unit time*, it will usually allow that branches have different durations. Recall from §3.5 that Markov models allow one to compute the probability that a branch ends in one state, given that it begins in another; the values of these *branch* transition probabilities are functions of the *duration* of the branch and the *instantaneous* probabilities of different changes. A given change will be more probable on a branch that lasts a long time than it is on a branch that has only a short duration.

If most of the models of molecular evolution used in phylogenetic inference ignore selection and assume that a given change on a branch has the same probability, regardless of which site one considers, how do these models differ? The Jukes–Cantor (1969) model contains a single adjustable parameter that represents the (instantaneous) probability of all change at all sites on all branches. The Kimura (1980) model has two parameters; it allows transversions and transitions to have different probabilities.[37] These models assume that the four nucleotides have the same expected frequencies throughout the tree. The Felsenstein (1981) model says that all substitutions on all branches have the same probability but allows that base frequencies may be unequal. All three of these models are special cases of the general time-reversible (GTR) model (Lanave et al. 1984; Taveré 1986; Rodriguez et al. 1990). As shown in Figure 4.23, the relation of logical implication links some of these models to others, just as was true in Figure 4.21. As before, the two intermediate models, Kimura

---

[37] Changes between *A* and *G* and between *C* and *T* are *transitions*; all other changes are transversions.

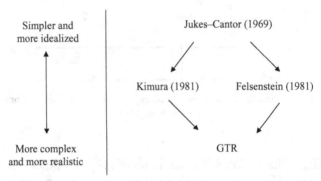

Figure 4.23 Four models of molecular evolution and their logical relationships (figure adapted from Swofford et al. 1996: 434).

(1981) and Felsenstein (1981) are not related in this way; neither is a special case of the other.

How are these different process models put to work in a likelihood assessment of phylogenetic hypotheses? Let's continue to use the example of humans, chimps, and gorillas. Assuming that the tree must be strictly bifurcating (i.e., that it contains no reticulations or polytomies), there are three possible rooted trees: $(HC)G$, $H(CG)$, and $(HG)C$. As noted earlier, none of these, by itself, confers a probability on the characteristics we observe. However, the same is true if we conjoin one of these genealogical hypotheses with one or another of the process models just described. The reason is that each process model contains at least one adjustable parameter. Until values for adjustable parameters are specified, we cannot talk about the probability of the data under different hypotheses. In short, the propositions that have well-defined likelihoods take the form of a conjunction that contains three conjuncts:

Tree topology & process model & specified values for the parameters in the model.

The parameters that describe the probabilities of different changes are examples of what statisticians call *nuisance parameters*. The reason for this name is not that biologists never take an interest in the values of these parameters; rather, the point is that when we are interested in comparing the likelihoods of different tree topologies, we are forced to deal with questions about the evolutionary process even though these are not the focus of our inquiry. Naturally, what is a nuisance parameter in one problem may be the subject of interest in another. Our present concern is testing tree topologies against each other; in Chapter 3, we considered different process models (for example, selection versus drift). In that setting, the tree topology might be thought of as a nuisance parameter.

To assess the likelihood of a three-conjunct conjunction that has the form just described, we first need to recall the very different approaches that Bayesianism and the Neyman–Pearson theory take to the problem of handling nuisance parameters (§1.3, §1.5). For a Bayesian, the likelihood of a tree topology is an *average*. There are many different process models that might be true and many different values that the parameters in a given model might have. The likelihood of $(HC)G$ reflects all of these:

$$Pr[\text{data} \mid (HC)\, G]$$

$$= \sum\nolimits_{i,j} Pr[\text{data} \mid (HC)G \ \& \ \text{Model } i \text{ and Parameter vaules } j]$$

$$\times Pr[\text{Model } i \ \& \ \text{parameter vaules } j \mid (HC)\, G].$$

If a given model M were known to be true (or if this is an assumption whose consequences one wishes to explore), this summation would simplify to

$$Pr[\text{data} \mid (HC)\, G] = \sum\nolimits_{j} Pr_M[\text{data} \mid (HC)G \ \& \ \text{parameter vaules } j]$$

$$\times Pr_M[\text{parameter vaules } j \mid (HC)\, G].$$

The subscript M on the probability function means that the probabilities are all assigned on the assumption that model M is true. The same sort of averaging would have to be undertaken for the other topologies under consideration, and then the average likelihoods of the different topologies could be compared.

Given a process model M that one is prepared to regard as true, the Neyman–Pearson method of handling nuisance parameters is very different. One isn't interested in averaging over all possible values; rather, one looks at the single setting of those parameters that makes the data most probable. For the topology $(HC)G$, the quantity of interest is

$$Pr\{\text{data} \mid L[(HC)\, G \ \& \ \text{model } M]\}.$$

Here "L[$(HC)G$ & model M]" denotes the likeliest member of [$(HC)G$ & model M]. The values of the parameters in M that maximize the likelihood of $(HC)G$ need not be the same as the ones that maximize the likelihood of other topologies.

	(HC)G	H(CG)	(HG)C
Jukes–Cantor	L[(HC)G & Jukes–Cantor]	L[H(CG) & Jukes–Cantor]	L[(HG)C & Jukes–Cantor]
Felsenstein 1981	L[(HC)G & F]	L[H(CG) & F]	L[(HG)C & F]
Kimura 1980	L[(HC)G & K]	L[H(CG) & K]	L[(HG)C & K]
GTR	L[(HC)G & GTR]	L[H(CG) & GTR]	L[(HG)C & GTR]

Figure 4.24 Conjunctions of the form "tree topology & process model" containing adjustable parameters; these are nuisance parameters in the context of making inferences about topologies. Frequentists set these at their maximum likelihood values, denoted by "L(process model & tree topology)."

Most statistical work in phylogenetic inference has been carried out within a frequentist, not a Bayesian, framework. The usual practice has been to adopt a single model of the evolutionary process and then compare topologies with the parameters in the model set at their maximum likelihood values. This amounts to making "horizontal" comparisons within a single row in Figure 4.24. When the trees are "specified in advance," biologists frequently seek to determine which of the three conjunctions has the highest likelihood, thus bypassing questions about which hypothesis is the null and what value of α should be chosen. However, other procedures (e.g., the SOWH test; see Felsenstein 2004: 371–2 for discussion) are sometimes used when the ML tree is compared with one that is less likely; here, the ML tree is regarded as the null hypothesis, and the question is whether an alternative tree is *significantly* less likely than it. We see here a pattern that often arises in frequentist practice; the statistical procedure is not determined by logical and mathematical relationships among data, hypotheses, and background assumptions but involves facts about what goes on in the mind of the investigator (recall the discussion of stopping rules in §1.6). Your treatment of (*HC*)*G*, *H*(*CG*), and (*HG*)*C* depends on whether you design your test before gathering data or design the test already knowing that (*HC*)*G* is the most likely tree. Bayesians and likelihoodists find it hard to understand why this difference should make a difference.

Frequentists also make vertical comparisons in Figure 4.23; here, you are not testing a topology; rather, you are testing different process models against each other, given the assumption that some topology is true. The typical procedure is to use the likelihood ratio test. The question is not which conjunction has the higher likelihood; we know in advance that models with a larger number of adjustable parameters will fit the data better, so likelihoods must increase as one moves from the Jukes–Cantor model to Kimura (1980) and then to GTR. Rather, the question is whether the likelihood of a more complex model is *sufficiently* greater than the likelihood of a simpler model to justify rejecting the simpler model. As noted in §1.5, this methodology has a frequentist justification only for nested models. It is possible to compare each of [$(HC)G$ & Felsenstein] and [$(HC)G$ & Kimura] with [$(HC)G$ & Jukes–Cantor], but one can't compare the first two with each other. Another property of the likelihood ratio test is that it can yield different answers depending on whether one starts with the simplest model and works up or starts with the most complex model and works down (§1.5).

These limitations of the Neyman–Pearson theory suggest that it may make sense to place phylogenetic inference within a model-selection framework.[38] In using AIC, or some other model-selection criterion, one obtains an ordered list, from best to worse, of conjunctive hypotheses, each of which has the form "genealogical hypothesis & process model." The Duhemian point continues to apply: In the first instance, what one is testing are the different conjunctions, not the genealogical hypotheses taken on their own. Still, one can reach inside these conjunctions and examine the conjuncts of interest in the following way. Suppose $(HC)G$ is the genealogical hypothesis that figures in the first five, or the first ten, or the first fifteen conjunctions at the top of the list. The larger this group of conjunctions is, the more we are entitled to conclude that the data favor $(HC)G$. In this case, $(HC)G$ is *robust* across variation in process model, and the more robust the better. But suppose that $(HC)G$ appears in the first, but not the second, of the conjunctions on this list and then appears in the third through twentieth entries. Since AIC provides a quantitative score for each conjunction, and not just an ordering of conjunctions, one can ask what the *average effect* is of shifting from one tree topology to another, within each of several process models. For example, perhaps AIC scores are on average improved by moving from $H(CG)$ to $(HC)G$.

What resources does Bayesianism have for testing tree topologies against each other across a range of possible process models? Just like frequentist work on phylogenetic inference, most Bayesian analyses have opted for a single process model and then compare topologies within the context of that one model; what makes the work distinctly Bayesian is that a prior distribution is employed for the values that the nuisance parameters in the model might have. But Bayesians also have started to consider

---

[38] Kishino and Hasegawa (1990) applied AIC to choice between tree topologies; see Posada and Crandall (2001) and Posada and Buckley (2004) for further discussion.

multiple process models within a model-selection framework (see, for example, Huelsenbeck et al. 2004). If one topology has a higher average likelihood than another for each of the process models one has considered, this shows that the result is *robust*; it does not depend on which of these process models one chooses. And if unanimity across models fails, the fact that BIC provides quantitative values for the average likelihoods of different conjunctions, and not just an ordering, becomes important. BIC can be used to evaluate the average likelihoods of conjunctions of the form (tree topology & process model M), and one can see what the *average effect* is of shifting from one tree topology to another across each of several process models.

Model-selection theory, whether it is Akaikean or Bayesian, provides the resources for statistically testing tree topologies against each other without requiring one to decide in advance which process model is true. Choosing between the two approaches requires one to consider the different goals that AIC and BIC have and involves the questions surveyed in Chapter 1 concerning whether various assumptions that go into the two procedures are reasonable. I won't repeat those points here, but I want to recall one theme. If the process models one is considering all contain idealizations, all are false, so there won't be much point to asking which of them has the highest probability of being true. A better paradigm is the goal of estimating predictive accuracy, of finding fitted models that are close to the truth.

What does cladistic parsimony assume about the evolutionary process?

What does the word "assume" mean in the question that is the title of this section? An example from outside science provides some guidance. Consider the two sentences

(P)   Jones is poor but honest

and

(A)   There is a conflict between being poor and being honest.

I hope it is clear that *P assumes* that *A* is true, but that *A* does not assume that *P* is true. Notice that *P entails A* – that is, if *P* is true, then *A* must also be true. However, *A* does not entail *P*; if there is a conflict between poverty and honesty, this says nothing about Jones and the characteristics he happens to have. This example points to a general fact about what it means to talk about the assumptions of a proposition:

If P assumes A, then P entails A.

To find out what a proposition assumes, you must look for conditions that are *necessary* for the proposition to be true, not for conditions that *suffice* for the proposition's truth.

When are likelihood and parsimony ordinally equivalent?

Given this clarification of what an assumption is, we can turn to the question of what it means to talk about the assumptions that are involved in using cladistic parsimony to infer tree topologies. What parsimony assumes about the evolutionary process are the propositions that must be true if parsimony is to be a legitimate method of phylogenetic inference. But what does "legitimate" mean? There are a number of choices to consider. For example, one might demand that a legitimate phylogenetic method be statistically consistent – that it converge on the true phylogeny as the number of observations is made large without limit. We will consider this idea in the next section. Another interpretation – the one I want to explore now – maintains that parsimony is a legitimate method precisely when it is *ordinally equivalent* with likelihood. This idea is easy to understand by considering the Fahrenheit and Centigrade scales of temperature. These are ordinally equivalent, meaning that for any two objects, the first has a higher temperature in Fahrenheit than the second, precisely when the first has a higher temperature in Centigrade than the second. The two scales induce the same ordering of objects. For parsimony and likelihood to be ordinally equivalent, the requirement is that

(OE)   For any phylogenetic hypotheses H_1 and H_2, and for any data set D, H_1 provides a more parsimonious explanation of D than H_2 does precisely when $Pr_M(D \mid H_1) > Pr_M(D \mid H_2)$.

The subscript M in the likelihood terms is a reminder of the Duhemian point that phylogenetic hypotheses do not confer probabilities on data, save in the context of a process model. In fact, it is misleading to talk of parsimony and "likelihood" being, or failing to be, ordinally equivalent. Rather, the question is whether likelihood when implemented by an assumed process model M is or is not ordinally equivalent with parsimony.

This may be true for some process models and false for others. I am interested in (OE) as a device for exploring the legitimacy of parsimony because I think that likelihood is a good measure of the degree to which evidence favors one hypothesis over another. However, (OE) could be employed in the opposite direction – by someone who believes that cladistic parsimony is legitimate and wants to see whether likelihood (when implemented by using some process model M) can be justified in terms of parsimony.

Viewed from the vantage point of likelihoodism, our question concerning the assumptions that parsimony makes about the evolutionary process comes to this: Which propositions about evolution must be true if (OE) is correct? More specifically, which model or models of the evolutionary process must be true if parsimony and likelihood are to coincide in their evaluation of how data bear on competing tree topologies? One mathematical result that needs to be understood in this context is Felsenstein's (1973, 1979) demonstration that likelihood and parsimony are ordinally equivalent when all probabilities of change in character state are very small. Many biologists have taken this result to show that parsimony "assumes" that evolutionary change is very improbable, but the result shows nothing of the kind. Felsenstein's result provides a *sufficient* condition for ordinal equivalence, not a *necessary* condition. The logical relationships involved here are as follows:

(E) Felsenstein's model (in which changes are very improbable)
→ ordinal equivalence → assumptions of parsimony   -

Felsenstein did not demonstrate the following very different relationship:

Ordinal equivalence → low probability of change

The first link in the chain of entailments shown in (E) is Felsenstein's result, and it does throw light on what parsimony assumes – or rather, on what parsimony does *not* assume. We can see from (E) that any assumptions that parsimony makes must be entailed by Felsenstein's model. This provides the following partial test for whether parsimony assumes that this or that proposition is true (Sober 2005):

- If a proposition is entailed by the Felsenstein's model, it *may or may not be* an assumption that parsimony makes.
- If a proposition is not entailed by Felsenstein's model, it is *not* an assumption that parsimony makes.

This test has some interesting consequences. For example, Felsenstein's model does not require neutral evolution. Though change in character state must be improbable, there is no demand that a change in one direction must have the same probability as a change in the opposite direction. Parsimony, therefore, does not assume neutrality. Nor does Felsenstein's model include the requirement that a change from one state to another has the same probability on all branches of the tree. Parsimony, therefore, does not assume that branches are homogeneous.

Tuffley and Steel (1997) report a result that throws further light on the question of what parsimony assumes. The model they describe involves *no common mechanism*. There is no requirement that change is improbable and none that different traits on the same branch evolve according to the same rules or that the same trait on different branches must have the same probability of changing. However, all traits evolve by neutral evolution. Their result can be represented in the same way that (E) represents Felsenstein's:

Tuffley and Steel's model of no-common-mechanism → ordinal equivalence → assumptions of parsimony

The two bullet points above that describe the partial test for parsimony's assumptions can be applied to the Tuffley–Steel result; if a proposition is not entailed by their model, then it is not an assumption of parsimony. One consequence is that *parsimony does not assume that change is improbable*. This goes contrary to what many biologists have claimed.

The partial test described here has an obvious limitation; it can demonstrate that this or that proposition is *not* an assumption required by cladistic parsimony, but it cannot demonstrate that a given proposition *is* one of parsimony's presuppositions. However, the criterion of ordinal equivalence suggests a second procedure that goes beyond the partial test. If a model of the evolutionary process entails that parsimony and likelihood are *not* ordinally equivalent, then parsimony assumes that that model is false. This approach to the problem is both simpler and more powerful than the partial test. Felsenstein and Tuffley and Steel each produced *general* results; notice that the criterion of ordinal equivalence describes *any* two topologies and *any* data set. The idea now on the table simply requires an *example* – one in which parsimony and likelihood disagree when likelihood is implemented by a given model.

We considered an example of this sort in §3.3 in connection with the problem of reconstructing the character states of ancestors. In a single branch in which a descendant *D* is observed to have quantitative character

state $D = x$, the most likely assignment of character state to its ancestor
(A) is $A = x$ if drift is the process at work, but this will not be the most
likely estimate if there is directional selection pushing the lineage towards
an optimal trait value O where $O \neq x$. Using the ordinal equivalence
criterion as our guide, we may conclude that parsimony assumes that the
process at work in the lineage is *not* directional selection of the kind just
described. For other examples in which this line of reasoning is pursued,
see Sober (2002c).

The quest to discover parsimony's presuppositions could be set to one
side if a model of the evolutionary process could be presented that
everyone grants is plausible and that suffices to induce ordinal equivalence. The demonstration of sufficiency would not, of course, show that
parsimony *assumes* that this model is true. However, people prepared to
grant that the model is true will thereby have reason to conclude that
parsimony is legitimate (when judged by the criterion of ordinal
equivalence). *They* assume the model is true, even if *parsimony* does not,
and that will suffice to justify parsimony in their eyes. Unfortunately, no
such solution will be available for those who think that tractable models
of the evolutionary process inevitably contain idealizations and so are false.

One reason it is difficult to keep a clear head when the assumptions
of parsimony are discussed is that the word "assumptions" gets used in
different ways. We talk about the assumptions that *people* make, the
assumptions that go into *mathematical proofs* and the assumptions that a
proposition requires. People who use this or that method of inference may
make different assumptions or none at all, but that is a matter of psychology, not logic. Biologists sometimes assert that they make no assumptions
about the evolutionary process when using parsimony to infer phylogenetic
relationships, and this remark may be correct as an item of autobiography.
But for those interested in the logic of phylogenetic inference, the biographical remark is beside the point. The assumptions that go into a
mathematical proof are the stated propositions from which various the-
orems are shown to follow. The proof of Felstenstein's mentioned earlier
assumes that changes in character state have low probabilities of occurring;
this assumption is part of a set of assumptions that *suffices* for parsimony to
have a likelihood rationale. I hope it is clear that the correctness of the
proof does not show that *parsimony* assumes that evolutionary change is
improbable. The assumptions in a proof *suffice* to guarantee that a theorem
is true; they may or may not be *necessary* (and usually they are not). To put
the point somewhat paradoxically: The assumptions made in a proof may
or may not be assumptions of the proposition proved.

Much remains to be learned about parsimony's presuppositions. Felsenstein (1973, 1979) and Tuffley and Steel (1997) use the frequentist procedure for handling nuisance parameters. What connection can be established between parsimony and likelihood when nuisance parameters are handled in a Bayesian fashion? And what if AIC, or some other model-selection criterion, is used to rank tree topologies across multiple-process models?

Statistical consistency

Although the criterion of ordinal equivalence describes one gold standard that has been used to judge the legitimacy of parsimony, there has been a great deal of discussion of another requirement that a method of phylogenetic inference might be asked to satisfy. This is the demand that a method must be statistically consistent, meaning that as the number of data increases, it becomes more and more certain that the method will reconstruct the true tree. Felsenstein (1978) initiated this approach, describing a circumstance in which MP is statistically inconsistent and claiming that some types of ML inference are statistically consistent. Many biologists extracted from Felsenstein's paper the lesson that ML is the better procedure – it passes a test that MP fails.

Felsenstein's argument is based on the simple example shown in Figure 4.25. Characters are assumed to be dichotomous, and all evolve according to the same rules. Changes from 0 to 1 have probability p on the two branches shown and probability q on the two others; changes from 1 to 0 are impossible. The root of the tree is in state 0 for each

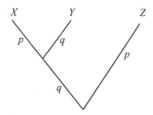

Figure 4.25 The example described in Felsenstein (1978) in which parsimony can converge on the incorrect tree as more and more data are consulted. The probability of a character's changing from state 0 to state 1 is p on the two branches shown and q on the two others. The root is assumed to be in state 0. Changes from 1 to 0 are impossible. Parsimony will be statistically inconsistent precisely when $p \gg q$.

character. With respect to any character, the probability of this tree's evolving $X = 1$, $Y = 1$, and $Z = 0$ (the 110 pattern, for short) is

$$[q + (1 - q)pq]\,(1 - p)$$

while the probability of its evolving the 101 pattern is

$$[(1 - q)p\,(1 - q)]p.$$

Parsimony interprets a character exhibiting the 110 pattern as evidence favoring $(XY)Z$ over $(XZ)Y$; it interprets a character with the 101 pattern as evidence with the opposite significance. Parsimony also judges the 111, 000, 001, 010, and 100 character distributions to be uninformative because the three rooted trees are equally parsimonious explanations of each.

   If $(XY)Z$ is the true tree and the model of trait evolution is the one just described, when is it more probable that parsimony will choose $(XY)Z$ rather than $(XZ)Y$ based on the data that the true tree generates? This is probable precisely when

$$[q + (1 - q)pq](1 - p) > [(1 - q)p\,(1 - q)]p,$$

which simplifies to

$$p^2 < q(1 - pq)/(1 - q).$$

For each value of q, there are values of p that satisfy this inequality and values that violate it. These are shown in Figure 4.26. If $p \gg q$, the law of

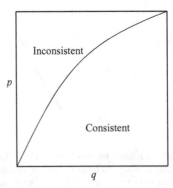

Figure 4.26   The tree in Figure 4.25 is in the "Felsenstein zone" when $p \gg q$. In this case, parsimony will converge on a false tree as more and more data are drawn.

large numbers guarantees that the probability approaches unity that the 101 pattern will occur more frequently than the 110 pattern as more and more data are gathered. In this unhappy circumstance, the more data you consult, the more certain you can be that parsimony will say that the false tree $(XZ)Y$ is better supported than the true tree, $(XY)Z$. Figure 4.26 indicates that the absolute values of p and q aren't relevant; parsimony can be inconsistent even when p and q are both small; a low probability of change on each branch does not suffice for parsimony to converge on the truth.

Farris (1983) responded to Felsenstein's argument by pointing out that the model of evolution used in the example is unrealistic. The failure of consistency in this case does not show that parsimony will often or ever be inconsistent when real data sets are analyzed. Felsenstein did not reply to this criticism by insisting that a method should be statistically consistent no matter what the underlying evolutionary process is like. Rather, he said that one can't simply assume that parsimony will be consistent when a more realistic model is used to generate the data; one must demonstrate that this is so. For Felsenstein, a method of inference must be consistent when applied to real data, so using parsimony on real data requires the assumption that the process model he constructed in his example is false. His point is that parsimony makes substantive assumptions about the evolutionary process.

Biologists were probably not surprised by the performance of parsimony in Felsenstein's (1978) example; if homoplasies are more probable than homologies, it is to be expected that parsimony will be misleading. What might have been more striking was Felsenstein's claim that "it can be shown quite generally that the maximum likelihood estimation procedure has the property of consistency" (1978: 408); in support Felsenstein cites his 1973 paper. However, Felsenstein then notes that when the number of parameters increases indefinitely as more characters are added, "maximum likelihood methods are particularly prone to lack of consistency" (1978: 409). Felsenstein's remark prefigures Tuffley and Steel's (1997) result. In their no-common-mechanism model, each new character has its own suite of transition probabilities, one parameter for each change that can take place on each branch. Within this model, the number of parameters does not hold steady as the size of the data set increases. Likelihood and parsimony are ordinally equivalent within this model, and both will fail to be consistent if the process generating the data obeys the no-common-mechanism model (for some settings of the parameters in that model). This means that an investigator using ML in conjunction with the no-common-mechanism model (but not knowing what the true values of the parameters are in the model) can be systematically misled.

A sufficient condition for the statistical consistency of ML was demonstrated by Chang (1996); he considers a model in which all sites follow the same rules of evolution, so increasing the size of the data set does not involve introducing new parameters that need to be estimated. However, this demonstration shows that the statistical consistency of ML is not the trump card it might first have appeared to be. This is because demonstrating that likelihood is statistically consistent involves examining what is true when the model generating the data is the same as the model used to implement the maximum-likelihood inference. As noted before, "ML" is not a single method. Rather, there is a family of methods that might be called "ML using model X," where the members of the family differ as to which process model gets used. Demonstrating that ML using model X is statistically consistent means showing that

For any vector v of values that the parameters in model X might have, Pr(ML using model X chooses tree topology $T \mid T$ is true & $X(v)$ is true & $X(v)$ generates n items of data) approaches unity as n approaches infinity.

Here $X(v)$ denotes the result of setting the adjustable parameters in model X at the values given by v and ML using model X means using the model but not knowing what the true values are of the parameters in the model. So what does it mean for ML using model X to be statistically consistent? It just means that if nature obeys model X and you use model X to compare tree topologies, you can be increasingly certain of finding the true topology as ever-larger data sets are considered. The point to notice is that there can be no general guarantee that ML using model X will reconstruct the true tree if the evolutionary process is governed by a *different* process model (Steel and Penny 2000: 840). That an ML method can fail to converge on the true topology is not just an abstract possibility; for example, Gaut and Lewis (1995) showed that ML can be inconsistent if the modeler assumes that rates are the same across sites while the underlying reality is that they are not.

So the fact that ML using model X can be shown to be statistically consistent does not guarantee that you'll converge on the true tree if you use this method of inference on the ever-larger data sets that *nature* provides. Once again, the fact that tractable models of the evolutionary process contain idealizations comes into play. If the underlying process is more complex than any model that a statistician will be willing to touch with a stick, the guarantee of statistical consistency for tractable models boils down to a very hypothetical assertion: If this idealized model were true (which it is not), then using that model in the context of ML inference would converge on the true phylogeny. Perhaps De Niro had a point when he replied to Charles Grodin's counterfactual by saying "but you're not my accountant" (§3.3).

Concluding comments on phylogenetic inference

Since the early 1970s, a dispute has raged between defenders of ML and defenders of MP. The former group has mainly embraced a frequentist statistical philosophy, with Bayesian methods entering the arena only more recently (see, for example, Rannala and Yang 1996 and Huelsenbeck et al. 2001).

The main criticism that has been leveled against the statistical approach to phylogenetic inference is that it requires one to accept at the outset a single model of the evolutionary process. This is a troubling requirement, since biologists who don't know what the true tree topology is for a set of taxa usually also fail to know which processes governed the evolution of the traits that those taxa exhibit. Although this criticism has had some merit in the past, there is nothing intrinsic to the statistical approach that forces a single model to be used. Rather, a model-selection framework, whether it is Bayesian or Akaikean, permits tree topologies to be compared across a range of possible process models.

The main criticism that has been leveled against cladistic parsimony is that it makes implausible assumptions about the evolutionary process. It may seem "obvious" that parsimony assumes that evolution proceeds parsimoniously – that it generates few homoplasies and that a lineage that begins in one state and ends in another gets from start to finish by way of the smallest possible number of changes. These allegations are frequently expressed, but they are nothing more than bald assertions until an argument is supplied that shows that they are true. The first step in assessing them is to separate *necessity* from *sufficiency*. Felsenstein (1973, 1979) and Tuffley and Steel (1997) constructed models of the evolutionary process that each *suffice* to ensure that the most parsimonious tree will be the tree with the highest likelihood. These two models force parsimony and likelihood to *agree*. However, Felsenstein's model is *incompatible* with the no-common-mechanism model of Tuffley and Steel. If parsimony assumes that *both* are true, it has lapsed into contradiction. In fact, parsimony does not assume that either model is true. It does not assume that changes in character state are very improbable, and it does not assume that neutral evolution is true. To discover what parsimony does assume about the evolutionary process, one needs to find process models in which parsimony and likelihood *disagree*. Friends of likelihood will see such cases as casting doubt on parsimony, but friends of parsimony will conclude that these are cases in which the process model is mistaken. Either way, cladistic parsimony turns out to involve assumptions about the evolutionary process.

Conclusion

I won't try to summarize the preceding four chapters, but I do want to describe some of their main themes.

The lazy way to test a hypothesis H is to focus on one of its possible competitors H_0, claim that the data refute H_0, and then declare that H is the only hypothesis left standing. This is an attractive strategy if you are fond of the hypothesis H but are unable to say what testable predictions H makes. I have nothing against laziness per se. If H_0 really is the only alternative to H, and if H_0 really does deductively entail some observation statement O that turns out to be false, then it really does follow that H_0 is false and H is true. The examples discussed in this book do not exhibit this tidy pattern.

It is false that the only alternative to intelligent design is gradual evolution by natural selection in an infinite population (§2.16), though it is true that this particular hypothesis does entail that fitness valleys cannot be crossed. Suppose we consider, instead, an evolutionary model that says that valleys *can* be crossed. If we want to test this evolutionary hypothesis against the hypothesis of intelligent design, the question we need to consider is contrastive: Does the evolutionary hypothesis or the intelligent-design hypothesis make the observations more probable? Intelligent-design proponents can't leave their "theory" in the background in the hope that evolutionary theory will shoot itself in the foot. Rather, they need to describe what intelligent-design theory asserts beyond the one-sentence slogan that an intelligent designer made this or that feature; in particular, they need to say what their "theory" predicts.

Precisely the same considerations apply to testing selection against drift. It is no good dismissing drift on the grounds that it says that the complex and useful trait we are studying is very improbable. Even if

Pr(observations | drift) is low, that is not enough. The value of Pr(observations | natural selection) also needs to be assessed. The hypothesis of natural selection can take many concrete forms, and what will in fact be tested are specific formulations of that idea, not the generic idea unalloyed. This point holds regardless of whether the observations consist of phenotypic data or molecular data (§3.9). Fortunately, there is a lot more to the theory of natural selection than a one-sentence slogan.

Although examples like Royall's valet (§1.4) are enough to discredit probabilistic *modus tollens*, there are two additional arguments against that dubious inference principle. The first derives from the fact that repeatedly testing a probabilistic theory will drive its likelihood lower and lower (§1.3). A theory T may successfully predict each of 1,000 observations $(O_1, O_2, \ldots, O_{1,000})$ – for example, by saying of each of these observations that $Pr(O_i \mid T) = 0.99$ – and still the value of $Pr(O_1 \ \& \ O_2 \ \& \ \ldots \ \& \ O_{1,000} \mid T)$ is tiny. Probabilistic *modus tollens* is a recipe for interpreting success as failure. The second point about probabilistic *modus tollens* derives from the discussion of common ancestry (CA) and separate ancestry (SA) in Chapter 4. If species X and Y are observed to share trait T and the trait obeys the Markov model described in §3.5 and §4.4, then

$$\frac{Pr(X \ \& \ Y \text{ have trait } T \mid \text{CA})}{Pr(X \ \& \ Y \text{ have trait } T \mid \text{SA})}$$

gets larger as its numerator gets smaller. The reason this happens is that as the value of the numerator shrinks, the value of the denominator shrinks even more. This is an interesting illustration of why the support for a hypothesis should not be measured by seeing how much it probabilifies the observations.

WHAT ARE THE ALTERNATIVE HYPOTHESES WITH WHICH A GIVEN HYPOTHESIS COMPETES?

Probabilistic *modus tollens* (and significance tests) aside, the three statistical philosophies discussed in Chapter 1 all understand evidence contrastively: For an observation O to be evidence *for* the hypothesis H, it must be evidence *against* some alternative to H. The philosophies differ over what that contrasting alternative must be like.

Bayesian philosophers of science see each hypothesis as competing with its own negation. The Bayesian criterion is that O confirms H if and only if $Pr(O \mid H) > Pr(O \mid notH)$. Sometimes this framework is unproblematic, as in the case of testing the hypothesis that your patient has tuberculosis against the hypothesis that he does not (§1.2). But if *H* is a scientific theory like the general theory of relativity, then *notH* will be the catchall hypothesis that covers all possible alternatives to that theory, even ones that have not yet been formulated. Likelihoodists and frequentists both think that testing such catchalls is impossible. The likelihoodist solution to this problem is to restrict testing to the evaluation of statistically simple hypotheses. Frequentists think that likelihoodism is too austere. Even though the frequentists' likelihood ratio test (§1.5) and model-selection criteria such as AIC (§1.7) are very different, each tries to show how evidence can be brought to bear on (some) statistically composite hypotheses.

TESTABILITY

In §2.14, I tried to formulate a criterion of testability that avoids the problems that attach to earlier proposals, including Popper's concept of falsifiability. A first step in this project is to separate the epistemic concept of testability from the semantic concept of meaningfulness. Perhaps a sentence is meaningful only if its negation is as well (Hempel 1965), but it is far from clear that a sentence is empirically testable only if its negation is. Bayesians think the testability of *H* and the testability of *notH* go hand in hand, since $Pr(H \mid O) > Pr(H)$ if and only if $Pr(notH \mid O) < Pr(notH)$. Likelihoodists and frequentists are committed to the view that some hypotheses are testable even though their negations are not.

The hypothesis that an intelligent designer produced the vertebrate eye can always be fleshed out so that the resulting proposition logically entails exactly the features of the organ that you observe. If you observe that the eye has features *F*, you can easily construct the hypothesis that an intelligent designer produced the vertebrate eye and endowed it with features *F* (§2.9). The very same point applies to the hypothesis that natural selection produced the vertebrate eye. Indeed, the point is very general: The claim that *gravity*, or *the casting of hexes*, is responsible for the vertebrate eye also can be fleshed out in a way that precisely captures the details of the eye that you observe. A Bayesian might point out that these beefed-up hypotheses have lower prior and posterior probabilities than the more modest claims that inspire them. But it also is true that these logically stronger claims have likelihoods of unity. The evidence therefore cannot provide Bayesian disconfirmation of these stronger hypotheses, and, given a few modest assumptions, the Bayesian is forced to concede that all these hypotheses are confirmed by the observation that the eye has features *F* (§1.2).

This is obviously an unsatisfactory situation, but that does not mean that the Bayesian claims just mentioned are false. Rather, the point is that there must be more to the epistemology of testing hypotheses about intelligent design, natural selection, gravity, and hexes than this. The remedy I have suggested was developed within a likelihood framework and eschews epistemological holism (Sober 2004c). Rather than testing the conjunction "an intelligent designer created the vertebrate eye and wanted the eye to have features *F*" as a whole by looking just at the observation that the eye has features *F*, we instead might try to test the first conjunct – that an intelligent designer created the eye – by using the observation that the eye has features *F*. To do this, we need to determine whether there are *independently justified* auxiliary propositions that can be added to this modest hypothesis that allow it to confer some probability on what we observe. This strategy is widely used in the sciences. Eddington didn't invent assumptions about the earth, sun, and other celestial bodies that would allow the general theory of relativity to fit the eclipse data he obtained; rather, he obtained independently justified information about those objects that permitted the theory to make a prediction about the eclipse. The inability of intelligent-design "theory" to follow this protocol is central to my argument that the theory is untestable (§2.12).

The legitimacy and importance of the demand for independently justified auxiliary propositions is not unique to likelihoodism and Baye- sianism. When a frequentist uses a model-selection criterion like AIC, various assumptions must be true for the criterion to have the operating characteristics that give it a claim on our attention. Akaike's (1973) proof that AIC is an unbiased estimator of predictive accuracy relies on various normality and uniformity of nature assumptions (§1.7). If we use AIC to score a number of competing models in part because we think that AIC is unbiased, we need some assurance that these assumptions (or others that entail the same result) are true of the problem at hand. It had better be possible to decide this matter prior to forming an opinion as to which models are more predictively accurate and which are less.

PLURALISM ABOUT THE CONCEPT OF EVIDENCE?

I am prepared to be a Bayesian on Monday, Wednesday, and Friday, a likelihoodist on Tuesday, Thursday, and Saturday, and a model selectionist on Sunday. When values for likelihoods and priors can be defended,

Bayesianism is fine. When likelihoods are defensible, but priors are not, likelihoodism makes sense. This shift reflects a change in subject (§1.1). Bayesianism is an answer to Royall's (1997) second question (What should I believe?) whereas likelihoodism is an answer to his first (What does the evidence say?). The fact that different questions have different answers is hardly surprising; this does not deserve to be called "evidential pluralism." In fact, Bayesianism and likelihoodism use *the same* concept of evidence, whose central principle is the law of likelihood.

The shift from Bayesianism and likelihoodism to a model-selection criterion like AIC is more profound. This change also reflects a change in the question asked, but it is not on Royall's list. Rather than asking whether H_1 has a higher probability of being true than H_2, or whether the evidence favors the hypothesis that H_1 is true over the hypothesis that H_2 is true, the Akaike framework asks whether model M_1 will be more predictively accurate than model M_2, or, equivalently, whether the fitted model $L(M_1)$ is closer to the truth (as measured by Kullback–Leibler distance) than is the fitted model $L(M_2)$. If all the models being considered contain idealizations, then none of them is true, but this does not mean that scientists are wrong to take an interest in them.

The move from truth to predictive accuracy as an inferential goal is important, but it does not mean that a new concept of evidence has been put on the table. In §1.7, I briefly contemplated the possibility that AIC scores (or the scores produced by some other model-selection criterion) might count as evidence that bears on hypotheses about the predictive accuracies of different models. Perhaps AIC scores are to predictive accuracy as thermometer readings are to temperature. When this is true, Bayesians and likelihoodists have no reason to dismiss AIC as a frequentist construct, nor is a new concept of evidence needed to interpret AIC scores. The law of likelihood suffices. The question remains, of course, of how generally AIC and the law of likelihood fit together in this way.

There is another way to narrow the gap between Bayesianism and AIC. I have contrasted the Bayesian goal of ascertaining the probability that a model is true with the Akaikean goal of discovering the model's expected degree of predictive accuracy. But the formalism of Bayes' theorem allows any H you please to be considered. H can be the statement that temperature and pressure are linearly related in a pressure cooker, but it also can be the proposition that LIN is a predictively accurate model of the relationship of those two quantities. This means that there is no *formal* reason why Bayesians can't consider predictive accuracy as a goal; they can do so by addressing the question of which theories have the highest probability of being predictively accurate, or which have the highest expected degree of predictive accuracy. Does that mean that there is no substantive difference between Bayesianism and the Akaike framework? A difference that remains is the Bayesian point that computing posterior probabilities requires information about prior probabilities. In contrast, when frequentists attempt to estimate a model's expected degree of predictive accuracy, given the data at hand, they think they can do this without needing to have a prior expectation on which to build.

A different question about evidential pluralism arises when we consider rules of inference that are applicable only to a delimited scientific subject matter. The main example considered in this book is cladistic parsimony. Although cladists have sometimes resisted the suggestion that the justification of this inference principle depends on the correctness of a probabilistic model of the evolutionary process, they generally have sought to justify the use of cladistic parsimony in terms of wider considerations. Some have appealed to Popperian concepts of parsimony and corroboration (Eldredge and Cracraft 1980; Wiley 1981), others to the concept of explanatory power (Farris 1994). But there is another take on the question of what justifies cladistic parsimony. This is the view that cladistic parsimony is *sui generis*: It stands on its own and does not need to be justified in terms of anything else. This position engenders a form of evidential pluralism according to which cladistic parsimony is the right way to interpret evidence when phylogenetic hypotheses are evaluated, but other concepts of evidence are needed for hypotheses on other subject matters. Few have thought of cladistic parsimony in this way; I have not.

TWO KINDS OF PARSIMONY

In his book *Experience and Prediction*, Reichenbach (1938) distinguished two concepts of simplicity, which he terms "inductive" and "descriptive." The first was so called because it is epistemically relevant; inductively simpler hypotheses, Reichenbach thought, are apt to make more accurate predictions. The second is merely aesthetic and of no epistemic relevance. Two logically equivalent sentences express the same proposition; they can't differ in their evidential support or predictive accuracy because they say the same thing. However, one of those sentences may be long and cumbersome and the other short and elegant; if so, they have the same inductive simplicity though they may differ in their descriptive simplicity.

I too have discussed two concepts of simplicity (or parsimony), but both fall in Reichenbach's first category. The literature in statistics on model selection equates the complexity of a model with the number of adjustable parameters it contains. This is not a syntactic feature of the string of symbols that is used in some language to express the model (§1.7). Parsimony in this sense is relevant to estimating how predictively accurate a model will be, or, equivalently, how close to the truth a fitted model is, though model-selection criteria disagree as to exactly what weight parsimony should be assigned. The second concept of parsimony is specifically biological; I have called it *cladistic* or *phylogenetic* parsimony. The cladistic parsimony of a phylogenetic tree depends on the data; this is one respect in which it differs from model-selection parsimony. A tree is parsimonious in this sense if it requires few changes in character state to generate the data attaching to the tree's tips. This is not a syntactic feature of the sentence expressing a proposition but is a property of the proposition itself, no matter how it is expressed. Philosophers have often complained that it isn't clear what makes one theory more parsimonious than another. It is gratifying that scientists have isolated two kinds of parsimony whose meanings are clear.

The question of their justification is another matter. I have looked at cladistic parsimony through the lens of models of the evolutionary process, asking under what circumstances the more parsimonious of two hypotheses has the higher likelihood (§3.3, §4.7) or the higher probability (§3.11). I have been interested in determining how cladistic parsimony on the one hand and likelihood and probability on the other *fit together*; one can explore this question without needing to decide which is more fundamental. When parsimony and a probability model of the evolutionary process agree on the ordering of hypotheses, probabilists may conclude that the model justifies parsimony, but cladists will think that parsimony justifies the model. Since they *agree* about which hypotheses are better and which are worse, there is no practical urgency to deciding in which direction the justification flows. Only when parsimony and a probability model *disagree* about the ordering of hypotheses does the difference in these approaches take on a practical importance. If you believe the probability model, you have reason to doubt the authority of parsimony. On the other hand, if you think that parsimony is authoritative, you must reject the probability model. Cladists will embrace the latter option, but in doing so they must abandon the claim that parsimony makes no assumptions about the evolutionary process beyond the idea of common ancestry. In this sense, cladistic parsimony has an empirical justification if it has a justification at all.

Perhaps the simplest problem that cladistic parsimony addresses is the inference of a single ancestor's character state from the observed character state of one or more of its descendants. Consider a star phylogeny in which all tip species have the same character state. What could be more intuitive than the conclusion that the character state exhibited by these descendants is evidence that their most recent common ancestor was in the same state? If the character in question is dichotomous, the Markov model described in §3.5 entails a *backwards inequality*; if D is a descendant of A, $Pr(D = 1 \mid A = 1) > Pr(D = 1 \mid A = 0)$, regardless of whether drift or selection is the process at work and regardless of what the lineage's (finite) duration is. The more parsimonious reconstruction of the ancestor's character state is also the one with the higher likelihood, and the likelihood ratio of $A = 1$ to $A = 0$ is larger the larger the number of descendants that all are in state *1*. The situation is different for a continuous trait; here the backwards inequality – that $Pr(D = i \mid A = i) > Pr(D = i \mid A = j)$ for all $i \neq j$ – holds when drift is the process at work, but not when selection is pushing the lineage towards an optimum different from i. Parsimony clearly has biological presuppositions in this case. In fact, the same is true for dichotomous traits, but seeing this point requires one to move from the simple example of a star phylogeny to the more complicated case of a bifurcating tree (§3.11). The use of parsimony to reconstruct the character states of ancestors depends on empirical assumptions even when traits are dichotomous.

The idea that the epistemic authority of cladistic parsimony rests on empirical assumptions also is true of model-selection parsimony, notwithstanding the body of mathematics that has been developed in connection with AIC and other model selection criteria. Akaike proved that AIC is an unbiased estimator of predictive accuracy, but, as noted above, the assumptions that Akaike used to derive his theorem are empirical in character (Forster and Sober 1994).

UNIFICATION

Unification is related to parsimony. In fact, the double meaning of the latter is reflected in two meanings that attach to the former. When we ask whether the wings of present-day birds derive from a single common ancestor or independently evolved thousands of times, we face a choice between two possible explanations that differ in their cladistic parsimony. The hypothesis that bird wings are a *homology* unifies the observations and the hypothesis of *homoplasy* disunifies them. With respect to model-selection parsimony, a model that explains multiple data sets by invoking the same set of n parameters is more unified than a model that assigns to each data set its own set of n parameters (§3.7). The law of likelihood cannot explain how the unified model could be better; the Akaike framework can.

Darwin thought that an important virtue of the theory of evolution by natural selection is its ability to unify diverse phenomena. Does the hypothesis of intelligent design do the same? What could unify nature more than the hypothesis that *all* natural phenomena (not just the complex adaptations of organisms) flow from the hand of God? Darwin disparaged this explanation, not by arguing that it is *false* but by suggesting that it is scientifically *empty*. What does "empty" mean in this context? It does not mean that the statement asserts nothing; atheists and theists are disagreeing about *something*, the testability theory of meaning (§2.14) notwithstanding. A better interpretation is that the intelligent-design hypothesis is empty because it makes no predictions. What, then, becomes of the virtue of unification that Darwin claimed for his theory? If the competing theory is *empty*, what's the point of praising Darwin's theory for being *unified*? The answer comes not from pitting Darwin's theory against intelligent design but from seeing Darwin's theory as one of several *biological* alternatives. Common ancestry should be tested against the hypothesis of separate origination. And natural selection should be tested against drift, phylogenetic inertia, and other evolutionary alternatives. This is what biologists routinely do.

Darwin embraced the most unifying of all phylogenetic hypotheses about the genealogy of present life forms when he opted for CA1 – the claim that all present-day living things and all the fossils that now exist trace back to a single common ancestor (§4.2). With respect to his view of the evolutionary process, Darwin was something of a pluralist. He says in *The Origin of Species* that "natural selection has been the main but not exclusive means of modification" ([1859] 1964: 6), a remark that Gould and Lewontin (1978) embraced in their attack on adaptationism. Darwin invokes ancestral influence (aka phylogenetic inertia) as well as natural selection (§4.8) and, within the category of natural selection, he usually understands selection to mean *individual* selection, though he occasionally thinks that *group* selection has been important, as when he discusses human morality or the barbed stinger of the honeybee (Sober and Wilson 1998). Less pluralistic views of evolution are certainly possible – for example, a maximal adaptationism that assigns zero importance to ancestral influence and to group selection. And more pluralistic views are now on the table as well, thanks, for example, to the development of the neutral theory (Kimura 1983).

Philosophers have often viewed unification as a tie breaker: If different theories can each explain the data, the one that is most unifying is the one that is "best" (i.e., most probably true). One imperfection in this formula is that it sweeps past the issue of fit to data, asking only whether or not a theory "can" explain the observations. The problem is that fit to data is a quantitative, not a dichotomous, concept, and a quantitative assessment of goodness of fit is indispensable when evolutionary models are compared. But goodness of fit is not the only virtue of models, at least if the Akaike framework is any guide. The contest between monistic and pluralistic conceptions of the evolutionary process needs to be understood as a problem in model selection.

"TESTING EVOLUTIONARY THEORY"

Creationists often talk of "testing evolutionary theory," and biologists sometimes talk this way as well. The context of their remarks sometimes reveals which specific proposition the authors have in mind, but often this is not the case. It is important to recognize that the phrase "evolutionary theory" is too vague when the subject of testing is broached. There are a number of propositions that evolutionary biologists take seriously. The first step should be to specify which of these is to be the focus.

Physicists standardly draw a distinction between laws on the one hand and initial (and boundary) conditions on the other. The general theory of relativity, by itself, makes no predictions about when eclipses will occur and what features they will have. However, if auxiliary propositions about the features of various celestial bodies are added, the laws plus these auxiliaries do generate testable predictions. The distinction between laws and initial conditions also is important in evolutionary biology. The "laws of motion of populations" are general statements that are conditional in form. They say that if a population has a given set of properties at time t_1 and is subject to this or that evolutionary process, then it has various probabilities of exhibiting different properties at time t_2. These laws make no predictions until initial conditions are specified. Duhem's thesis (§2.12) applies to evolutionary biology no less than it applies to physics, though it, of course, needs to be understood probabilistically.

This point reveals a yawning incompleteness in questions such as "What is the probability that organisms with the intelligence of human beings would exist now, according to evolutionary theory? Which initial conditions should we consider?" Different choices yield different answers.

We might conditionalize on the start of the universe 17 billion years ago, or on the start of the Earth about 4.5 billion years ago, or on the splitting of the lineages leading to humans and chimps some 6 million years ago or on the evolution of anatomically modern humans, about 200,000 years ago, or on the exodus from Africa around 50,000 years ago. There are other choices as well. We could even consider the probability that there is human intelligence *now*, conditional on the fact that there was human intelligence *yesterday*. When asked what the probability is of an event, you should reply, "Conditional on which initial conditions?" But even when a starting point is identified, one can't assume that evolutionary biology has in its pocket an estimate of the probability of any current fact that happens to strike one's fancy. I would say that the values of the conditional probabilities just contemplated, except perhaps the last one, are *unknown*.

Consider a chain of events that contains 1,000 links – E_1, E_2, ..., $E_{1,000}$ – and suppose that a later event occurs only if all the events before it did. Even if $Pr(E_{i+1} \mid E_i) = 0.99$ (for each i), it still will be true that $Pr(E_{1,000} \mid E_1)$ is tiny. In this case, the farther back you start, the lower the probability of the end point you now observe. This relationship between time and probability holds because $E_{1,000}$ can be achieved via just one path. Suppose, instead, that there are many pathways from E_1 to $E_{1,000}$ and that each has a very small probability of occurring. Now it is possible that $Pr(E_{1,000} \mid E_1)$ is large. There is less path dependence in this case. Gould's (1989) thesis that evolutionary outcomes are radically contingent endorses the first probability model; Conway Morris's (2003) antithesis – that the level of contingency is much lower – endorses the second. The fact that the evolutionary process is probabilistic does not settle which of these is the right way to think about the existence of humanlike intelligence or of any other trait. In fact, we should not assume that one of these two patterns is always right and the other is always wrong. Maybe some outcomes involve a high degree of sensitivity to initial conditions while others are more robust (Sober 2003a).

KNOWABILITY AND DEEP TIME

Time is often the enemy of knowability. In a singly connected causal chain extending from E_1 to E_2 to ... to $E_{1,000}$, the *information-processing inequality* holds true. The present state of the system provides more evidence about the recent past than it does about the distant past (Sober and Steel 2002). Think of a lineage in which a dichotomous trait is subject to drift. If you observe that the lineage is now in the state $D = 1$, what evidence does this provide about the state of the lineage when it began? If the lineage had a brief duration, you have strong evidence that the ancestor was in the state $A = 1$. But if the lineage had a longer duration, the evidence for this conclusion weakens. With infinite time, you have no information at all. An optimist will see good news in this fact: With finite time, the state of the descendant always provides *some* evidence concerning the state of the ancestor. A pessimist will see bad news here: We have only *very weak* evidence about deep time. As with glasses half full and glasses half empty, both are right.

Is there any process that can stand in the way of the information-destroying march of time? Consider a process that tends to keep a lineage in the state in which it began. If it begins in state 1, it has a high probability of remaining in state 1, and if it begins in state 0, that is where it probably will remain. Drift does not do this, and neither does unconditional selection for state 1 or unconditional selection for state 0. What does the trick is selection for the majority trait, or, more generally, selection in which there are two adaptive peaks. In this instance, $D = 1$ is evidence for $A = 1$, even when time is infinite. In §4.5, I explored Darwin's claim that adaptive similarities provide scant evidence about common ancestry, but useless similarities provide more. Darwin was right when there is one adaptive peak but wrong when there are several. The fact that all Eukaryotes use the same genetic code in their nuclear genes is evidence for their common ancestry, even if that code turns out to be optimal.

POPULATION THINKING AND TREE THINKING

Ernst Mayr (1976) suggests that Darwin's greatest conceptual achievement was to push aside essentialist (or "typological") assumptions and replace them with population thinking. Mayr distinguishes these two frameworks in terms of how they view the variation in traits that is found in a population. For the populationist, variability is "real"; it is the engine that drives evolutionary processes. For the essentialist, variability is a superficial distraction that must be "seen through" in order to perceive a deeper underlying uniformity. The essentialist grants that human beings vary but holds that the important thing to grasp is *human nature*, which is a set of properties that all human beings share. According to Mayr, populationists see variability as a cause whereas essentialists see variability only as an effect – as something that needs to be explained but that is not itself explanatory.

Mayr's population thinker views populations as *objects*, subject to their own laws of motion (Sober 1980). The rich system of models found in population biology attests to the fertility of this mode of thought. A modest example was described at the start of Chapter 3 in the context of considering why polar bears have fur that averages 10 centimeters in length. I represented the population as a point on a line with probabilistic forces impinging on it that might move it to the left or to the right. Drift is a random walk on the line; a small step to the left has the same probability as a small step to the right. Selection is a biased walk under the influence of a probabilistic attractor (Figure 3.1). In models such as these, one doesn't worry about the individual organisms that make up the population any more than one worries about the individual molecules in a gas when one uses the kinetic theory of gases. This feature of evolutionary modeling is something that the founder of American pragmatism, Charles Sanders Peirce, noticed in Darwin's theory shortly after the publication of *The Origin of Species*:

Mr. Darwin proposed to apply the statistical method to biology. The same thing has been done in a widely different branch of science, the theory of gases. Though unable to say what the movements of any particular molecule of gas would be on a certain hypothesis regarding the constitution of this class of bodies, Clausius and Maxwell were yet able, eight years before the publication of Darwin's immortal work, by the application of the doctrine of probabilities, to predict that in the long run such and such a proportion of the molecules would, under given circumstances, acquire such and such velocities [. . .] and from these propositions were able to deduce certain properties of gases, especially in regard to their heat-relations. In like manner, Darwin, while unable to say what the operation of variation and natural selection in any individual case will be, demonstrates that in the long run they will, or [would], adapt animals to their circumstances. (Peirce 1934: 226)

The same analogy was a repeated motif in R. A. Fisher's thinking about natural selection. Here is what he says in his paper on dominance:

The investigation of natural selection may be compared to the analytic treatment of the Theory of Gases, in which it is possible to make the most varied assumptions as to the accidental circumstances, and even the essential nature of the individual molecules, and yet to develop the general laws as to the behavior of gases, leaving but a few fundamental constants to be determined by experiment. (Fisher 1922a: 321–2)

Eight years later, he returns to this point in discussing his *fundamental theorem of natural selection*:

It will be noticed that the fundamental theorem [...] bears some remarkable resemblances to the second law of thermodynamics. Both are properties of populations, or aggregates, true irrespective of the nature of the units which compose them; both are statistical laws; each requires the constant increase of a measurable quantity, in the one case the entropy of a physical system and in the other the fitness [...] of a biological population.   (Fisher 1957: 39)

Fisher's remarks highlight the fact that population thinking does not require one to deny that organisms or molecules have "essential natures"; the point is that there are higher-level population laws that hold regardless of whether they do.

   Robert O'Hara (1988, 1998) sees a second pervasive pattern in evolutionary biology. In addition to population thinking, there is *tree thinking*. Besides viewing populations as objects buffeted by forces, evolutionary biologists understand current populations as descendants tracing back to common ancestors. This is not an idle detail but a fact that plays a central role in testing theories about the evolutionary process. Phylogenetic trees are constantly used to test selection against drift (§3.2, §3.9) and can also be used to test selection against inertia (§3.10). If assumptions about phylogenetic trees are used to test hypotheses about the causes of character evolution, and assumptions about character evolution are used in inferring phylogenetic trees, is this interpenetration of pattern and process cause for alarm? There is no vicious circularity here, in part because *one* set of traits can be used to infer a phylogeny, which then is used to test process theories about *other* traits that evolve on that tree. In addition, models that combine phylogenies and process hypotheses can be tested as conjunctions, with different conjunctions being compared with each other, as described in §4.7.

   O'Hara ends his 1998 paper with an interesting historical detail. Whewell's 1847 book, *The Philosophy of the Inductive Sciences*, places biological systematics with mineralogy among the "classificatory sciences," not in the category that Whewell called "the palaeontological sciences," which is where he located geology and comparative philology. The paleontological sciences aim at historical reconstruction; Whewell's attention to them was prompted in part by Lyell's revolutionary ideas about geology. Darwin, who took Lyell's geology as a model for the kind of theory he wanted to develop, "in effect took systematic biology out of the classificatory sciences and placed it squarely among the palaeontological sciences" (O'Hara 1998: 327). For Whewell, explaining why gold is yellow is the same type of task as explaining why tigers have stripes. In what sense did Darwin drive these projects apart? After all, every object has its history – a lump of gold no less than an organism – so what is especially historical about "the historical sciences?" One difference is that tigers are genealogically related to each other, whereas lumps of gold (mostly) are not. If the processes governing the evolutionary process were sufficiently powerful, it would not matter what the character state was in which the tiger lineage began. But tigers don't just *have* a history; their present features show the imprint of that history. The tree of life is not a neutral backdrop on which evolutionary portraits are painted. Rather, the branches of the tree are the pathways along which the traits of ancestors influence the traits of their descendants.

Bibliography

Akaike, H. (1973) "Information Theory as an Extension of the Maximum Likelihood Principle," in B. Petrov and F. Csaki (eds), *Second International Symposium on Information Theory*, Budapest: Akademiai Kiado, pp. 267–81.

Alexander, R. M. (1996) *Optima for Animals*, Princeton, NJ: Princeton University Press.

Allman, J., Rosin, A., Kumar, R., and Hasenstaub, A. (1998) "Parenting and Survival in Anthropoid Primates: Caretakers Live Longer," *Proceedings of the National Academy of Science USA* 95: 6866–9.

Anscombe, F. J. (1954) "Fixed-Sample-Size Analysis of Sequential Observations," *Biometrics* 10: 90–100.

Antonovics, J. and Van Tienderen, P. H. (1991) "Ontoecogenophyloconstraints? The Chaos of Constraint Terminology," *Trends in Ecology and Evolution* 6: 166–8.

Arbuthnot, J. (1710) "An Argument for Divine Providence, Taken from the Constant Regularity Observ'd in the Births of Both Sexes," *Philosophical Transactions of the Royal Society* 27: 186–90.

Armitage, P. (1975) *Sequential Medical Trials*, 2nd edn, New York: John Wiley & Sons.

Backe, A. (1999) "The Likelihood Principle and the Reliability of Experiments," *Philosophy of Science* 66: S354–61.

Baum, D. and Larson, A. (1991) "Adaptation Reviewed: A Phylogenetic Methodology for Studying Character Coevolution," *Systematic Zoology* 40: 1–18.

Behe, M. (1996) *Darwin's Black Box*, New York: Free Press.

(2001) "Reply to My Critics:" "A Response to Reviews of Darwin's Black Box: The Biochemical Challenge to Evolution," *Biology of Philosophy*, 16(5) 683–707.

(2005) "Design for Living," *New York Times*, February 7, p. A27.

(2006) "Whether Intelligent Design Is Science: A Response to the Opinion of the Court in Kitzmiller vs Dover Area School District," Special Report, Center for Science and Culture, Discovery Institute. Available online at http://www.discovery.org/scripts/viewDB/filesDB-download.php?command=download&id=697. (Accessed April 21, 2007.)

Borel, E. (1913) "Mécanique Statistique et Irréversibilité," *Journal of Physics*, 5(3): 189–96.

Bradshaw, W. and Holzapfel, C. M. (1996) "Genetic Constraints to Life-History Evolution in the Pitcher-Plant Mosquito, *Wyeomyia smithii*," *Evolution* 50: 1176–81.

Brooke, J. (2003) "Darwin and Victorian Christianity," in J. Hodge and G. Radick (eds), *The Cambridge Companion to Darwin*, Cambridge: Cambridge University Press, pp. 192–213.

Brown, F. B. (1986) "The Evolution of Darwin's Theism," *Journal of the History of Biology* 19: 1–45.

Burkhardt, F. (1993) *The Correspondence of Charles Darwin*, Volume VIII: *1860*, Cambridge: Cambridge University Press.

Burnham, K. and Anderson, D. (1998) *Model Selection and Inference: A Practical Information-Theoretic Approach*, New York: Springer.
(2002) *Model Selection and Multimodel Inference: A Practical Information-Theoretic Approach*, 2nd edn, New York: Springer.

Burt, A. (1988) "Comparative Methods Using Phylogenetically Independent Contrasts," *Oxford Surveys in Evolutionary Biology*, Oxford: Oxford University Press, pp. 33–53.

Butler, M. and King, A. (2004) "Phylogenetic Comparative Analysis: A Modeling Approach for Adaptive Evolution," *American Naturalist* 164: 683–95.

Cairn-Smith, A. G. (1982) *Genetic Takeover and the Mineral Origins of Life*, Cambridge: Cambridge University Press.

Carnap, R. (1947) *Meaning and Necessity*, Chicago, Ill.: University of Chicago Press.
(1950) *Logical Foundations of Probability*, Chicago, Ill.: University of Chicago Press.

Carroll, S. (2005) *Endless Forms Most Beautiful: The New Science of Evo Devo*, New York: Norton.

Cartwright, N. (1994) *Nature's Capacities and their Measurement*, Oxford: Oxford University Press.

Cavalcanti, A. and Landweber, L. (2004) "Genetic Code," *Current Biology* 14: R147.

Chang, J. (1996) "Inconsistency of Evolutionary Tree Topology Reconstruction Methods When Substitution Rates Vary Across Characters," *Mathematical Biosciences* 134: 189–215.

Cheverud, J. M., Dow, M. D., and Leutenegger, W. (1985) "The Quantitative Assessment of Phylogenetic Constraints in Comparative Analyses: Sexual Dimorphism in Body Weight among Primates," *Evolution* 39: 1335–51.

Clifford, W. K. (1999) "The Ethics of Belief," in *The Ethics of Belief and Other Essays*, Amherst, NY: Prometheus Books. First published 1877.

Conway Morris, S. (2003) *Life's Solution: Inevitable Humans in a Lonely Universe*, Cambridge: Cambridge University Press.

Cook, R. and Cockrell, B. (1978) "Predator Ingestion Rate and its Bearing on Feeding Time and the Theory of Optimal Diets," *Journal of Animal Ecology* 47: 529–47.

Courant, R. and Robbins, H. (1959) *What is Mathematics?* Oxford: Oxford University Press.

Coyne, J. and Orr, A. (2004) *Speciation*, Sunderland, Mass.: Sinauer.

Creath, R. (1991) "Every Dogma Has Its Day," *Erkenntnis* 35: 347–89.

Crick, F. (1968) "The Origin of the Genetic Code," *Journal of Molecular Biology* 38: 367–79.

Crow, J., Budowle, B., Erlich, H., Lederberg, J., Reeder, D., Schumm, J., Thompson, E., Walsh, P., and Weir, B. (2000) "The Future of Forensic DNA Testing: Predictions of the Research and Development Working Group," *National Institute of Justice*, NCJ 183697.

Crow, J. and Kimura, M. (1970) *An Introduction to Population Genetics Theory*, Minneapolis, Minn.: Burgess.

Cunningham, C. W., Omland, K. E., and Oakley, T. H. (1998) "Reconstructing Ancestral Character States: A Critical Appraisal," *Trends in Ecology and Evolution* 13: 361–6.

Da Costa, N. C. and French, S. (2003) *Science and Partial Truth: A Unitary Account to Models and Scientific Reasoning*, Oxford: Oxford University Press.

Darwin, C. (1876) *The Variation of Animals and Plants under Domestication*, 2nd edn, New York: D. Appleton. First published 1868.

—— (1887) *The Life and Letters of Charles Darwin*, ed. F. Darwin, London: Murray, Vols I–III.

—— (1958) *The Autobiography of Charles Darwin, 1809–1882, with original omissions restored*, ed. N. Barlow, London: Collins.

—— (1959) *On the Origin of Species: A Variorum Edition*, ed. M. Peckham, Philadelphia, Pa.: University of Pennsylvania Press.

—— (1964) *The Origin of Species*, London: Murray; Cambridge, Mass.: Harvard University Press. First published 1859.

Dauben, J. W. (1994) "Topology: Invariance of Dimension," in I. Grattan-Guinness (ed.), *Companion Encyclopedia of the History and Philosophy of the Mathematical Sciences*, London and New York: Routledge, pp. 939–49.

Dawid, P. (2002) "Bayes' Theorem and Weighing Evidence by Juries," in R. Swinburne (ed.), *Bayes's Theorem*, Oxford: Oxford University Press, pp. 71–90.

Dawkins, R. (1986) *The Blind Watchmaker*, New York: Norton.

—— (2006) *The God Delusion*, Boston, Mass.: Houghton Mifflin Co.

Dembski, W. (1998) *The Design Inference*, Cambridge: Cambridge University Press.

—— (2004) *The Design Revolution: Answering the Toughest Questions about Intelligent Design*, Downers Grove, Ill.: InterVarsity Press.

Dennett, D. (1978) *Brainstorms: Philosophical Essays on Mind and Psychology*, Cambridge, Mass.: MIT Press.

—— (1987a) "Intentional Systems in Cognitive Ethology: The 'Panglossian Paradigm' Defended," in *The Intentional Stance*, Cambridge, Mass.: MIT Press, pp. 237–86.

—— (1987b) "True Believers," in *The Intentional Stance*, Cambridge, Mass.: MIT Press, pp. 13–42.

(1995) *Darwin's Dangerous Idea*, New York: Touchstone.

(2006) *Breaking the Spell: Religion as a Natural Phenomenon*, New York: Viking.

Descartes, R. (1985) *The Philosophical Writings of Descartes*, trans. J. Cottingham, R. Stoothoff, and D. Murdoch, 2 vols, Cambridge: Cambridge University Press.

Desmond, A. and Moore, J. (1991) *Darwin: The Life of a Tormented Evolutionist*, New York: Times Warner.

Diaconis, P. and Mosteller, F. (1989) "Methods of Studying Coincidences," *Journal of the American Statistical Association* 84: 853–61.

Diaz-Uriarte, R. and Garland, T. (1996) "Testing Hypotheses of Correlated Evolution Using Phylogenetically Independent Contrasts: Sensitivity to Deviations from Brownian Motion," *Systematic Biology* 45: 27–47.

Dick, S. (1996) *The Biological Universe: The Twentieth-Century Extraterrestrial Life Debate and the Limits of Science*, Cambridge: Cambridge University Press.

Doolittle, W. F. (2000) "The Nature of the Universal Ancestor and the Evolution of the Proteome," *Current Opinion in Structural Biology* 10: 355–8.

Doolittle, W. F. and Bapteste, E. (2007) "Pattern Pluralism and the Tree of Life Hypothesis," *Proceedings of the National Academy of Sciences USA* 104: 2043–9.

Draper, P. (1989) "Pain and Pleasure: An Evidential Problem for Theists," *Noûs* 23: 331–50. Reprinted in D. Howard-Snyder (ed.), *The Evidential Argument from Evil*, Bloomington, Ind.: Indiana University Press, 1996, pp. 12–29.

(1996) "The Skeptical Theist," in D. Howard-Snyder (ed.), *The Evidential Argument from Evil*, Bloomington, Ind.: Indiana University Press, pp. 175–92.

Duhem, P. (1954) *The Aim and Structure of Physical Theory*, Princeton, NJ: Princeton University Press. First published in\French 1914.

Dye, J. (1988) "Supernatural Speech and Biological Books," *History of Philosophy Quarterly* 3: 257–72.

Earman, J. (1986) *A Primer on Determinism*, Dordrecht: Reidel.

(1992) *Bayes or Bust: A Critical Examination of Bayesian Confirmation Theory*, Cambridge, Mass.: MIT Press.

(2000) *Hume's Abject Failure: The Miracles Argument*, New York: Oxford University Press.

Eaton, T. H. (1960) "The Aquatic Origin of Tetrapods," *Transactions of the Kansas Academy of Science* 63: 115–20.

Eddington, A. (1928) *The Nature of the Physical World*, Cambridge: Cambridge University Press.

(1939) *The Philosophy of Physical Science*, Cambridge: Cambridge University Press.

Edwards, A. (1972) *Likelihood*, Cambridge: Cambridge University Press.

Edwards, J. L. (1989) "Two Perspectives on the Evolution of the Tetrapod Limb," *American Zoologist* 29: 235–54.

Eells, E. (1985) "Problems of Old Evidence," *Pacific Philosophical Quarterly* 66: 283–302.

Efron, B. and Morris, C. (1977) "Stein's Paradox in Statistics," *Scientific American* 236: 119–27.

Eldredge, N. and Cracraft, J. (1980) *Phylogenetic Patterns and the Evolutionary Process*, New York: Columbia University Press.

Escoto, B. (2004) "Philosophical Significance of Akaike's Theorem," Unpublished.

Falconer, D. and MacKay, T. (1996) *Introduction to Quantitative Genetics*. Edinburgh: Pearson.

Farris, J. S. (1994) "The Logical Basis of Phylogenetic Analysis," in E. Sober (ed.), *Conceptual Issues in Evolutionary Biology*, Cambridge, Mass.: MIT Press. First published 1983 in N. Platnick and V. Funk (eds), *Advances in Cladistics: Proceedings of the 2nd Annual Meeting of the Willi Hennig Society*, New York: Columbia University Press, pp. 7–36.

—— (1986) "On the Boundaries of Phylogenetic Systematics," *Cladistics* 2: 14–27.

Felsenstein, J. (1973) "Maximum Likelihood and Minimum-Step Methods for Estimating Evolutionary Trees from Data on Discrete Characters," *Systematic Zoology* 22: 240–9.

—— (1978) "Cases in which Parsimony and Compatibility Methods Can Be Positively Misleading," *Systematic Zoology* 27: 401–10.

—— (1979) "Alternative Methods of Phylogenetic Inference and their Inter-relationships," *Systematic Biology* 28: 49–62.

—— (1981) "A Likelihood Approach to Character Weighting and What It Tells Us About Parsimony and Compatibility," *Biological Journal of the Linnaean Society* 16: 183–96.

—— (1985) "Phylogenies and the Comparative Method," *American Naturalist* 125: 1–15.

—— (2004) *Inferring Phylogenies*, Sunderland, Mass.: Sinauer.

Feyerabend, P. (1974) *Against Method: Outline of an Anarchistic Theory of Knowledge*, Atlantic Highlands, NJ: Humanities Press.

Fisher, R. A. (1922a) "On the Dominance Ratio," *Proceedings of the Royal Society of Edinburgh* 42: 321–41.

—— (1922b) "On the Mathematical Foundations of Theoretical Statistics," *Philosophical Transactions of the Royal Society A* 222: 309–68.

—— (1956) *Statistical Methods and Scientific Inference*, London: Oliver and Boyd.

—— (1957) *The Genetical Theory of Natural Selection*, 2nd edn, New York: Dover.

—— (1959) *Statistical Methods and Scientific Inference*, 2nd edn, Edinburgh: Oliver & Boyd.

Fitelson, B. (1999) "The Plurality of Bayesian Measures of Confirmation and the Problem of Measure Sensitivity," *Philosophy of Science* 66: S362–78.

—— (2007) "Likelihoodism, Bayesianism, and Relational Confirmation," *Synthese* 156: 473–489.

Fitelson, B., Stephens, C., and Sober, E. (1999) "How Not to Detect Design: A Review of W. Dembski's *The Design Inference*," *Philosophy of Science* 66: 472–88.

Forster, M. (1986) "Statistical Covariance as a Measure of Phylogenetic Relationship," *Cladistics* 2: 297–317.

(1999) "Model Selection in Science: The Problem of Language Invariance," *British Journal for the Philosophy of Science* 50: 83–102.

(2000a) "Hard Problems in the Philosophy of Science: Idealisation and Commensurability," in R. Nola and H. Sankey (eds), *After Popper, Kuhn, and Feyerabend*, London: Kluwer, pp. 231–50.

(2000b) "Key Concepts in Model Selection: Performance and Generality," *Journal of Mathematical Psychology* 44: 205–31.

(2001) "The New Science of Simplicity," in A. Zellner, H. Keuzenkamp, and M. McAleer (eds), *Simplicity, Inference, and Modeling*, Cambridge: Cambridge University Press, pp. 83–119.

(2006) "Counterexamples to the Likelihood Theory of Evidence," *Mind and Machines* 16: 319–38.

(in press) "A Philosopher's Guide to Empirical Success," *Philosophy of Science.*

Forster, M. and Sober, E. (1994) "How to Tell When Simpler, More Unified, or Less *Ad Hoc* Theories Will Provide More Accurate Predictions," *British Journal for the Philosophy of Science* 45: 1–36.

(in preparation) "AIC Scores as Evidence: A Bayesian Interpretation."

Freeland, S., Knight, R., Landweber, L., and Hurst, L. (2000) "Early Fixation of an Optimal Genetic Code," *Molecular Biology and Evolution* 17: 511–18.

Frigg, R. and Hartmann, S. (2006) "Models in Science," in E. Zalta (ed.), *The Stanford Encyclopedia of Philosophy*. Available at http://plato. stanford. edu/archives/spr2006/entries/models-science. (Accessed May 5 2007.)

Gaut, B. S. and Lewis, P. O. (1995) "Success of Maximum Likelihood Phylogeny Inference in the Four Taxon Case," *Molecular Biology and Evolution* 12: 152–62.

Gehring, W. J. (2002) "The Genetic Control of Eye Development and Its Implications for the Evolution of the Various Eye-Types," *The International Journal of Developmental Biology* 46: 65–73.

Gillespie, J. H. (1986) "Natural Selection and the Molecular Clock," *Molecular Biology and Evolution* 3: 138–55.

Gilovich, T., Valone, R., and Tversky, A. (1985) "The Hot Hand in Basketball: On the Misperception of Random Sequences," *Cognitive Psychology* 17: 295–314.

Goldman, N. (1990) "Maximum Likelihood Inference of Phylogenetic Trees, with Special Reference to a Poisson Process Model of DNA Substitution and to Parsimony Analyses," *Systematic Biology* 39: 345–61.

Good, I. J. (1967) "On the Principle of Total Evidence," *British Journal for the Philosophy of Science* 17: 319–21.

Goodman, S. N. (1999) "Toward Evidence-Based Medical Statistics: 1. The P Value Fallacy," *Annals of Internal Medicine*, 130: 995–1004.

Gould, S. (1977) "The Misnamed, Mistreated, and Misunderstood Irish Elk," in *Ever Since Darwin*, New York: Norton, pp. 79–90.

(1980) *The Panda's Thumb*, New York: Norton.

(1989) *Wonderful Life*, New York: Norton.

(1991) "The Ant and the Plant," in *Bully for Brontosaurus*, New York: Norton, pp. 479–88.

Gould, S. and Lewontin, R. (1979) "The Spandrels of San Marco and the Panglossian Paradigm: A Critique of the Adaptationist Paradigm," *Proceedings of the Royal Society B* 205: 581–98.

Gould, S. and Vrba, E. (1982) "Exaptation: A Missing Term in the Science of Form," *Paleobiology* 8: 4–15.

Griffiths, A., Wessler, S., Lewontin, R., Gelbart, W., Suzuki, D., and Miller, J. (2005) *Introduction to Genetic Analysis*, New York: W. H. Freeman.

Grossman, J. (2007) "Statistical Inference: From Data to Simple Hypotheses," Unpublished.

Hacking, I. (1965) *The Logic of Statistical Inference*, Cambridge: Cambridge University Press.

—— (1975) *The Emergence of Probability*, Cambridge: Cambridge University Press.

—— (1987) "The Inverse Gambler's Fallacy: The Argument from Design. The Anthropic Principle Applied to Wheeler Universes," *Mind* 96: 331–40.

Hajek, A. (2003) "What Conditional Probabilities Could Not Be," *Synthese* 137: 273–323.

Hansen, T. (1997) "Stabilizing Selection and the Comparative Analysis of Adaptation," *Evolution* 51: 1341–51.

Hanson, N. R. (1969) *Perception and Discovery: An Introduction to Scientific Inquiry*, San Francisco, Calif.: Freeman-Cooper.

Harman, G. and Kulkarni, M. (2007) *Reliable Reasoning: Induction and Statistical Learning Theory*, Cambridge, Mass.: MIT Press.

Harvey, P. and Pagel, M. (1991) *The Comparative Method in Evolutionary Biology*, Oxford: Oxford University Press.

Hausman, D. (1992) *The Inexact and Separate Science of Economics*, Cambridge: Cambridge University Press.

Hempel, C. (1951) "The Concept of Cognitive Significance: A Reconsideration," *Proceedings of the American Academy of Arts and Sciences* 80: 61–77.

—— (1965a) "Empiricist Criteria of Cognitive Significance: Problems and Changes," in *Aspects of Scientific Explanation and Other Essays in the Philosophy of Science*, New York: Free Press, pp. 101–22.

—— (1965b) "Studies in the Logic of Confirmation," in *Aspects of Scientific Explanation and Other Essays in the Philosophy of Science*, New York: Free Press, pp. 3–46.

Hereford, J., Hansen, T., and Houle, E. (2004) "Comparing Strengths of Directional Selection: How Strong Is Strong?" *Evolution* 58: 2133–43.

Hesse, M. (1966) *Models and Analogies in Science*, Notre Dame, Ind.: University of Notre Dame Press.

Hick, J. (1978) *Evil and the God of Love*, rev. edn, New York: Harper and Row.

Himma, K. (2005) "The Application-Conditions of Design Inferences: Why the Design Argument Needs the Help of Other Arguments for God's Existence," *International Journal for the Philosophy of Religion* 57: 1–33.

Hitchcock, C. and Sober, E. (2004). "Prediction versus Accommodation and the Risk of Over-fitting." *British Journal for the Philosophy of Science*, 51: 1–34.

Hodge, M. J. S. (1987) "Natural Selection as a Causal, Empirical and Probabilistic Theory," in L. Krüger, R. Daston, and M. Heidelberg (eds), *The Probabilistic Revolution*, Vol. II, Cambridge, Mass.: MIT Press, pp. 233–70.
(1991) "The History of the Earth, Life, and Man: Whewell and Palaetiological Science," in M. Fisch and S. Schaffer (eds), *William Whewell: A Composite Portrait*, Oxford: Clarendon Press, pp. 255–89.
Hoover, K. (2003) "Nonstationary Time Series, Cointegration, and the Principle of the Common Cause," *British Journal for the Philosophy of Science* 54: 527–51.
Howard-Snyder, D. (ed.) (1996) *The Evidential Argument from Evil*, Bloomington, Ind.: Indiana University Press.
Howson, C. (2001) "The Logic of Bayesian Probability," in D. Corfield and J. Williamson (eds), *Foundations of Bayesianism*, Dordrecht: Kluwer Academic.
Howson, C. and Urbach, P. (1993) *Scientific Reasoning: The Bayesian Approach*, Peru, Ill.: Open Court.
Huelsenbeck, J., Larget, B., and Alfaro, M. (2004) "Bayesian Phylogenetic Model Selection Using Reversible Jump Markov Chain Monte Carlo," *Molecular Biology and Evolution* 21: 1123–33.
Huelsenbeck, J., Ronquist, F., Nielsen, R., and Bollback, J. (2001) "Bayesian Inference of Phylogeny and Its Impact on Evolutionary Biology," *Science* 294: 2310–14.
Hull, D. (2000) "Why Did Darwin Fail? The Case of John Stuart Mill," in R. Creath and J. Maienschein (eds), *Biology and Epistemology*, Cambridge: Cambridge University Press, pp. 48–63.
Hume, D. (1990) *Dialogues Concerning Natural Religion*, London: Penguin. First published in 1779.
James, W. (1897) "The Will to Believe," in *The Will to Believe and Other Essays in Popular Philosophy*, New York: Longmans Green & Co., pp. 1–31.
James, W. and Stein, C. (1961) "Estimation with Quadratic Loss," in J. Neyman (ed.), *Proceedings of the Fourth Berkeley Symposium on Mathematical Statistics and Probability*, Vol. I, Berkeley, Calif.: University of California Press, pp. 361–79.
Jeffrey, R. (1983) *The Logic of Decision*, 2nd edn, Chicago, Ill.: University of Chicago Press. First published 1965.
Johnson, D. (1995) "Statistical Sirens: The Allure of Nonparametrics," *Ecology* 76: 1998–2000.
Johnson, P. (1991) *Darwin on Trial*, Downers Grove, Ill.: InterVarsity Press.
(1997) *Defeating Darwinism by Opening Minds*, Downers Grove, Ill.: InterVarsity Press.
Jukes, T. and Cantor, C. (1969) "Evolution of Protein Molecules," in H. Munro (ed.), *Mammalian Protein Metabolism*, New York: Academic Press, pp. 21–132.
Justus, J. (2007) "The Search for a Formal Criterion of Empirical Significance: A Reconsideration," Unpublished.
Kadane, J. B., Schervish, M., and Seidenfeld, T. (1996), "Reasoning to a Foregone Conclusion," *Journal of the American Statistical Association* 91: 1228–35.
Kaplan, M. (1996) *Decision Theory as Philosophy*, Cambridge: Cambridge University Press.

Keynes, J. (1921) *A Treatise on Probability*, London: Macmillan.

Kimura, M. (1980) "A Simple Method for Estimating Evolutionary Rates of Base Substitutions through Comparative Studies of Nucleotide Sequences," *Journal of Molecular Evolution* 16: 111–20.

——— (1981) "Estimation of Evolutionary Distances Between Homologous Nucleotide Sequences," *Proceedings of the National Academy of Sciences USA* 78: 454–8.

——— (1983) *The Neutral Theory of Molecular Evolution*, Cambridge: Cambridge University Press.

Kingsolver, J. and Koehl, M. (1985) "Aerodynamics, Thermoregulation, and the Evolution of Insect Wings: Differential Scaling and Evolutionary Change," *Evolution* 39: 488–504.

Kishino, H. and Hasegawa, M. (1990) "Converting Distance to Time: Application to Human Evolution," *Methods in Enzymology* 183: 550–70.

Kitcher, P. (1983) *Abusing Science: The Case Against Creationism*, Cambridge, Mass.: MIT Press.

——— (1993) *The Advancement of Science: Science Without Legend, Objectivity Without Illusions*, New York: Oxford University Press.

——— (2001) "Real Realism: The Galilean Strategy," *Philosophical Review* 110: 151–98.

——— (2003) "Darwin's Achievement," in *In Mendel's Mirror*, New York: Oxford University Press, pp. 45–93.

Knight, R., Freeland, S., and Landweber, L. (2001) "Rewiring the Keyboard: Evolvability of the Genetic Code," *Nature Reviews: Genetics* 2: 49–58.

Koh, K., Evans J. M., Hendricks, J. C., and Sehgal, A. (2006) "A Drosophila Model for Age-Associated Changes in Sleep:Wake Cycles," *Proceedings of the National Academy of Sciences USA* 103: 13843–7.

Kolmogorov, A. N. (1950) *Foundations of the Theory of Probability*, New York: Chelsea.

Krebs, J. and Davies, N. (1981) *An Introduction to Behavioral Ecology*, Sunderland, Mass.: Sinauer.

Kreitman, M. (2000) "How to Detect Selection in Populations with Applications to the Human," *Annual Review of Genomics and Human Genetics* 1: 539–59.

Kuhn, T. (1962) *The Structure of Scientific Revolutions*, Chicago, Ill.: Chicago University Press.

Kyburg, H. (1970) "Conjunctivitis", in M. Swain (ed.) *Induction, Acceptance, and Rational Belief*, Dordrecht: Reidel, pp. 55–82.

Lanave, C., Preparata, G., Saccone, C., and Serio, G. (1984) "A New Method for Calculating Evolutionary Substitution Rates," *Journal of Molecular Evolution* 20: 86–93.

Land, M. (1984) "Molluscs," in M. Ali (ed.), *Photoreception and Vision in Invertebrates*, New York: Plenum, pp. 699–725.

Lande, R. (1976) "Natural Selection and Random Genetic Drift in Phenotypic Evolution," *Evolution* 30: 314–34.

——— (1985) "Expected Time for Random Genetic Drift of a Population between Stable Phenotypic States," *Proceedings of the National Academy of Science USA* 82: 7641–5.

Lang, C., Sober, E., and Strier, K. (2002) "Are Human Beings Part of the Rest of Nature?" *Biology and Philosophy* 17: 661–71.

Laplace, P. S. de (1951) *Philosophical Essay on Probabilities*, New York: Dover. First published in French 1829.

Lauder, G. (1996) "The Argument from Design," in M. Rose and G. Lauder (eds), *Adaptation*, New York: Academic Press, pp. 55–92.

Leroi, A. M., Rose, M. R., and Lauder, G. V. (1994) "What Does the Comparative Method Reveal About Adaptation?" *The American Naturalist* 143: 381–402.

Lewin, R. (1980) "Evolutionary Theory under Fire," *Science* 210: 883–7.

Lewis, P. (1998) "Maximum Likelihood as an Alternative to Parsimony for Inferring Phylogeny Using Nucleotide Sequence Data," in D. Soltis, P. Soltis, and J. Doyle (eds), *Molecular Systematics of Plants II*, Boston, Mass.: Kluwer, pp. 132–63.

(2001) "A Likelihood Approach to Estimating Phylogeny from Discrete Morphological Character Data," *Systematic Biology* 50: 913–25.

Lewontin, R. (1978) "Adaptation," *Scientific American* 239: 156–69.

Li, W. H. (1993) "So, What About the Molecular Clock Hypothesis?" *Current Opinion in Genetics and Development* 3: 896–901.

Li, W. H. and Tanimura, M. (1987) "The Molecular Clock Runs More Slowly in Man than in Apes and Monkeys," *Nature* 326: 93–6.

Lindley, D. V. (1957) "A Statistical Paradox," *Biometrika* 44: 187–92.

Lindley, D. V. and Phillips, L. D. (1976) "Inference for a Bernouilli Process (a Bayesian view)," *The American Statistician* 30: 112–19.

Littlewood, J. (1953) *A Mathematician's Miscellany*, London: Methuen.

Maddison, W. (1991) "Squared-Change Parsimony Reconstructions of Ancestral States for Continuous-Valued Characters on a Phylogenetic Tree," *Systematic Zoology* 40: 304–14.

Maynard Smith, J. (1978) "Optimization Theory in Evolution," *Annual Review of Ecology and Systematics* 9: 31–56.

Mayo, D. (1996) *Error and the Growth of Experimental Knowledge*, Chicago, Ill.: University of Chicago Press.

Mayr, E. (1976) "Typological versus Population Thinking," in *Evolution and the Diversity of Life*, Cambridge, Mass.: Harvard University Press, pp. 26–9. First published 1963.

(1988) *Towards a New Philosophy of Biology*, Cambridge, Mass.: Harvard University Press.

(2000) "The Biological Species Concept," in Q. Wheeler and R. Meier (eds), *Species Concepts and Phylogenetic Theory*, New York: Columbia University Press, pp. 17–29, 93–100, 161–6.

McDonald, J. and Kreitman, M. (1991) "Adaptive Protein Evolution at the Adh Locus in Drosophila," *Nature* 351: 652–4.

McMullin, E. (1985) "Galilean Idealization," *Studies in the History and Philosophy of Science* 16: 247–73.

Monod, J. (1971) *Chance and Necessity*, New York: Alfred Knopf.

Morgan, M. and Morrison, M. (eds) (1999) *Models as Mediators: Perspectives on Natural and Social Science*, Cambridge: Cambridge University Press.

Morgenbesser, S. (1960) "The Realist-Instrumentalist Controversy," in S. Morgenbesser, P. Suppes, and M. White (eds), *Philosophy, Science, and Method*, New York: Harcourt, Brace, and World, pp. 106–22.

Morris, H. (1980) *King of Creation*, San Diego, Calif.: CLP Publishers.

Mougin, G. and Sober, E. (1994) "Betting Against Pascal's Wager," *Noûs* 28: 382–95.

Nelson, P. (1996) "The Role of Theology in Current Evolutionary Reasoning," *Biology and Philosophy* 11: 493–517.

Neyman, J. and Pearson, E. S. (1933) "On the Problem of the Most Efficient Tests of Statistical Hypotheses," *Philosophical Transactions of the Royal Society, Series A* 231: 289–333.

Nielsen, R. (2005) "Molecular Signatures of Natural Selection," *Annual Review of Genetics* 39: 197–218.

Nilsson, D. (1989) "Vision Optics and Evolution," *Bioscience* 39: 298–307.

Nilsson, D. and Pelger, S. (1994) "A Pessimistic Estimate of the Time Required for an Eye to Evolve," *Proceedings of the Royal Society London, Series B, Biological Sciences* 256: 53–8.

Numbers, R. (1992) *The Creationists: The Evolution of Scientific Creationism*, New York: Knopf.

—— (2003) "Science Without God: Natural Law and Christian Beliefs: An Essay on Methodological Naturalism," in D. Lindberg and R. Numbers (eds), *When Science and Christianity Meet*, Chicago, Ill.: University of Chicago Press, pp. 265–86.

O'Hara, R. (1988) "Homage to Clio, or, Towards an Historical Philosophy for Evolutionary Biology," *Systematic Zoology* 37: 142–55.

—— (1998) "Population Thinking and Tree Thinking in Systematics," *Zoological Scripta* 26: 323–9.

Olby, R. (1985) *The Origins of Mendelism*, Chicago, Ill.: University of Chicago Press.

Oparin, A. (1953) *The Origin of Life*, New York: Dover.

Orr, A. (1996) "Darwin v. Intelligent Design (Again)," *Boston Review of Books*, Dec / June: 28–31.

Orzack, S. and Sober, E. (1993) "A Critical Examination of Richard Levins' 'The Strategy of Model Building in Population Biology'," *Quarterly Review of Biology* 68: 533–46.

—— (2001) "Adaptation, Phylogenetic Inertia, and the Method of Controlled Comparisons," in *Adaptationism and Optimality*, Cambridge: Cambridge University Press, pp. 45–63.

Ospovat, D. (1981) *The Development of Darwin's Theory: Natural History, Natural Theology, and Natural Selection 1838–1859*, Cambridge: Cambridge University Press.

Page, R. and Holmes, E. (1998) *Molecular Evolution: A Phylogenetic Approach*, Oxford: Blackwell.

Paley, W. (1809) *Natural Theology, or, Evidences of the Existence and Attributes of the Deity, Collected from the Appearances of Nature*, 12th edn, London: Rivington. First published 1802. Available online at http://darwin-online. org.uk/content/frameset?itemID=A142 & viewtype=text & pageseq=1.

Parker, G. (1978) "Searching for Mates," in J. Krebs and N. Davies (eds), *Behavioral Ecology: An Evolutionary Approach*, Oxford: Blackwell, pp. 214–44.

Parker, G. and Stuart, R. (1976) "Animal Behavior as a Strategy Optimizer: Evolution of Resource Assessment Strategies and Optimal Emigration Thresholds," *American Naturalist* 110: 1055–76.

Parzen, E. (1962) *Stochastic Processes*, San Francisco, Calif.: Holden-Day.

Patterson, C. (1981) "Significance of Fossils in Determining Evolutionary Relationships," *Annual Review of Ecology and Systematics* 12: 195–223.

Peirce, C. S. (1934) "The Fixation of Belief," in C. Hartshorne, P. Weiss, and A. Burks (eds), *The Collected Papers of Charles Sanders Peirce*, Cambridge, Mass.: Harvard University Press, Volume V, pp. 223–47. First published 1877.

Pennock, R. (1999) *Tower of Babel: The Evidence Against the New Creationism*, Cambridge, Mass.: MIT Press.

Penny, D. Foulds, L. R., and Hendy, M. D. (1982) "Testing the Theory of Evolution by Comparing Phylogenetic Trees Constructed from Five Different Protein Sequences," *Nature* 297: 197–200.

Plantinga, A. (1974) *The Nature of Necessity*, Oxford: Oxford University Press.

(1993) *Warrant and Proper Function*, Oxford: Oxford University Press.

(2000) *Warranted Christian Belief*, Oxford: Oxford University Press.

Popper, K. (1959) *Logic of Scientific Discovery*, London: Hutchinson.

(1976) *Unended Quest*, LaSalle, Ill.: Open Court.

(1978) "Natural Selection and the Emergence of Mind," *Dialectica* 32: 339–55.

Posada, D. and Buckley, T. (2004) "Model Selection in Phylogenetics: Advantages of the AIC and Bayesian Approaches," *Systematic Biology* 53: 793–808.

Posada, D. and Crandall, K. (2001) "Selecting the Best-Fit Model of Nucleotide Substitution," *Systematic Biology* 50: 580–601.

Provine, W. (1989) "Progress in Evolution and Meaning in Life," in M. Nitecki (ed.), *Evolutionary Progress*, Chicago, Ill.: University of Chicago Press.

Quine, W. (1953) "Two Dogmas of Empiricism," in *From a Logical Point of View*, Cambridge, Mass.: Harvard University Press, pp. 20–46.

Raddick, G. (2005) "Deviance, Darwinian-Style: A Review of A. Lustig, R. Richards, and M. Ruse's *Darwinian Heresies* (Cambridge: CUP, 2004)," *Metascience* 14: 453–7.

Rannala, B., and Yang, Z. (1996) "Probability Distribution of Molecular Evolutionary Trees: a New Method of Phylogenetic Inference," *Journal of Molecular Evolution* 43: 304–11.

Reichenbach, H. (1938) *Experience and Prediction*, Chicago, Ill.: University of Chicago Press.

(1956) *The Direction of Time*, Berkeley, Calif.: University of California Press.

Ridley, M. (1983) *The Explanation of Organic Diversity*, Oxford: Oxford University Press.

Robbins, H. (1970) "Statistical Methods Related to the Law of the Iterated Logarithm," *Annals of Mathematical Statistics* 41: 1397–1409.

Rodriguez, F., Oliver, J. L., Marin, A., and Medina, J. R., (1990) "The General Stochastic Model of Nucleotide Substitution," *Journal of Theoretical Biology* 142: 485–501.

Rowe, W. (1979) "The Problem of Evil and Some Varieties of Atheism," *American Philosophical Quarterly* 16: 335–341.

Royall, R. (1997) *Statistical Evidence: A Likelihood Paradigm*, Boca Raton, Fla.: Chapman and Hall.

Ruse, M. (1979) *The Darwinian Revolution: Science Red in Tooth and Claw*, Chicago, Ill.: University of Chicago Press.

(2000) *Can a Darwinian Be a Christian?* Cambridge: Cambridge University Press.

Russell, B. (1919) *Introduction to Mathematical Philosophy*, New York: The Macmillan Company; London: George Allen & Unwin, Ltd.

Sakamoto, Y., Ishiguro, M., and Kitagawa, G. (1986) *Akaike Information Criterion Statistics*, New York: Springer.

Salmon, W. (1984) *Scientific Explanation and the Causal Structure of the World*, Princeton, NJ: Princeton University Press.

Salvini-Plaven, L. and Mayr, E. (1977) "On the Evolution of Photoreceptors and Eyes," in M. Hecht, W. Sterre, and B. Wallace (eds), *Evolutionary Biology*, Vol. X, New York: Plenum, pp. 207–63.

Schlichting, C. D. and Pigliucci, M. (1998) *Phenotypic Evolution: A Reaction Norm Perspective*, Sunderland, Mass.: Sinauer.

Schwarz, G. (1978) "Estimating the Dimension of a Model," *Annals of Statistics* 6: 461–5.

Shanks, N. (2004) *God, the Devil, and Darwin: A Critique of Intelligent Design Theory*, Oxford: Oxford University Press.

Shapiro, L. and Sober, E. (2007) "Epiphenomenalism: The Do's and Don'ts," in G. Wolters and P. Machamer (eds), *Studies in Causality: Historical and Contemporary*, Pittsburgh, Pa.: University of Pittsburgh Press.

Shoesmith, E. (1987) "The Continental Controversy over Arbuthnot's Argument for Divine Providence," *Historia Mathematica* 14: 133–46.

Simon, H. A. (1981) "The Architecture of Complexity," in *The Sciences of the Artificial*, Cambridge, Mass.: MIT Press, pp. 193–229.

Skyrms, B. (1984) *Pragmatics and Empiricism*. New Haven, Conn.: Yale University Press.

(1996) *The Evolution of the Social Contract*, Cambridge: Cambridge University Press.

Snyder, L. J. (2006) *Reforming Philosophy: A Victorian Debate on Science and Society*, Chicago, Ill.: University of Chicago Press.

Sober, E. (1980) "Evolution, Population Thinking, and Essentialism," *Philosophy of Science* 47: 350–83.

(1984) *The Nature of Selection*, Cambridge, Mass.: MIT Press. Second edition published 1993 Chicago, Ill.: University of Chicago Press.

(1988) *Reconstructing the Past: Parsimony, Evolution, and Inference*, Cambridge, Mass.: MIT Press.

(1989) "Independent Evidence About a Common Cause," *Philosophy of Science* 56: 275–87.

(1990) *Core Questions in Philosophy: A Text with Readings*, New York: Macmillan.

(1993a) "Mathematics and Indispensability," *Philosophical Review* 102: 35–58.

(1993b) *Philosophy of Biology*, Boulder, Col.: Westview Press.

(1994) "Progress and Directionality in Evolutionary Theory," in J. Campbell (ed.), *Creative Evolution?!* Boston, Mass.: Jones and Bartlett, pp. 19–33.

(1998) "Three Differences Between Evolution and Deliberation," in P. Danielson (ed.), *Modeling Rationality, Morality, and Evolution*, Oxford: Oxford University Press, pp. 408–22.

(1999a) "Modus Darwin," *Biology and Philosophy* 14: 253–78.

(1999b) "Testability," *Proceedings and Addresses of the American Philosophical Association* 73: 47–76.

(2000a) "Evolution and the Problem of Other Minds," *Journal of Philosophy* 97: 365–86.

(2000b) "Quine's Two Dogmas," *Proceedings of the Aristotelian Society*, Supplementary Volume 74: 237–80.

(2001) "Venetian Sea Levels, British Bread Prices, and the Principle of the Common Cause," *British Journal for the Philosophy of Science* 52: 1–16.

(2002a) "Intelligent Design and Probability Reasoning," *International Journal of Philosophy of Religion* 52: 65–80.

(2002b) "Instrumentalism, Parsimony, and the Akaike Framework," *Philosophy of Science* 69: S112–23.

(2002c) "Reconstructing Ancestral Character States: A Likelihood Perspective on Cladistic Parsimony," *The Monist* 85: 156–76.

(2003a) "It Had to Happen: A Review of Simon Conway Morris' *Life's Solution in a Lonely Universe*," *New York Times*, November 30, Book Review Section, p. 18.

(2003b) "Two Uses of Unification," in F. Stadler (ed.), *The Vienna Circle and Logical Empiricism: Vienna Circle Institute Yearbook 2002*, Dordrecht: Kluwer, pp. 205–16.

(2004a) "The Contest Between Likelihood and Parsimony," *Systematic Zoology* 53: 6–16.

(2004b) "The Design Argument," in W. Mann (ed.), *The Blackwell Guide to the Philosophy of Religion*, Oxford: Blackwell, pp. 117–47.

(2004c) "Likelihood, Model Selection, and the Duhem–Quine Problem," *Journal of Philosophy* 101: 1–22.

(2004d) "A Modest Proposal: A Review of John Earman's *Hume's Abject Failure: The Miracles Argument*," *Philosophy and Phenomenological Research* 118: 487–94.

(2005) "Parsimony and its Presuppositions," in V. Albert (ed.), *Parsimony, Phylogeny, and Genomics*, Oxford: Oxford University Press.

(2007a) "Intelligent Design Theory and the Supernatural: The 'God or Extra-Terrestrials' Reply," *Faith and Philosophy* 24: 72–82.

(2007b) "Sex Ratio Theory, Ancient and Modern: An 18th Century Debate about Intelligent Design and the Development of Models in Evolutionary Biology," in J. Riskin (ed.), *Genesis Redux: Essays on the History and Philosophy of Artificial Life*, Chicago, Ill.: University of Chicago Press, pp. 131–62.

(2008a) "Empiricism," in S. Psillos and M. Curd (eds), *The Routledge Companion to the Philosophy of Science*, London: Routledge.

(2008b) "Evolutionary Theory and the Reality of Macro-Probabilities," in E. Eells and J. Fetzer (eds), *Probability in Science*, Chicago, Ill.: Open Court.

Sober, E. and Orzack, S. (2003) "Common Ancestry and Natural Selection," *British Journal for the Philosophy of Science* 54: 423–37.

Sober, E. and Steel, M. (2002) "Testing the Hypothesis of Common Ancestry," *Journal of Theoretical Biology* 218: 395–408.

Sober, E. and Wilson, D. S. (1998) *Unto Others: The Evolution and Psychology of Unselfish Behavior*, Cambridge, Mass.: Harvard University Press.

Sorenson, R. (2001) *Vagueness and Contradiction*, Oxford: Clarendon Press; New York: Oxford University Press.

Spirtes, P., Glymour, C., and Scheines, R. (2001) *Causality, Prediction, and Search*, Cambridge, Mass.: MIT Press.

Stace, C. (2000) "Cytology and Cytogenetics as a Fundamental Taxonomic Resource for the 20th and 21st Centuries," *Taxon* 49: 451–77.

Stanford, K. (2006) *Science, History, and the Problem of Unconceived Alternatives*, Oxford: Oxford University Press.

Stanley, S. (1979) *Macroevolution: Pattern and Process*, San Francisco, Calif.: W. H. Freeman.

Steel, M. and Penny, D. (2000) "Parsimony, Likelihood, and the Role of Models in Molecular Phylogenetics," *Molecular Biology and Evolution* 17: 839–50.

(2004) "Two Further Links between *MP* and *ML* under the Poisson Model," *Applied Mathematics Letters* 17: 785–90.

Sterelny, K. and Griffiths, P. (1999) *Sex and Death: An Introduction to the Philosophy of Biology*, Chicago, Ill.: University of Chicago Press.

Stone, M. (1977) "An Asymptotic Equivalence of Choice of Model by Cross-Validation and Akaike's Criterion," *Journal of the Royal Statistical Society B* 39: 44–7.

Sugiura, N. (1978) "Further Analysis of the Data by Akaike's Information Criterion and the Finite Corrections," *Communications in Statistics, Theory and Methods* A7: 13–26.

Swift, J. (1984) *Gulliver's Travels*, London: Herbert Press. First published 1726.

Swinburne, R. (1968) "The Argument from Design," *Philosophy* 43: 199–212.

(1990) "The Limits of Explanation," in D. Knowles (ed.), *Explanation and Its Limits*, Cambridge: Cambridge University Press, pp. 177–93.

Swofford, D., Olsen, G., Waddell, P., and Hillis, D. (1996) "Phylogenetic Inference," in D. Hillis, C. Moritz, and B. Marble (eds), *Molecular Systematics*, 2nd edn, Sunderland, Mass.: Sinauer, pp. 407–514.

Takeuchi, K. (1976) "Distribution of Informational Statistics and a Criterion of Model Fitting," (in Japanese), *Suri-Kapaku* (Mathematical Sciences) 153: 12–18.

Taveré, S. (1986) "Some Probabilistic and Statistical Problems on the Analysis of DNA Sequences," *Lectures on Mathematics in the Life Sciences* 17: 57–86.

Tuffley, C. and Steel, M. (1997) "Links Between Maximum Likelihood and Maximum Parsimony under a Simple Model of Site Substitution," *Bulletin of Mathematical Biology* 59: 581–607.

Turelli, M. (1988) "Population Genetic Models for Polygenic Variation and Evolution," in B. Weir, E. Eisen, M. Goodman, and G. Namkoong (eds), *Proceedings of the Second International Conference on Quantitative Genetics*, Sunderland, Mass.: Sinauer, pp. 601–18.

Van Fraassen, B. (1982) "The Charybdis of Realism: Epistemological Implications of Bell's Inequality," *Synthese* 52: 25–38.

Van Inwagen, P. (1993) *Metaphysics*, Boulder, Col.: Westview Press.

Venn, J. (1866) *The Logic of Chance*, New York: Chelsea.

Wagner, C. (2004) "Modus Tollens Probabilized," *British Journal for the Philosophy of Science* 55: 747–53.

Wake, D. B., Roth, G., and Wake, M. H. (1983) "On the Problem of Stasis in Organismal Evolution," *Journal of Theoretical Biology* 101: 211–24.

Wald, A. (1947) *Sequential Analysis*, New York: John Wiley & Sons.

Walsh, D., Lewens, T., and Ariew, A. (2002) "The Trials of Life: Natural Selection and Random Drift," *Philosophy of Science* 69: 452–73.

Wardrop, R. (1999) "Statistical Tests for the Hot-Hand in Basketball in a Controlled Setting," Technical Report, Department of Statistics, University of Wisconsin, Madison. Available online at http://hot-hand. behaviouralfinance. net/Ward99. pdf. (Accessed 4 February 2007.)

Whewell, W. (1833) *Astronomy and General Physics Considered with Reference to Natural Theology*, Philadelphia, Pa.: Carey, Lea and Blanchard.

(1847) *The Philosophy of the Inductive Sciences, Founded Upon Their History*, Vol. I, London: J. Parker.

Wiley, E. (1981) *Phylogenetics: The Theory and Practice of Phylogenetic Systematics*, New York: Wiley.

Williams, G. C. (1966) *Adaptation and Natural Selection*, Princeton, NJ: Princeton University Press.

Williamson, T. (1994) *Vagueness*, London and New York: Routledge.

Wilson, E. O. (1975) *Sociobiology: The New Synthesis*, Cambridge, Mass.: Belknap Press of Harvard University Press.

Woese, C. (1998) "The Universal Ancestor," *Proceedings of the National Academy of Sciences USA* 95: 6854–9.

Woodward, J. (2003) *Making Things Happen*, Oxford: Oxford University Press.

Wright, L. (1976) "Functions," *Philosophical Review* 85: 70–86.

Wykstra, S. (1984) "The Humean Obstacle to Evidential Arguments from Suffering: On Avoiding the Evils of 'Appearance'," *International Journal for Philosophy of Religion* 16: 73–93.

Yoccoz, N. (1991) "The Use, Overuse, and Misuse of Significance Tests in Evolutionary Biology and Ecology," *Bulletin of the Ecological Society of America* 32: 106–11.

Young, R. (1985) *Darwin's Metaphor: Nature's Place in Victorian Culture*, Cambridge: Cambridge University Press.

Yule, G. U. (1926) "Why Do We Sometimes Get Nonsensical Relations Between Time Series? A Study of Sampling and the Nature of Time Series," *Journal of the Royal Statistical Society* 89: 1–64.

Zuckerkandl, E. and Pauling, L. (1965) "Evolutionary Divergence and Convergence in Proteins," in V. Bryson and H. J. Vogel (eds), *Evolving Genes and Proteins*, New York: Academic Press, pp. 97–166.

Index

acceptance
and rejection 5, 70, 75–6, 77, 79, 107
prudential vs. evidential 6–7, 109
see also action, belief
acheulean tools 332
acting troupe 231, 307
action 3–7, 65, 107
Adam, Y-chromosome 272
Adams, E. 104–7
adaptation
adaptationism 297, 312, 361
adaptive characters 261
adaptive peaks 157, 212–15, 293–352, 364
agnosticism 186
see also atheism, theistic evolutionism
Akaike, H. *see* Akaike Information Criterion
Akaike Information Criterion 82–96, 98, 108, 153, 239, 245, 356, 360
and frequentism 102–4
and instrumentalism 96–8
and phylogenetic inference 341
see also model selection
Alexander, R. M. 205
Alfaro, M. 342
alignment 290
Allman, J. 227–9, 234
ancestor
inferring character state of 207–12, 253–8, 359–60
meaning of 268
versus relative 204
Anderson, D. 70, 72, 91, 95, 98
Anscombe, F. J. 76
Antonovics, J. 192
appendix 305
Aquinas, St. T. *see* design argument
Arbuthnot, J. 117–18, 227
arch and keystone 162
Armitage, P. 77

assumptions 252–3, 342, 344
see also auxiliary assumptions
atheism 112
see also agnosticism, theistic evolutionism
auxiliary assumptions 151, 153, 333, 356, 362
and design argument 142
and testing for selection 201
need for independent justification 143, 151
see also De Niro fallacy

Backe, A. 78
backwards inequality 216, 246, 279, 301, 360
see also Markov model
Bacon, F. 109, 113
Bapteste, E. 268
Baum, D. 203, 255
Bayes, T. 8
Bayes' Theorem 8–9, 11–13, 35, 41, 258
Bayesian Information Criterion 92
see also Model Selection
Bayesianism 2–3, 8, 11–13, 15–17, 18, 20–4, 32–5, 37, 42, 46, 48–9, 53, 64, 65, 66, 70–1, 78, 81, 90, 337, 341, 351, 355
and common ancestry 275
and logic 30–2
and statistics 30–2
and stopping rules 72–8
and testability 46
objections to 24–30
see also experimental design and bayesianism
Beatty, J. 281
beetle 205
Behe, M. 128, 154, 168, 182
see also design argument, irreducible complexity
belief 3–7, 5, 8, 32, 39–40, 56
see acceptance
Bentley, R. 116
Bertrand's Paradox 28
see indifference, principle of
Big Ben 134, 135

biogeography 324, 330
biological species concept 268
birthday fallacy 115, 177, 331
Borel, E. 116
breakfast 43–4, 139
breast cancer 169
Breeder's equation 195, 196
Brooke, J. 109
Brouwer, L. E. J. 100
Brown, F. B. 186, 187
Brownian motion 194, 252
Buckley, T. 341
Burkhardt, F. 186
Burnham, K. 70, 72, 95, 98
Burt, A. 222, 251
Butler, M. 194

Cairn-Smith, A. G. 162
cake slicing 27, 55
Cantor, C. 284, 337
Cantor, G. 100
Cape Verde 329
Carnap, R. 76
Carroll, S. 320
catch-all hypotheses 28–9, 30–2, 37, 47, 78
   see also bayesianism
causal Markov condition 231
Cavalcanti, A. 289
Chang, J. 350
characters
   coding 310
   dichotomous versus continuous 287, 325
   genetic and phenotypic 290
   individuation of 136
   weighting of 295
   see also adaptation, drift
Cheverud, J. M. 252
Christianity 187
chromosome number 285
Cicero 117
Clifford, W. K. 7, 188
Cockrell, B. 205
coincidences 104–7
combination lock 123
common ancestry 124
   and common causes 292
   and fossil evidence 318–24
   and LUCA (last universal common ancestor)
      270
   and parsimony 314–18
   and similarity 266, 291, 311
   and tests of selection 239
   as control 252
   biogeographical evidence 324
   evidence from matching 277, 293

   meaning of 268
   versus separate ancestry 274
common cause, principle of 232, 278, 307
competitive exclusion 282
composite versus simple hypotheses 31, 47, 66,
      68, 70–1
conditional probability, definition of 9
confirmation 15–17, 32, 34–5
   degree of 16–17, 33
   see also bayesianism, evidence
consistency, statistical 24, 90–1, 347–51
contexts of discovery and justification 185
contingency of evolutionary process 363
controlled comparison in testing selection
      against inertia 248
Conway Morris, S. 66, 363
Cook, R. 205
correlation
   as evidence for common ancestry 306
   spatial and temporal 233
   see also backwards inequality
cosmological argument 186
Cottingham, J. 147
Courant, R. 100
Coyne, J. 281
Cracraft, J. 358
Crandall, K. 341
creationism 51, 274, 318, 329
   and politics 184–5
   and wedge strategy 184
   as empty 110, 361
   testability of 130, 141–7, 189
   testing 353
   versus theistic evolutionism 112
   see also intelligent design, irreducible
      complexity
Crick, F. 289
Crow, J. 52, 237
curve-fitting 95
   see simplicity

Da Costa, N. C. 80
Darwin, C.
   on adaptation 261, 297, 361
   on biogeography and common ancestry 324,
      330
   on evidence for common ancestry 265,
      266
   on fossils 324
   on function-switching 161
   on group selection
   on imperfect adaptation 128
   on intelligent design 109–12
   on origin of life 276
   on randomness 125

on selection explaining variation 226
on similarity and common ancestry 297, 313
on space-time principle 326
on theism 186
on vestigial organs 324
Dauben, J. W. 100
Dawkins, R. 50−1, 112, 123
Davies, N. 191
decision theory 6
see also action
default reasoning 245
see also Assumptions
Dembski, W. 51, 168
De Niro fallacy 202, 208
Dennett, D. 112, 188
Descartes, R. 146
Desjardins, E.
design argument 113
  and Darwin's theory 154
  and Hume 139−41, 169−70, 170
  and imperfect adaptations 127
  and model selection 180−2
  and Paley's stone 147
  and Paley's watch 119
  and problem of evil 164−7, 186
  Aquinas 114−15, 177
  as a likelihood argument 125, 141, 189
  as an analogy argument 139−40
  as an inductive sampling argument 140, 167−77, 168
  existence and attributes of the designer 140, 167
  necessity versus high probability 115
  see also irreducible complexity
Desmond, A. 187
deterministic theories 156
Diaconis, P. 104
dimension of a model 100
dispersal 326, 331
Doolittle, F. 253, 268, 273
Doyle, A. C. 57
Draper, P. 166
drift 192, 296, 345
  and test of common ancestry 301
  as random walk 193
  phenotypic and genetic 197
  testing 235
Duhem, P. 57, 144, 333
dungfly 205
Durham, W. 119
Dye, J. 171

Earman, J. 13, 29, 43, 156
Eaton, T. H. 244

Eddington, A. 16, 28−30, 29, 37, 41, 47, 57, 76, 116, 132
Edwards, A. 38, 51, 244
Eells, E. 13
Efron, B. 66
Eldredge, N. 358
Einstein, A. *see* theory of relativity
epicureanism 116, 122, 124, 137, 155
estimates versus estimators 66−7, 102
  see also likelihood error
  observational 57
  probabilities 58
  types of 58, 63
Escoto, B. 94
essentialism 364
Eve, mitochondrial 272
evidence 1−7, 32, 45−6, 52, 65, 73, 77−8, 107
  absence of 323
  and acceptance 5, 56, 58, 63, 64
  and prediction 294
  and the special consequence condition 316
  and time 311, 363
  concepts of 356
  contrastive character of 32, 52, 61, 116, 131, 149, 152, 190, 227, 267, 314
  principle of total 41−6, 46, 53, 63−4, 73, 89, 94, 105, 134, 136−9, 225, 289, 290, 308
  strength of 302, 304
  strengthening and weakening 43−5
  see also likelihood, law of
evil 52, 165, 187
  see design argument
evo-devo 213
evolution, micro and macro 182
evolutionary theory
  as unifying 361
  testing of 362
  see also common ancestry, drift, selection
expectation 19, 21−2, 84, 86, 123
experimental design
  and bayesianism 75
  and data interpretation 65, 78
  and frequentism 74
eye architectures 212

Falconer, D. 195
fallacy of affirming the consequent 129
falsifiability 49, 129
  and evolutionary theory 130
  and probability statements 130
  see also Modus Tollens
Farris, J. 256, 334, 349, 358
Felsenstein, J. 246, 252, 337, 338, 340, 344, 347, 351

Feyerabend, P. 152
fine-tuning 76
finite population size 157
    see also drift
Fisher, R. A. 9, 35, 45, 49, 53−8, 61, 130, 365,
    366
fishing 76
Fitelson, B. 16, 37, 51, 168
fitness function 157, 194, 196, 212
    valleys in 214
footprints on beach 324
forensic tests 52
Forster, M. 84−5, 87, 88, 91, 93, 94, 96−164,
    101, 103, 307, 360
fossils
    as ancestors or relatives 204, 324
    as evidence for common ancestry 318−24
    see also ancestor
free will 166
Freeland, S. 289, 313
French, S. 80
frequency data 24−7
frequentism 2−3, 7, 30, 31, 32, 48, 48−9, 53,
    58−72, 79, 87, 102
    and stopping rules 72−8
    versus bayesianism and likelihoodism 42
Frigg, R. 80
functions and teleology 114, 115, 134

Galapagos 329
Galileo 176
gases, theory of 365
Gassendi 147
Gaut, B. S. 350
Gehring, W. J. 213
gene transfer, lateral 272
genealogies, reticulate versus tree 269
    see also phylogenetic trees
genetic code 289, 312
    and common ancestry 289, 364
genetic fallacy 185
gill slits 305
Gillespie, J. H. 239
Gilovich, T. 96
Glymour, C. 231
goals of inference 93, 96
God's work and word 109, 113
Goldman, N. 257
Good, I. J. 46
Goodman, S. N. 73
Gossett, W. S. 57
Gould, S. 127−8, 143, 144, 244, 261, 285,
    361, 363
gradualism 320
Gray, A. 186

Griffiths, A. 160
Griffiths, P. 254
Grossman, J. 35

Hacking, I. 32, 45, 51, 57, 65, 78, 110, 119,
    138
Hajek, A. 39
Hansen, N. R. 152
Hansen, T. 194, 196, 250
Hartmann, S. 80
Harvey, P. 194, 244, 248
Hasegawa, M. 341
Hausman, D. 80, 144
heap, paradox of 288
Hempel, C. 149, 150, 316
Hereford, J. 196
heredity 298
heritability 195, 282, 325
    see also backwards inequality; Breeder's
    equation
Herschel, J. 124
Hesse, M. 80
Hick, J. 165
Himma, K. 121
Hodge, M. J. S. 109, 330
Holmes, E. 235, 238, 240, 241
Holmes, Sherlock 57
homology 283
homoplasy 283
honeybee's stinger 361
Hooker, J.D. 111
Hoover, K. 235
hot hands 96
Houle, E. 196
Howard-Snyder, D. 166
Howson, C. 31, 53−5, 56, 72, 74, 75
Huelsenbeck, J. 342, 351
Hull, D. 124
Hume, D. 43, 126, 139−41, 166, 169, 186

idealization 80, 81, 91, 96, 144, 156
identifiability of models 90
imperfect adaptations
    and Paley 128
    and natural selection 159, 213−14
independent contrasts, Felsenstein's method of
    252
indifference, principle of 21, 27−8, 306
induction 20−4, 140
    eliminative 57
    rules of 20−4, 25, 27
    see also design argument
inertia and stasis 244, 250
information processing inequality 305, 363
insect wings 50

instrumentalism 97
　see also realism and instrumentalism
intelligence, evolution of 363
intelligent design 51
　see also creationism; irreducible complexity
interleaving 100
intolerance of traits 146
　see also irreducible complexity
inverse gambler's fallacy 138
irreducible complexity 154
　and epistasis 163
　and fitness functions 158
　and four legged horses 160
　and function switching 161
　and the arch 162
　see also wine-bottle problem

James, W. (philosopher) 7, 188
James, W. (statistician) 66
Jeffrey, R. 12, 78
Johnson, D. 81
Johnson, P. 154, 185
Jukes, T. 284, 337
Justus, J. 149

Kadane, J. B. 77
Kaplan, M. 5
Kepler, J. 113
Keynes, J. 121
Kimura, M. 236, 237, 239, 337, 340, 361
King, A. 194
Kingsolver, J. 161
Kishino, H. 341
Kitcher, P. 110, 144, 176
Knight, R. 289
Koehl, M. 161
Koh, K. 229
Kolmogorov, A. N. 9, 39–41
Krebs, J. 191
Kreitman, M. 235, 240, 241, 243, 316
Kuhn, T. 13, 152
Kullback, S. see Kullback–Leibler Distance
Kulback–Leibler distance 98, 101
Kyburg, H. 5

Lanave, C. 337
Land, M. 213
Lande, R. 194, 196, 214
Landweber, L. 289
Lang, C. 226, 230
lanugo 304
Laplace, P. S. de 20–4, 21, 27
Larget, B. 342
Larson, A. 203, 255
Lauder, G. 254

laws 112
Leibler, R. A. see Kullback–Leibler distance
Leroi, A. M. 231
Lewin, R. 243
Lewis, P. 334, 350
Lewontin, R. 261, 361
Li, W. H. 238
life
　number of start-ups 276
　origin of 51, 111
　see also common ancestry
likelihood 9–11, 18, 25, 30–2, 32–5, 78
　and common causes 278, 281
　and evidence 14
　and nested models 83
　and phylogenetic trees 333–4
　and statistical consistency 349
　average versus maximum 28, 31, 70–1,
　　92–3, 99, 102–3, 239, 339–42
　definition of 9–10, 35
　law of 32–5, 35–8, 46, 52, 55, 56, 62–3,
　　63, 66, 76, 77, 103, 105, 108, 121, 147,
　　166, 198, 233, 284, 294, 299, 310, 354,
　　357
　maximum-estimate 23–4, 25, 65–6, 81, 83,
　　90, 91, 309
　of selection and drift 198
　principle 35
　ratio 32, 43, 45–6, 46, 52, 63, 64, 66, 73,
　　75–8
　ratio test 66, 71–2, 89–90
　versus posterior probability 120, 255
likelihoodism 3, 32–5, 35, 46, 48, 49, 52, 55,
　　64, 65, 66, 79, 81
　and Duhem's Thesis 144
　and interpretation of AIC scores 102
　and stopping rules 72–8
　objections to 35–41, 46–8
　versus bayesianism and frequentism 37, 42
Lindley, D. V. 72
Littlewood, J. 104
longevity, sex difference in 226
lottery
　models of 105
　paradox 5, 50
LUCA (last universal common ancestor) 270
　see also common ancestry
Lyell, C. 127

MacKay, T. 195
Maddison, W. 209, 256, 257
Markov model 215, 246, 300, 305, 337, 354,
　　360
Maynard Smith, J. 191, 205
Mayo, D. 76

Mayr, E. 213, 214, 245, 268, 364
McDonald, J. 240, 241, 316
McDonald—Kreitman test 240, 316
McMullin, E. 80, 144
Mendelism 26, 26—41
model selection 79, 226, 228, 335, 341, 362
  and coin tossing 177—80
  and intelligent design 177—84
  Bayesian 92
  see also Akaike information criterion; natural
    selection; phylogenetic inference
models
  and character evolution 335
  averaging 95
  fit to data 85
  fitted versus unfitted 98
  in logic 80
  LIN and PAR 67—72, 79, 83—90, 90, 93—4,
    100—1
  nested 69, 71, 83, 89, 93
  of no common mechanism 345
  parameters in 67, 69, 79, 81, 83, 85, 90,
    99—102
  time reversible 337
modus tollens 49—50
  probabilistic 49—53, 52, 53, 57, 105, 129,
    192
  see also falsifiability
molecular clock 236
molecular data on drift versus selection 235
monkeys and typewriters 116, 122
Morgan, M. 80
Morris, J. 187
Morris, H. 51
Morrison, M. 80
Mosteller, F. 104
Mougin, G. 6

natural selection
  and common ancestry 221, 264
  and dichotomous traits 217
  and explaining variation 191, 219—26, 262
  and imperfection 127
  and perfect adaptations 159
  and valley crossing 214
  as biased walk 194
  as non-random 123
  chronological test of 253
  frequency dependent 298, 299, 303, 312
  fundamental theorem of 365
  intensity of 195, 196
  models of 193—7, 215—17, 247
  of groups 245
  of species 281
  response to 195

versus artificial selection 188
versus drift, testing 193, 194, 353
natural theology 118—20, 124
naturalism, methodological 111
Nelson, P. 128
neutral evolution *see* drift
Newtonian theory 29, 32, 37, 47, 48, 57, 132
Neyman, J. *see* Neyman—Pearson testing
Neyman—Pearson testing 7, 49, 58—78, 79, 81,
    96, 102, 238, 339
  and stopping rules 72
  see also significance tests
Nielsen, R. 235
Nilsson, D. 213, 214
no-designer-worth-his-salt 126—8
  see Panda's thumb
null hypothesis 60—2, 69, 71, 73—8, 79, 80—1
Numbers, R. 111

O'Hara, R.J. 366
observation
  absolute versus relative theory neutrality 153
  as theory laden 152
observation selection effect 76
Oparin, A. 276
optimality model of when to give up 205
optimum
  as attractor 194
  inferring 202—7
ordinal equivalence
  of definitions of confirmation 16
  of parsimony and likelihood 343
Ornstein—Uhlenbeck model 194—5, 220, 222
Orr, A. 163, 281
Orzack, S. 212, 219, 243, 316
Ospovat, D. 191

P-value 54, 57, 78
Page, R. 235, 238, 240, 241
Pagel, M. 194, 244, 248
Paley, W. 118—20, 155
  his design argument as a likelihood argument
    121
  on evil 164
  on imperfect adaptation 128
  your head's pointing in the direct in which
    you step 120
  see also design Argument
panda's thumb 127—8
paradox of the heap *see* heap, paradox of
parameters
  counting numbers of 99—102
  nuisance 338
parasitic wasp 186
Parker, G. 205, 211

parsimony 250
  and group selection 245
  and inferring ancestral trait values 209, 255,
    261
  and likelihood 209, 359
  and probability 256
  and statistical consistency 347
  assumptions of 345, 351
  cladistic/phylogenetic 207, 332−4, 358
  model selection 207
  two types of 359
Parzen, E. 215
Pascal, B. 6−7
Patterson, C. 204
Pauling, L. 290
Pearson, E. S. *see* Neyman−Pearson testing
Pelger, S. 214
Pennock, R. 144
Penny, D. 314−18, 350
Phillips, L. D. 72
phylogenetic inertia 243
phylogenetic inference 293−352
phylogenetic trees 264, 333, 366
Pigliucci, M. 192
Plantinga, A. 134, 167, 188
polymorphisms and fixed differences 240
polynomials 70−1, 82, 90
polyploidy 286, 320
Popper, K. 49, 83, 129, 130, 358
Posada, D. 341
pragmatics 41, 94−5
prediction 79, 80, 82, 84, 95, 296, 362
predictive accuracy 84−5, 86, 87, 88, 90, 91, 95
  see Akaike Information Criterion
  see model selection
principle of common cause *see* common cause,
  principle of
principle of indifference *see* indifference,
  principle of
principle of total evidence *see* evidence, principle
  of total
prior probability 358
  improper 24
  objectivity of 24−8
  of nested models 93
  swamping of 25
probability
  conditional 38
  density 21−2, 27
  interpretations of 12, 49
  objective versus subjective 40, 47
  of reconstruction of ancestral character states
    255−8
  unconditional 39−40
  unconditional of observations 29−30

updating 11−13, 12, 50
pseudo-processes 324

quantum mechanics 231
Quine, W. 144

Raddick, G. 127
random versus biased processes 122, 287, 365
Rannala, B. 351
realism and instrumentalism 96−9
  see also idealization
redundancy 134
  see also irreducible complexity
Reichenbach, H. 20−4, 24, 25, 150, 185,
    231, 278, 307, 324, 358
relative rates test 238
reliability 17, 42
Ridley, M. 245
Rinard, S. 148
Robbins, H. 100
Rodriguez, F. 337
Rosales, A.
Rowe, W. 166
Royal Society 109, 116
Royall, R. 3, 8, 32, 46, 51, 51−2, 60,
    62−3, 65, 77, 107, 302, 354, 357
Ruse, M. 109, 112
Russell, B. 202

Sakamoto, Y. 86, 87
Salmon, W. 278, 324
Salvini-Plaven, L. 213, 214
Scheines, R. 231
Schervish, M. 77
Schlichting, C. D. 192
Schwarz, G. 92
screening-off 255
Seidenfeld, T. 77, 78
semantics versus epistemology 12, 49, 149
  see also probability, interpretations of
sex ratio, Arbuthnot on 117−18, 227
Shanks, N. 144
Shapiro, L. 195, 324
Shoesmith, E. 117
significance tests 49, 53−8, 61, 79, 130
  and choice of level of significance 54, 69,
    74, 76
  and sample size 56
  and stopping rules 72
  rejection versus evidential interpretations
    54−5, 56
similarity
  adaptive 297, 302
  deleterious 303
  neutral 298, 306

similarity (cont.)
  overall 294
  see also common ancestry
Simon, H. A. 123
simplicity 81−2, 83, 85, 88, 90, 97, 179
  and fit trade-off 86
  inductive and descriptive 358
  see also parsimony
Skyrms, B. 150
smoking 226, 247, 248
Snyder, L. J. 109
Sorenson, R. 5
species 226
Spirtes, P. 231
St. Petersburg Paradox 78
  see also expectation
Stanford, K. 97
Stanley, S. 281
Steel, M. 271, 305, 314, 345, 347, 349, 350,
  351, 363
Stein, C. 66
Stephens, C. 51
Sterelny, K. 254
Stone, M. 88
stopping rules 72−8
Strier, K. 226, 230
Stuart, R. 205, 211
subfamily problem 93−5 see Akaike Information
  Criterion
Sugiura, N. 95
Swofford, D. 337, 338
Swift, J. 116, 124
synapomorphies 266

Takeuchi, K. 95−164
Taveré, S. 337
teleology see also functions and teleology
testability 129, 148−54, 355
  and creationism 130, 141−7, 189
  and evolutionary theory 189
  and logical positivism 149
testing, contrastive character of 32, 52, 61, 116,
  131, 149, 152, 190, 227, 267, 314, 353,
  354
tetrapods 244
theft versus honest toil 202
theistic evolutionism 110, 112
theodicy 165
theory of relativity 16, 26−7, 28−31, 32, 37,
  41, 47, 48, 57, 132
topological invariance 100
trait see character
transformation series 284
tree thinking 366

truly large numbers, law of 104, 348−51
truth versus predictive accuracy as inference
  goals 80−1, 97
Tuffley, C. 345, 347, 349, 351
Turelli, M. 198
turkey baldness 226

unbiased estimators 86−7, 92
unification 106, 110, 226, 228, 360
  see also common cause, principle of;
    parsimony; simplicity
uniformity of nature 87, 177
Urbach, P. 53−5, 56, 72, 74, 75
useful and intolerant traits 133, 147
  see also irreducible complexity
utilitarianism 65

vagueness 5 see also heap, paradox of
validity 1, 25, 49, 50, 53, 129
values and ethics 7, 60, 78
Van Fraassen, B. 231
Van Inwagen, P. 137
Van Tienderen, P. H. 192
variation, among and within species 219−26,
  229
Vrba, E. 261
vulture 226

Wagner, C. 53
Wake, D.B. 250
Wald, A. 78
Walsh, D. 195
Wardrop, R. 96
Wedgewood, E. 187
Whewell, W. 109−10, 366
Wiley, E. O. 358
Williams, G. C. 261
Williamson, T. 5
Wilson, D. S. 92, 361
Wilson, E. O. 244
wine bottle problem 135−6, 146
  see also irreducible complexity
witness testimony 42−3
Wittgenstein, L. 90
Woese, C. 268, 273
Woodward, J. 231
Wright, L. 115
Wykstra, S. 167

Yang, Z. 351
Young, R. 191
Yule, G. U. 233

Zuckerkandl, E. 290

Printed in the United States
By Bookmasters